A Philosophical History of Western Medicine

Michael Boylan

A Philosophical History of Western Medicine

 Springer

Michael Boylan
Department of Philosophy
Marymount University
Arlington, VA, USA

ISBN 978-3-031-97805-0   ISBN 978-3-031-97806-7   (eBook)
https://doi.org/10.1007/978-3-031-97806-7

© The Editor(s) (if applicable) and The Author(s), under exclusive license to Springer Nature Switzerland AG 2025

This work is subject to copyright. All rights are solely and exclusively licensed by the Publisher, whether the whole or part of the material is concerned, specifically the rights of translation, reprinting, reuse of illustrations, recitation, broadcasting, reproduction on microfilms or in any other physical way, and transmission or information storage and retrieval, electronic adaptation, computer software, or by similar or dissimilar methodology now known or hereafter developed.
The use of general descriptive names, registered names, trademarks, service marks, etc. in this publication does not imply, even in the absence of a specific statement, that such names are exempt from the relevant protective laws and regulations and therefore free for general use.
The publisher, the authors and the editors are safe to assume that the advice and information in this book are believed to be true and accurate at the date of publication. Neither the publisher nor the authors or the editors give a warranty, expressed or implied, with respect to the material contained herein or for any errors or omissions that may have been made. The publisher remains neutral with regard to jurisdictional claims in published maps and institutional affiliations.

This Springer imprint is published by the registered company Springer Nature Switzerland AG
The registered company address is: Gewerbestrasse 11, 6330 Cham, Switzerland

If disposing of this product, please recycle the paper.

To Vivian Nutton, whose writing, correspondence, and friendship have helped me develop my ideas on medicine and its evolution

Preface

In the ancient Greek world, a book titled *On Nature* was very common. Some say that this interest separated philosophy from religion as its own discipline in the West. Among various candidates, Thales is often brought forward in his theory of water as the principle (*arche*) of nature (*phusis*). As time went by, the ways that *nature* could be understood expanded, too. Because of this and because the scope of any text must, by necessity, be limited, I have chosen to further narrow the scope of this book by focusing upon Western Medicine in the ancient and the modern eras. At times, medicine is understood via other natural compartments such as biology, chemistry, and physics. As these are relevant to the study of Western Medicine, selected forays into other scientific departments will be conducted with a focus: how do they illumine our philosophical understanding of the development of Western Medicine.

The book will begin with the earliest texts that bring forth concepts relevant to Western Medicine. These include early poetry and pre-Socratic philosophy. Then we will proceed to Plato, Aristotle, and the Hippocratic writers. Various intermediate figures will be examined as we continue to Galen and the stopping point in one portion of our history.

Then, we will examine a critical turning point in scientific understanding from the seventeenth century to the twenty-first century. The object of this excursion will be to create a structure by which we can unite the way we have variously understood nature and its effect upon medicine in the West. It will be the contention of this book that there are three important unifying approaches to viewing nature—particularly when examining medicine. The first of these concerns the ontological question of *what nature essentially is*. It will be the contention of this author that from the beginnings to the present era, we understand nature as being connected with how we *come to know it*, its epistemology. One way we come to know nature (and thereby ascribe its essence *to be*) is to see it as a set of material relationships that bring about material results that are qualitatively (later quantifiably) understood. A second way we come to know nature is via ascription of complex material problems to the supernatural/non-natural—first to the divine, and then later to mathematic modeling. Thus, an examination into nature is an examination of the divine, initially. This evolved in the seventeenth century to another extra-empirical arena: mathematics. This new depiction of nature$_2$ has continued until the present day in general science and the way that medicine had incorporated technology in its method and practice.

A third way we come to know nature is via ascription of complex material problems to pragmatic solutions but backing away from generalizing about these origins because of skepticism of their general, necessary applicability—some form of epistemological antirealism.

It will be a contention of this book that these three forms of understanding the ontology of nature (through epistemology) will provide one context of how nature is presented to practitioners and the general public.

I use the term "nature" (with a small "n") to denote individual natures and "Nature" (with a capital "N") to denote group characteristics. When a concept applies to both groups, I will use "n/Nature" to make the reference.

Two other contexts of nature must also be highlighted. The first is aimed at practitioners and why they might be attracted to one causal account over another. I will contend that an important component to most of the pivotal figures I set out is that they are aesthetically attracted to one account over another due to a sense of *elegance* in the preferred account. I take "elegance" to be comprised of three principal factors: (a) *completeness,* (b) *coherence*, and (c) *simplicity* (which I take to be ontological simplicity—meaning that they believe that nature is basically well-organized and this well-formed organizational structure provides feedback to those who search for natural meaning). The more theory$_1$ is seen to be structured in such a way that it is complete, efficiently coherent, and has wide explanatory power within a structurally/conceptually simple presentation, the more attractive it is over a competitor, theory$_2$.

When the structure of a theory contains these three components to a greater degree, we often judge it to be *elegant* in its presentation—the final characteristic in judging one theory to be more desirable than another.

How we are to understand the n/Nature dynamic can also be understood by what I call the "maker/artifact distinction." This is a leitmotif throughout the book but is more forceful among certain practitioners.[1] It can affect the way we understood causation and what should be recognized by those under the realm of Nature. It can be both the foundation for precautionary reason and an underpinning of environmentalism.

Arlington, VA, USA   Michael Boylan

---

[1] Though the "maker/made" dynamic is a minor dimension of this treatise, I have written about this further in Michael Boylan, "What is 'Nature' and Why Should We Care?" in Michael Boylan, ed. *Environmental Ethics* 3rd ed. (Oxford: Wiley-Blackwell, 2022): 15–34.

Acknowledgments

As in any project of this kind there are many to thank. I would first like to thank Vivian Nutton for comments on this manuscript and those in the past. Then, there is Ralph M. Rosen whose detailed comments on this manuscript and my "blood book" have been very useful. Then there is Philip van der Eijk whose comments on this manuscript and my seminar at Humboldt University have helped me make my direction more coherent. I'd also like to mention earlier comments by Paul Demont and Geoffrey Lloyd on early versions of the themes that have emerged within the book.

And finally, thanks to the support of my family: Rebecca, Arianne, Seán, and é.

Competing Interests The author has no competing interests to declare that are relevant to the content of this manuscript.

Acknowledgments

As in any project of this kind there are many to thank. I would first like to thank Vivian Nutton for comments on this manuscript and those in the past. Then, there is Ralph M. Rosen whose detailed comments on this manuscript and my "blood book" have been very useful. Then there is Philip van der Eijk whose comments on this manuscript and my seminar at Humboldt University have helped me make my direction more coherent. I'd also like to mention earlier comments by Paul Demont and Geoffrey Lloyd on early versions of the themes that have emerged within the book.

And finally, thanks to the support of my family: Rebecca, Arianne, Seán, and é.

Competing Interests The author has no competing interests to declare that are relevant to the content of this manuscript.

Contents

Part I The Ancient Western World

1 The Beginnings 3
   Homeric Poetry and Nature 4
   A Shift to Material *Arche*: Thales and Empedocles 7
   Dueling Principles: Heraclitus 20
      The Role of Logos 21
      The Role of the Senses.................................... 21
      The Metaphysical Characteristics of Nature 22
      Dual Explanations 24
   Numbers: Pythagorean Mathematical Realism v. Parmenides'
   and Zeno's Anti-Realism....................................... 28
      Pythagoras 28
      Parmenides and Zeno..................................... 32
   Summary of All Pre-Socratics 44

2 The Hippocratic Contribution............................... 45
   Introduction.. 46
   Basic Science Applied to Medicine 48
      Psuche and the Role of External Air and Other External Factors ... 48
      The Roles of Interior Factors Causing Disease 52
   The Practice of Medicine 53
      The Fundamentals: Diagnosis, Prognosis, Treatment 53
      The Ethical Dimension of Practice: The Oath 62

3 Plato's *Timaeus* 69
   Plato's General Philosophy of Science.......................... 70
   Plato's Non-systematic Insights into Biomedicine 75

4	**Aristotle**	81
	Aristotle's Biomedical Applications	87
	Conclusion	98
5	**Galen's Resolution of *Nature* in Medicine**	99
	Introduction	100
	Galen and His Contemporaries	102
	A Brief Selection of Key Galenic Texts	111
	Conclusion	116

Part II The Medieval, Modern, and Contemporary Worlds

6	**The New Paradigm: Changes Leading up to the Seventeenth Century and Beyond**	121
	Introduction	121
	Transition in the Islamic World	122
	Transition in the European World	127
	The European Tradition and the Causes for Change	134
	The Genetic Order for Change	134
	The Logical Order for Change	139
	Argument 6.1: The Synthetic *A Priori* in Nature	151
	Argument 6.2: Time and Space in the Transcendental Aesthetic	152
	Argument 6.3: The Categories of the Understanding	153
	Argument 6.4: The Schematism of the Categories	153
	Conclusion	158
7	**Evolution, Germ Theory, and Their Consequences**	161
	Introduction	161
	Evolutionary Theory	163
	Teleology	163
	Germ Theory	176
	Introduction	176
	Conclusion	184
8	**Nature Split: Big Nature and Little Nature**	187
	Introduction	187
	Big Nature	190
	Special and General Relativity	193
	Logicist Thesis	196
	Little Nature	199
	Logical Empiricism and Reductionism	205
	The Unity of Science and Reductionism	206
	Linguistic Expressions and Meaning	207
	Big and Little Nature in Medicine	207
	20th Century Medicine	208
	Big Nature in Medicine	212
	Conclusion	215

9	**Technology and Nature: A Marriage or a Divorce?**	217
	Introduction	217
	Little Nature in General Science	218
	Big Nature in General Science: Cosmological Choices	222
	Loop Quantum Gravity	224
	String Theory	225
	Changes in the Philosophy of Science	226
	Technology and Medicine	230
	AI, Technology, and Modern Surgery	232
	Public Health—Basic Principles	233
	Public Health—Pandemics	233
	Conclusion to Pandemics	242
	An Assessment of Technology and the Mission of Medicine	243
	Conclusion	243
Appendix		245
Bibliography		261
Index		287

Introduction

Abstract: This chapter introduces the reader to the general background of the principal theoretical constructs that will be used to explore *what* nature *is* and how accounts of it can be understood and evaluated. The venue will be the ancient Greek world concerning science and how it is, generally, to be understood. These same distinctions will also be used in the second part of the book, but in a different guise.

Keywords: Symmetry; Simplicity; Elegance; Completeness; Coherence; Maker/artifact dynamic; Aristotle; Ancient Greek science; Plato; Pre-Socratics

This essay will further introduce the reader to the four principal interpretative concepts in this monograph: (a) the tripartite understanding of n/Nature; (b) the symmetry/simplicity/elegance criteria; (c) the completeness/coherence criteria; and (d) the maker/artifact dynamic. These four distinctions will be used in the rest of the book to explain methodology and biomedical practice in this select historical survey. Since the basis of the theoretical foundations of this inquiry takes place in the ancient Greek and Roman worlds, we will first set out these distinctions within this context.

One of the most important set of questions that philosophy should address concerns *nature*. In the ancient Greek World, it was a very common topic for analysis as was also the case in the ancient Chinese World.[1] In these civilizations, nature was something exterior to one (an environment which was composed of various sorts of supporting *non-living entities* such as aspects of the earth, air, fire, and water as well as *living entities* like plants, animals, other humans). This sense of nature was often characterized by types. But there was also another sense of nature referring to what it meant to be an individual human.[2] In this volume I will distinguish between these two

---

[1] For an overview of how nature was used to support theories of ethics and justice, see: Michael Boylan, *Natural Human Rights: A Theory* (New York and Cambridge: Cambridge University Press, 2014): chapters 2 and 3.

[2] For more on the dynamics of the community v. the individual perspective on nature, see: Michael Boylan, "What is Nature and Why should I Care?" in Michael Boylan, ed. *Environmental Ethics,*

understandings by using the small case "n" for individual natures and the capital "N" for the supporting cast of environmental living and non-living entities set out as types. This distinction is roughly the same as the contemporary type-token distinction.[3]

For readers in the twenty-first century a study of nature, focusing upon the history medicine and its intellectual environment within the Western tradition, is of interest because of its connection to the history of science that allows us a more nuanced understanding of the activities of science today as well as how we should think about these discussions and what they mean to our community worldview going forward into the future.[4]

Three Senses of Nature

To begin our journey let us turn to Aristotle, who introduces our discussion in *Posterior Analytics* II.1 "There are four questions we can ask about the [n/Natural] objects we know (*epistametha*): the fact (*hoti*); the reasoned fact or cause (*dioti*); the question about the manner of existence of the entity (*ei esti*); and the question of definition or essence (*ti esti*)."[5] In the context of this essay I'd contextualize this series of concerns as ultimately searching for a definition of nature that can be the basis of our discussion. But "definition" is last on Aristotle's list. Why is that? It is because the definition must be the result of experience (*empeiria*) that itself allows one to have confidence in her beliefs about what the fact *actually is* (*hoti*) and its structure and modalities within various contexts (*ei esti*) in order to speculate on the causal structure that underlies its operation (*dioti*). As has been the case in the past, this author is greatly influenced by Aristotle's approach to understanding n/Nature.[6]

In the first phase of the book, Chaps. 1 and 2 the view of nature among the Pre-Socratics[7] and Plato is rather general and leans toward metaphysics as it forms a

---

3rd ed. (Malden, MA and Oxford: Wiley Blackwell, 2022): Ch. 2. For another set of essays to support this thesis, see: G.E.R. Lloyd, Jingyi Jenny Zhao, Qiaosheng Dong, eds. *Ancient Greece and China Compared* (Cambridge: Cambridge University Press, 2018).

[3] The type-token distinction is variously represented. My use of the distinction owes its source to Paul Grice, 'Utterer's Meaning and Intentions' (1969) 78 *Philosophical Review* 147–177 and Willard Van Quine, *Quiddities: An Intermittently Philosophical Dictionary*. (Cambridge, MA: Harvard University Press, 1987).

[4] For a discussion of the various versions of community worldview, see: Boylan (2014): chapter 6.

[5] I am following W.D. Ross's text *Aristotle's Prior and Posterior Analytics* (Oxford: Oxford University Press, 1949). My translation is of: 89b 23–25: Τὰ ζητούμενά ἐστιν ἴσα τὸν ἀριθμὸν ὅσαπερ ἐπιστάμεθα. ζητοῦμεν δὲ τέττρα, τὸ ὅτι, τὸ διότι, εἰ ἔστ, τί ἐστιν.

[6] See particularly my books that explore Aristotle: *Method and Practice in Aristotle's Biology* (Lanham, MD and London: UPA/Rowman and Littlefield, 1983) and *The Origins of Ancient Greek Science: Blood—a philosophical Study* (New York and London: Routledge, 2015): ch. 3, and "Mechanism and Teleology in Aristotle's Biology" *Apeiron* 15.2 (1981): 96–102; "The Place of Nature in Aristotle's Biology" *Apeiron* 19.1 (1985): 126–139.

[7] For a good overview of supernaturalism among the ancient Greeks, see: Andrew Gregory, "The Ancient Greeks and the Supernatural" in G. Arabatzis ed. *Studies in Supernaturalism* (Berlin: Logos Verlag, 2009): 11–38.

broad overarching context of personal and community worldview.[8] In particular, these early forays into natural philosophy are characterized by their adoption of what I will term Nature$_1$ or a depiction of Nature as material in at least the *hoti* and the *dioti*. This would be a transition from Nature$_2$—the depiction of Nature as a province of the gods. Nature$_2$ is often the choice of the common person who faces the many questions of the strong forces about them and chooses to posit very strong divine figures to take over the management of these affairs.[9] There are some advantages of turning over the running of n/Nature to the gods (Olympian and pre-Olympian—not to mention regional gods).[10] This is because they are assumed to be a pure instance of *dioti*. It is their power that makes it so. They act as *demiourgos*, a fashioner or creator. When we think of the act of creation from the human perspective we think of the creator as being above that which s/he creates. There are several reasons for this. First, the artifact would not exist (*hoti*) save for the activity of this or another fashioner. Second, the *manner of existence (ei* esti) that the artifact achieves is largely due to the fashioner. Finally, if the fashioner is displeased with their efforts, then they might decide to destroy it and start again (another instance of *ei esti*). Because of these dynamics, it will be the contention of this treatise that we can identify a "maker-artifact" or "maker-made" dynamic[11] in which from the point of view of the artifact, the maker should be regarded as *good* (in at least a functional sense).[12]

---

[8] I use these terms *personal worldview* and *community worldview* technically via the normative imperatives I set out concerning these—see Michael Boylan, *A Just Society* (Lanham, MD and Oxford: Rowman and Littlefield, 2004): chapts. 2 and 6 and in Michael Boylan, *Natural Human Rights: A Theory* (New York and Cambridge: Cambridge University Press, 2014): ch. 6.

[9] For example, *gaia* (from *ge*) gets personified as earth; *aeolus* and *aether* refer to the winds and the heavens; *boreas* cold wind; *helios* (the sun—a primal source of fire—abstractly while *hephaestus*, is applied fire for technological purposes); *pontos* (pre-Olympian son of *gaia*) and the Olympian *poseidon*, god of the sea, storms, and earthquakes. For a good introduction to these and other figures, see: Edith Hamilton, *Timeless Tales of Gods and Heroes* (New York: Grand Central Publishing, rpt. 2011[1964]).

[10] A discussion of these dynamics can be found in Walter Burkert, *Greek Religion* (Cambridge, MA: Harvard University Press, 1987).

[11] Though the "maker/made" dynamic is a minor dimension of this treatise, I have written about this further in Michael Boylan, "What is 'Nature' and Why Should We Care?" in Michael Boylan, ed. *Environmental Ethics* 3rd ed. (Oxford: Wiley-Blackwell, 2022): 15–34. I list this dynamic in the "Introduction" because it is due to this causal hierarchy, that we revere n/Nature and give it status within our theories of science—so much so that from the ancient Greek world to the eighteenth century those who investigated the foundations of general science and medicine were called "natural philosophers."

[12] Of course, from the point of view of any self-regarding, sentient being, the functional will also include a normative element since without the creator the individual, herself, would not be. And if the individual regards herself as "good" then so also would be the entity responsible for her existence (on the principle of heritability of existing properties). In modern genetics, there is a similar dynamic when offspring consider their biological parents—See Michael Boylan and Kevin Brown, *Genetic Engineering: Science and Ethics on the New Frontier* (Upper Saddle River, N.J.: Prentice Hall, 2002).

A third sense of Nature, Nature$_3$, arises when the epistemology of the practitioner and his cadre of followers adopt *antirealism*. Under this understanding of Nature, no one can ever have knowledge (certainty). Instead, one is bound by a sense of *skepsis* that requires continual exploration that will never obtain closure. Since action requires that we have, at least, rules of thumb, the antirealist advocate of Nature$_3$ will satisfy himself with what has happened in the past as a possible indicator of what will occur in the future—not as a certainty, but only as grounds for a reasonable guess.[13]

The last major principle that will prove to be important in this presentation is the depiction of the operation of Nature (*dioti*). Most natural philosophers (perhaps because of the maker-artifact dynamic) seek to construct their versions of Nature's *dioti* in terms of elegance (simplicity and novel explanation) and symmetry (systemic balance), which give rise to an aesthetic appreciation. Nature's structure (*hoti*) and its mode of being (*ei esti*) give rise to orderly execution that is, in principle, knowable and possibly predictable (*dioti*). This construction comes together so that when its essence (*ti esti*) is known, the result is appreciated via an aesthetic judgment (elegance and symmetry).

These are the principal concepts that will be set out in this volume. However, before beginning it is important to make one important note on methodology. Since many primary texts will be quoted in making these arguments, it is key to understand the way this author understands primary texts and how they might be interpreted.[14] In Classics (and in most modern literature disciplines) the attitude toward texts is that: (a) they are static entities that reveal themselves only via literal examination without much inferential interpretation; and (b) that the use of counterfactuals and other deductive logical devices are to be eschewed.[15]

My methodology follows Boylan (2019) and this allows one to employ *als-ob* devices that can test meaning via exploration of various embedded implications within the text through hypotheticals and other logically connected outcomes (sometimes less than tight deductive inferences).

---

[13] In modern philosophy this stance is best enunciated by David Hume that because the future does not necessarily resemble the past, there can be no necessary inductive explanations. See: David Hume, *An Enquiry Concerning Human Understanding,* Edited by Eric Steinberg (Indianapolis, IN: Hackett, 1977 [1748]): IV.2, p. 22 ff.

[14] I set out the argument for this use of texts within the discipline of philosophy in Michael Boylan, *Fictive Narrative Philosophy: How Fiction Can Act as Philosophy* (New York and London: Routledge, 2019): ch. 7.

[15] On the use of counterfactuals in philosophy I am thinking here of Nelson Goodman in *Fact, Fiction, and Forecast* 4th ed. (Cambridge, MA: Harvard University Press, 1979): part I. I should note that Ralph M. Rosen mentioned to me that he thought (in support of me) that this conception of the static text is not supported by most contemporary classicists—especially in the history of medicine. I should also note that several classics researchers believe that I am being a bit harsh on the classical audience here harkening back to an earlier era. However, in 2018 in a keynote address I was questioned on this very point.

Introduction

For an example of this methodology let us consider an example concerning the three senses of Nature outlined above in the realm of ancient Greek philosophy concerning *phusis* (Nature) in its three modes.

Primary Texts on **phusis** *(Nature)*

Nature$_1$: Aristotle, *Generation of Animals*: 762a 9-18: *phusis*-1 (the materially based view of Nature).[16]

> All [living organisms] which neither produce side-shoots[17] nor make 'honeycombs'[18] reproduce by spontaneous generation;[19] and all which arise in this manner whether on land or in the water come to be formed, as can be seen, to the accompaniment of putrefaction and admixture of rainwater:[20] as the sweet ingredients are separated off into the principle which is taking form, that which remains over assumes a putrefying aspect. Nothing, however, is formed by a process of concoction: the putrefaction and the putrefied matter are a residue[21]

---

[16] Ὅσα δὲ μήτε παραβλαστάνει μήτε κηριάζει, τούτων δὲ πάντων ἡ γένεσις αὐτόματός ἐστιν. πάντα δὲ τὰ συνιστάμενα τὸν τρόπον τοῦτον καὶ ἐν γῇ καὶ ἐν ὕδατι φαίνεται γινόμενα μετὰ σήψεως καὶ μιγνυμένου τοῦ ὀμβρίου ὕδατος· ἀποκρινομένου γὰρ τοῦ γλυκέος εἰς τὴν συνισταμένην ἀρχὴν τὸ περιττεῦον τοιαύτην λαμβάνει μορφήν. Γίνεται δ' οὐθὲν σηπόμενον ἀλλὰ πεττόμενον· ἡ δὲ σῆψις καὶ τὸ σηπτὸν περίττωμα τοῦ πεφθέντος ἐστίν.

οὐθὲν γὰρ ἐκ παντὸς γίνεται, καθάπερ οὐδ' ἐν τοῖς ὑπὸ τῆς τέχνης δημιουργουμένοις, οὐθὲν γὰρ ἂν ἔδει ποιεῖν· νῦν δὲ τὸ μὲν ἡ τέχνη τῶν ἀχρήστων ἀφαιρεῖ, τὸ δ' ἡ φύσις. (This text follows The Leipzig edition, VIII edited and translated into German by H. Aubert and Fr. Wimmer (Leipzig: W. Engelmann, 1860).

[17] "Slideshots" (*parablastano*) are meant to encompass "budding." This is a biological form of reproduction that does not involve sexual interaction. The text seems to suggest that the source of this sort of reproduction is solely the province of the male. However, it might be questioned, that if only the male is concerned with this sort of reproduction, why Nature (which does nothing in vain) even created females for these species. Note that Theophrastus uses this term for plant reproduction (considered to be asexual), *HP* 1.2.6.

[18] "Honeycombs" (*keriazo*) are perhaps the eggs of Gastropods (Peck) or the spawn of purple-fish (*porphura*).

[19] "Spontaneous generation" (*genesis automatos*) note the use of *automatos*. This is one of the two technical words that Aristotle uses for chance (the other being *tuche*—cf. *Physics* II. 4–6). In the case of *tuche* we have the intervention of art (*techne*) for the desired effect. In the case of *automaton*, it is the intervention of nature—like the wind blowing a cat off its feet. This is *phusis*$_1$. It is from material causes that come from nature and not from the conscious activity of a human agent—such as a physician.

[20] The material components of the mixture (*mignumenou*) are rain water (*ombriou* that is *hudatos*—emphasizing the material property of "wet", cf. "hot, cold, wet, and dry") and fermented or putrefied (*sepseos*) matter. Fermented matter within the digestion system of humans creates a transformation *pepsis*. See Boylan (1982). This is thus a mysterious, but materially based process (cf. *pettomenon*).

[21] *Perittoma* is also the word that Aristotle technically uses for the human female menses that is the material cause for human reproduction, see Boylan, (1984).

of that which has been concocted, for no creature's formation[22] uses up the *whole* of the material, any more than in the case of objects fashioned by the agency of art, otherwise there would be no need to make anything at all, whereas what happens in actual fact is that the useless material[23] is removed in the one case by art and in the other by Nature.[24] [English Translation by A.L. Peck, (Loeb).]

As we can see by this text and the footnotes, I have set out Aristotle here as trying to account for living organisms (animals, for the most part) that do not seem to engage in sexual reproduction. One direction he could have gone is to say that they were due to Nature$_2$ (*phusis*-2). This would be a rather easy approach because it would involve in saying that in cases in which no observable sexual reproduction occurs, that it was due to the agency of the gods. But Aristotle does not take this approach. Nor does he take the approach of saying that the offspring occur without any causation. Though this group of organisms are said to occur spontaneously, there are still material factors that enter into this result. This is an important point in the methodology of understanding Nature: there are *no uncaused events*. This is a version of the Leibniz's principle of sufficient reason.[25] It is an important first-level principle to accept when exploring the *hoti* and *dioti* of Nature.

However, it is important to note that the word *phusis* is only used in the last word of the selection. But this does not detract from my analysis according to the terms of interpretation set out earlier. If it is required for us to insert *phusis* into the text to make sense of the discourse, then this is permissible as an instance of enthymeme (which employs suppressed premises necessary to bring about the logical inference within the conclusion). The logical use of enthymeme (*en-thumos*) "in the spirit of the argument that is assumed in order to create a valid deductive inference," is used (though often not mentioned) through much of philosophy (East and West) in its historical presentation.[26]

In the next example we turn to Plato's *Republic* 444d 3-6.[27]

---

[22] *Demiourgoumenois* emphasizes the active role of nature-1 & 2 in the process of bringing forth a spontaneously generated entity. The use of this particular word implies a creative, active role for Nature, cf. Plato's use of the same term in the *Timaeus* for the creation of the world. This is indeed the agency of art (*techne*).

[23] *Achreston* or useless material is taken out of the process. Whereas *katamenia* is a residue that has potentiality, *dunameis*.

[24] So here, again, we have the dichotomy (*men ...de*) of some clearly controllable or recognizable material process (the *techne*) v. the still material, but unknowable causal force of *phusis*-1.

[25] Gottfried Wilhelm Freiherr von Leibniz, "On Geometrical Method and the Method of Metaphysics" in *Discourse on Metaphysics,* ed. Albert R. Chandler, trans. George Montgomery (LaSalle, IL: Open Court, 1924 [1686]).

[26] For a discussion of *enthymeme* as it is used in logical deductive argument, see: Michael Boylan, *The Process of Argument: An Introduction,* 3rd edition (New York and London: Routledge, 2020).

[27] Ἔστι δὲ τὸ μὲν ὑγίειαν ποιεῖν τὰ ἐν τῷ σώματι κατὰ φύσιν καθιστάναι κρατεῖν τε καὶ κρατεῖσθαι ὑπ' ἀλλήλων, τὸ δὲ νόσον παρὰ φύσιν ἄρχειν τε καὶ ἄρχεσθαι ἄλλο ὑπ' ἄλλου. Plato, *Republic*: 444d 3–6—*Phusis*-1. (This text follows: John Burnet, *Platonis Opera,* vol. 4 (Oxford: Clarendon Press, 1902).

Introduction

> To produce health is to establish the parts of the body in a relation of mastering,[28] and being mastered by, one another that is according to nature,[29] while to produce sickness is to establish a relation of ruling and being ruled by, one another that is contrary to nature. [Translation by Allan Bloom].[30]

In the realm of medicine from the Hippocratic writers onward, health is seen from one perspective as a struggle between material forces that either work according to the rules of Nature$_1$ or against them. These materially based rules dictate that if you follow the program (the emerging four-humor account), then your body will be following the material paradigm of Nature and you will be healthy. Go against this material paradigm and sickness will result.

Nature$_2$: From Plato, *Phaedo*: 118a7-8 we encounter an example of *phusis$_2$* (the account of Nature that relies upon reference to the gods).[31]

> Crito, we owe a cock to Asclepius;[32] make this offering to him and do not forget. [English translation by Michael Boylan]

As suggested in footnote 29, it is unclear what the sacrifice of the cock to the god of health is supposed to procure. At any rate, though the word *phusis* does not occur in the passage it is clear that some sense of one's hygienic well-being is put to the authority of a god and not material nature. As per my interpretative guidelines mentioned above, this will still count as an instance of appealing to the gods in a causal instance involving health and thus under the depiction of Nature$_2$.

A second example of Nature$_2$ comes from the Hippocratic Writer, *The Sacred Disease;* I.1-3—*Phusis*-2.[33]

---

[28] REPUBLIC PASSAGE—*PHUSIS$_1$*. A critical idea of "mastering" and "being mastered" is through the physical properties.

[29] *Kata phusin* and *para phusin* are critical concepts in biomedicine. Though this is presented as *phusis*-1, one would be inclined to ask what structured the natural order to be just this way. This level-two question has only two answers: (a) by Divine forces (*phusis*-2) or (b) by accidental forces—such as Democritus and Leucippus. This passage, being from Plato, one must assume the former. This means that advocates of *phusis*-1 accounts might also be accepting a broader context of *phusis*-2 accounts, as well.

[30] Allan Bloom, *The Republic of Plato* (New York: Basic Books, 1968), cf. "To produce health is to establish the components of the body in a natural relation of control and being controlled, one by another, while to produce disease is to is to establish a relation of ruling and being ruled contrary to nature." (Translation by G.M.A. Grube, rev. C.D.C. Reeve, in *Plato: Complete Works,* ed. John M. Cooper, associate editor D.S. Hutchinson (Indianapolis, IN: Hackett, 1997).

[31] Ὦ Κρίτων, ἔφη, τῷ Ἀσκληπιῷ ὀφείλομεν ἀλεκτρυόνα ἀλλὰ ἀπόδοτε καὶ μὴ ἀμελήσητε.
(This text follows John Burnet, *Platonis Opera*, vol. 1 (Oxford: Clarendon Press, 1900). Plato, *Phaedo*: 118a7-8).

[32] PHAEDO—Whether this is to procure health or a good passing, it still involves invocation of a deity associated with health.

[33] Περὶ τῆς ἱερῆς νούσου καλεομένης ὧδ᾽ ἔχει. οὐδέν τί μοι δοκεῖ τῶν ἄλλων θειοτέρη εἶναι νούσων οὐδὲ ἱερωτέρη, ἀλλὰ φύσιν μὲν ἔχει καὶ πρόφασιν. τὰ λοιπὰ νουσήματα ὅθεν γίνεται. *Sacred Disease*. I. ll.1-3. (This text follows Émile Littré, *Hiippocrate, Oeuvres Complètes*, 10 vol. (Paris: J. B. Billière, 1839–1861).

I am about to discuss the disease called 'sacred.' It is not, in my opinion, any more divine or more sacred than other diseases,[34] but has a *phusis*[35] which also comes-to-be in the remaining sickness(es). [Translation by Michael Boylan English after W.H.S. Jones.]

As noted above, epilepsy presents itself through grand mal seizures which are so sudden and remarkable that they seem to be sui generis in the realm of health. This led to this condition being termed a "sacred disease." Though the term *phusis* does not occur in this sense during the passage, it still implied and falls under my rubric for interpretation. When *phusis* does occur, it is as a contrast to the invocation of the divine. As noted, this appears to be a contrast between Nature$_2$ in the first part of the passage and Nature$_1$ —here set out as the alternative by the Hippocratic Writer. This conflict suggests that within certain instances some individuals (physicians?) might assert that the presentation of epilepsy is due to intervention of the gods—for some reason or for no reason—while other individuals (at least the Hippocratic Writer of *The Sacred Disease*) contend that the wrong invocation of nature has occurred by the Nature$_2$ advocates and that the materially based Nature$_1$ is really at play here.

Nature$_3$: The third form of Nature exhibits the skeptical or anti-real perspective (*phusis$_3$*). An example of this can be found in Sextus Empiricus, *Outlines of Pyrrhonism*[36] I. 8-9

Skepticism is an ability,[37] or mental attitude, which opposes 'appearances to judgements'[38] in any way. [English translation by R. G. Bury (Loeb)]

In this passage the biomedical writer, Sextus Empiricus, orients the disposition toward Nature as one which exhibits an epistemological need to be leery of *appearances*. Why is this? Historically, the antirealists believe that Truth is very complicated and beyond the ken (empirical perceptions + judgments about those perceptions) of ordinary people. The ancient skeptics continually want more information and thus keep their options open because they see various avenues of interpretation. For the skeptic the branding phrase is "it's complicated; tell me more." Such a maxim creates an attitude that rarely accepts epistemological closure.

---

[34] THE SACRED DISEASE: Note that this author is not an advocate of *phusis-2,* but is commenting on those who are. This is not unusual for epilepsy seems quite inexplicable—cf. Cassandra's link to Apollo in the *Iliad* and the *Agamemnon*.

[35] *Phusis* here is presumably *phusis-*1. Thus, we have an implied *phusis-*2 *being* set against a direct invocation of *phusis-*1. This means that advocates of *phusis-*2 might also accept *phusis-*1 accounts, as well.

[36] Sextus Empiricus, *Outlines of Pyrrhonism* I. 8–9.
Ἔστι δὲ ἡ σκεπτικὴ δύναμις ἀντιθετικὴ φαινομένων τε καὶ νοουμένων καθ' οἱονδήποτε τρόπον. (This text follows R.G. Bury's Loeb edition: *Outlines of Pyrrhonism*, ed. and tr. R.G. Bury (Cambridge, MA: Harvard University Press, 1989).

[37] SEXTUS EMPIRICUS. *Dunamis* here refers to a capacity. It is possible to view nature as not entirely what we see. This "capacity" rests upon a critical attitude about causal connections we empirically experience.

[38] *Antithetike phainomenon te kai nooumenon*. It is not causally necessary to move from appearances to judgments. This leads to doubt—an essential property of epistemological anti-realists.

A second example of *Phusis₃* concerns Zeno of Elea.³⁹

> This reasoning, further, enables us to meet those who, in the terms of Zeno's argument, ask whether it is true that you must always go half-way to a point before you get there, and there is always a half-way point between the last half-way point that you have reached and the point itself you are making for, and so you can never get there, because you would have to pass through an infinite⁴⁰ number of points. [English Translation by P.H. Wickstead and F. M. Cornford (Loeb)]

> The same method should also be adopted in replying to those who ask, in the terms of Zeno's argument, whether we admit that before any distance must be traversed, that these half-distances are infinite in number and that it is impossible to traverse distances infinite in number. [English Translation by R. P. Hardie and R. K. Gaye (Oxford)]

These two translations of the same passage make essentially the same claim that (a) either it is true [that motion exists as we experience it] or (b) we have to admit that when we quantify motion along a geometric grid, that certain clearly held arithmetic principles exist that explain the geometry (which is closest to our empirical input).⁴¹ Thus, under our lived (geometrical) experience we uncritically accept what our senses present to us: viz., motion. If arithmetic is a derivative notion to geometry, then we can discuss parsing a line segment (by halves, in this case). Without any additional apparatus, when one is forced to add ($\frac{1}{2} + \frac{1}{4} + 1/8 + 1/16 + 1/32 + \ldots$) she is never going to get a sum of 1. This is what drives Zeno's Achilles paradox. Moderns see this sequence as an instance of a derivative that can be solved by calculus integration.⁴² Most ancients had no method to make this move.⁴³

Be that as it may, for the average educated Greek, Zeno's paradoxes of motion could not be logically refuted. They tended toward the Paramedian "One." But such a position is contrary to our experience of events in the world. Thus, Zeno's arguments encourage the adoption of a skeptical attitude toward what we perceive. Naïve empiricism requires that we accept as fundamental what we perceive as the

---

³⁹ Τὸν αὐτὸν δὲ τρόπον ἀπαντητέον καὶ πρὸς τοὺς [5] ἐρωτῶντας τὸν Ζήνωνος λόγον, [καὶ ἀξιοῦντας] εἰ ἀεὶ τὸ ἥμισυ διιέναι δεῖ, ταῦτα δ' ἄπειρα, τὰ δ' ἄπειρα ἀδύνατον διεξελθεῖν, Aristotle's *Physics* 263a 4–6 [On Zeno]
(this text follows W.D. Ross, *Physica* (Oxford: Clarendon Press, 1951).

⁴⁰ ZENO. What drives Aristotle's skepticism of Zeno's argument is its reliance upon the *apeiron*. (Though there is some controversy on this by those who supported Anaximander. But the *apeiron* in this context would make motion (that we all observe) to be impossible. Thus, in the tradition of epistemological anti-realists, Zeno would put into questions the empirical grounding of knowledge of the natural world, *phusis₃*.

⁴¹ It is generally accepted that within Greek mathematical thinking *geometry* was thought to be primary with *arithmetic* being derivative. For discussions of this position, see: Arpad Szabo, *The Beginnings of Greek Mathematics* (Dordrecht: D. Reidel, 1978), Jacob Klein, *Greek Mathematical Thought and the Origin of Algebra* (New York: Dover reprint, 1992), and Edward A. Maziarz and Thomas Greenwood, *Greek Mathematical Philosophy* (New York: Frederick Ungar, 1968).

⁴² See Gregory Vlastos who took this position in "Zeno's Race Course" *Journal of the History of Philosophy* 4.2 (1966): 95–108.

⁴³ Ian Mueller offers such a possibility within the realm of ancient mathematics via "mathematical exhaustion," see: Ian Mueller, *Philosophy of Mathematics and Deductive Structure in Euclid's Elements*" (Cambridge, MA: M.I.T. Press, 1981): 230–236.

ground of a developing realistic epistemology.[44] Thus, Zeno's sort of argument is very threatening to epistemological realists.

It will be the position of this book that to properly understand what those practicing natural philosophy in the Western Tradition understood as Nature, we have to make reference to these three categories. They will be instrumental in understanding arguments and developments in the accounts that are put forth.

The Symmetry/Elegance Criteria

A second key distinction that arose in natural philosophy in the ancient Greek world concerned an aesthetic understanding of and appreciation of the structuring of Nature. There are two ways that an observer of any object, p, can render aesthetic appreciation for p: qualitatively and quantitatively. Sometimes these can merge with one or the other with one being more prominent. In either case the epistemological dynamic is to *mentally* understand or judge an object on the basis of pre-determined understanding of form (*eidos*) from its shape (*morphe*). This qualitative judgment might be based upon how well the shape fulfills the supposed function of the form. When this presents itself to be excellent (*arete*) in the opinion of the observer, then the qualitative judgment contains not only a positive utility, but it is elegant.

Quantitative judgments are rather geometrically oriented. The geometrical figure can elicit interest in understanding how it is constructed and how it is measured. Let's look at some examples of each in ancient Greek science texts.

> Plato cites Pythagoras in the *Republic* 530d: It transpires, I said, that as the eyes are made for astronomy so the ears are made for harmony, and these are sister sciences, as the Pythagoreans say and we, Glaucon, agree.[45] [Translation by Kirk and Raven]

This passage emphasizes how astronomy (understood in mathematical terms)[46] connects with harmony (understood in musical terms) as united [as kindred]. Indeed, there is in mathematics (understood by the ancient Greeks with geometry as primary to arithmetic)[47] that which creates harmony which is known to be balanced symmetry. For Pythagoras, this balance is best known today through the Pythagorean

---

[44] Roderick Chisholm made such an argument in *Theory of Knowledge*, 2nd ed. (Englewood Cliffs, N.J.: Prentice Hall, 1977).

[45] Κινδυνεύει, ἔφην, ὡς πρὸς ἀστρονομίαν ὄμματα πέπηγεν, ὡς πρὸς ἐναρμόνοιν φορὰν ὦτα παγῆναι, καὶ αὗται ἀλλήλων ἀδελφαί τινες αἱ ἐπιστῆμαι εἶναι, ὡς ὅι τε Πυθαγόρειοί φασι καὶ ἡμεῖς, ὦ Γλαύκων, συγχωροῦμεν. (text follows John Burnet, Oxford).

[46] Since geometry combines the quantitative with the qualitative it is a perfect example here. For a discussion of this, see Daniel W. Graham, "The Geometry of the Heavens" in *Science before Socrates: Parmenides, Anaxagoras, and the New Astronomy* (Oxford: Oxford University Press, 2013): Ch. 7.

[47] Marvin J. Greenberg, *Euclidean and Non-Euclidean Geometries: Development and Histories*, 3rd ed. (New York: W.H. Freeman and Co., 1993): 6–19.

Theorem.[48] What the Pythagorean Theorem does is to show two important points about Nature: (a) that even irrational lines that cannot be represented by a whole number are nonetheless quantifiable exactly and (b) that the right triangle itself gives rise to a symmetry of complementary squares constructed from its sides.

Now in nature, side (y) might not be a whole number. Some could say: (a) that if it is not a whole number, then it could not be numerically described at all. This might count for an argument against numbers being able to describe everything. "And by the way of indicating this the Pythagoreans are accustomed sometimes to say, 'All things are like number.'" (tr. Kirk and Raven)[49] I take *note* here to mean Nature. Thus, every natural entity possesses and exhibits the properties of number. Pythagoras is on the side here of number being the proper descriptor of Nature. The theorem: $x^2 + y^2 = z^2$ is one sort of answer to this question. If geometry can be described arithmetically, then we are one step away from making geometry primary over arithmetic.[50] The Pythagorean Theorem requires a second-level arithmetical operation: square roots. Sometimes the square root is a whole number and sometimes it isn't. In cases of land taxation within states, if the area of one's property is not a whole number, then a dispute might arise.[51] This is because taxation was based upon the actual acreage one possessed. If one's property were not a proper rectangle, the square footage might not be a whole number. This would cause a dispute.

However, if arithmetic could set out a formula for an alternate number that would be between whole numbers, then some sort of compromise (probably favoring the state in taxes) could be achieved. Philosophically, this opens the door for a more gradated understanding of arithmetic, via the theory of exhaustion.[52] The point of importance from the point of view of symmetry and elegance is that even if the answer to measurement of side (z) is not whole, it is still determinate (given the theory of exhaustion). This lends an exactness to n/Nature.

A second walk away from the Pythagorean Theorem can be seen when the square root of (z) is a whole number. In this case the triangle in Fig. 1 can connect to other geometric figures to form a connection. For example, say side (x) is 3 units long and side (y) is 4 units long. Then side (z) would be 5 units long. This would allow for the construction of three rectangle squares (physically showing the squaring of each number) alongside each side of the triangle indicating the square of that side: 9 (3,

---

[48] For a discussion of the Pythagorean Theorem in the context of symmetry and elegance, see Nobel Prize Physicist Frank Wilczek, *A Beautiful Question: Finding Nature's Deep Design* (New York: Penguin, 2016): ch. 2.

[49] καὶ τοῦτο ἐμφαίνοντες οἱ Πυθαγορικοὶ ποτὲ μὲν εἰώθασι λέγειν τὸ "ἀριθμῷ δέ τε πάντ' ἐπέοικεν." Sextus, *adv.math.* vii, 94, (text follows Diels-Krantz, 1952).

[50] Most philosophers follow Dan Garber in assigning this complete project role to Descartes in his analytic geometry, see: Daniel Garber, *Descartes' Metaphysical Physics* (Chicago: University of Chicago Press, 1992): ch. 2.

[51] Some of these issues are raised in: Thorolf Christensen, Dorothy J. Thompson, Katelijn Vandorpe trs. *Land and Taxes in Ptolemaic Egypt: An Edition, Translation, and Commentary for the Edfu Land Survey (P. Haun. IV 70)* (Cambridge: Cambridge University Press, 2017).

[52] Mueller, 230–236.

Fig. 1 The symmetry of
the Pythagorean Theorem

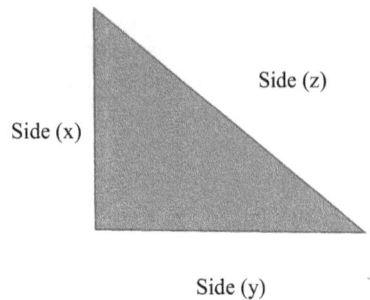

3, 3), 16 (4, 4, 4, 4), 25 (5, 5, 5, 5, 5). This would connect the triangle to rectangular squares and demonstrate that what is just set out connects to other figures. Such *connection* is another exhibition of symmetry as it hints at larger second-level connections in Nature.

These second-level connections that suggest a pattern to nature were also attributed to Pythagoras as being understood via *phusis*$_2$ and musically via harmony.

> What is the Oracle at Delphi? The *tetractus* which is the *harmony* which the Sirens sing is what is the finest. (my tr. after Kirk and Raven)[53]

Harmony in music understood via arithmetic/geometry via the *tetractus* (a four-tiered figure starting at the base with four points and then ascending to three, two, and one). This forms a triangle, a geometrical figure based upon arithmetic. For this to be connected to musical harmony is to make a connection between arithmetic/geometry and Nature via musical harmony. This is an example of *phusis* via elegance.[54]

In the history of the philosophy of science such connections have sometimes also been referred to as *simplicity* because the account offers a unified presentation in which the natural entities presented seem to be a part of a grand presentation that is unified.[55] Both arithmetic via its perceived completeness and coherence and geometry through its axiomatic approach[56] satisfy these requirements. "Well-ordered"

---

[53] τετρακτύς ὅπερ ἐστιν ἡ ἁρμονία, ἐν ᾗ αἱ Σειρῆνες. τὰ δὲ τί μάλιστα. Iamblichus, *Vita Pythagorae* 82 (DK 58c4).

[54] Any of the three senses of Nature (*phusis*) can exhibit elegance and symmetry as a positive evaluative characteristic.

[55] In the language of the simplicity literature in the history and philosophy of science, this amounts to both linguistic simplicity (the unifying manner of some particular presentation) and ontological simplicity (the *real* connections between natural entities in Nature). It does not deal with epistemological simplicity as that is relative to the observer and that some intricate connections may be difficult to ferret but once done, make sense and have wide explanatory power. For an excellent discussion of these points, see two books by Elliott Sober, *Simplicity* (Oxford: The Clarendon Press, 1975) and *Ockham's Razors: A User's Manual* (New York and Cambridge: Cambridge University Press, 2015).

[56] Mueller, see his comparison to the projects of Euclid and Hilbert in Chap. 1.

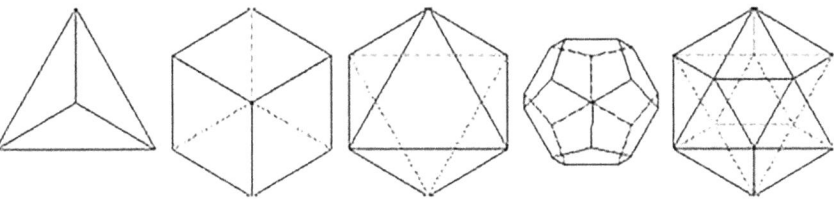

Fig. 2 Five regular platonic solids—tetrahedron, cube, octahedron, dodecahedron, icosahedron (Plato, *Timaeus*, 31a, ff.)

and "simple" will be taken to be similar concepts and both are related to *elegance* and *symmetry*.

Another example of this occurs in Plato's presentation of the five regular solids in the *Timaeus* 54b 7-55c 6. In this passage, Plato wants to assert that the underlying explanation behind the four elements (earth, air, fire, and water—from Empedocles[57]) geometrical properties. For example, triangles are sharp on all sides and might be involved with cutting results—like fire. This is not too different from the atomists who posited similar underlying geometrical explanation.[58]

The five regular solids are comprised with squares and triangles (Fig. 2).[59]

- The tetrahedron has four vertices and three faces coming together at each vertex
- The cube has six square faces (each of which could be constructed by two equilateral triangles) eight vertices, and three faces coming together at each vertex
- The octahedron has eight triangular faces, six vertices, and four faces coming together at each vertex
- The Dodecahedron has twelve pentagonal faces, twenty vertices, and three faces coming together at each vertex
- The icosahedron has twenty triangular faces, twelve vertices, and five faces coming together at each vertex

This represents *what* the five regular solids are (*ti esti*) but not *why* there are five and *only* five (*ei esti*). To answer this question, we must go back to our basis: the triangle. The way these triangles meet their vertices increases as we move from the tetrahedron to the octahedron, and then to the icosahedron[60] from three, four, to five triangles coming together. The next question to ask is what would happen if we moved six triangles together? This would form a "sixth regular solid." But unfortunately, this cannot be achieved in three-dimensional Euclidean Geometry. When we make this move, we have to get rid of the third dimension of *depth*. So, in order to

---

[57] Aetius I, 3, 20 and Aristotle, *Met.* 985a 31–33.

[58] Aetius I, 3, 18 (DK 68 A47) suggests that the properties of atoms depended upon their size and shape while Epicurus added 'weight.'

[59] Of course, Plato emphasizes triangles in this passage. Obviously, a square can be constructed by the conjunction of two identical right triangles.

[60] The cube and the dodecahedron will be either four or five depending on how you understand the underlying shapes to be united: the square and the pentagon.

go beyond "five" means creating two-dimensional depictions. For purposes of three-dimensional figures, five triangles fixed together in this way is the maximum.[61] Plato got it just right.

So, the existence of there being five and only five regular solids constructed on these limitations is an instance of symmetry that reveals to those who honor symmetry as a key factor in assessing scientific truth. We will see later how Johannes Keppler uses this symmetrical insight to speculate on the orbits of the planets in our solar system. Let us call this understanding of symmetry "mirror image symmetry." The reason we use the word "mirror" is that what happens on the right is the same structure as on the left with a "left-right" adaptation—just like a mirror.

The strength of this symmetry came from mathematics, particularly geometry. From Thales (one of the earliest recorded mathematical astronomers) through his student, Anaximander[62] there was an application of geometry to generate a theory of a celestial sphere above the earth that moved regularly for the stars and in a little more complicated way for the "wandering" planets.

This regularity put astronomy as a branch of mathematics/geometry which along with arithmetic and music represented this symmetrical branch of knowing.[63]

Another example of symmetry comes from biomedicine: Aristotle and the Hippocratic writers and Galen. In the case of Aristotle, we have many instances of his use of symmetry which sometimes refers to "proportionality" or "regular similarity" (*summetrias*) so that in the practice of biology one might observe a function in one sphere and connect it to another by analogy. For example, in discussing sex identification in conception theory Aristotle says (*G.A.* 767a 15-18):

> Male and female, then, differ generally with regard to each other in respect of the generation of male and female offspring on account of the causes which have been stated.[64] At the same time, they must stand in a right proportional relationship (*symmetrias*) to one another,[65] since everything that is formed either by art or by nature exists in virtue of some logical structuring. [my tr. after Peck][66]

Note here that Aristotle asserts that physiologically, males and female are quite a bit the same. The body parts that separate them must stand in some caused regular

---

[61] Wilczek, pp. 40–41.

[62] For a brief discussion of this, see Jeffrey Bennett, *Life in the Universe* (London: Pearson, 2017): 17–19.

[63] For a further discussion of this developing relationship see: David C. Lindbert, *The Beginnings of Western Science: The European Scientific Tradition in Philosophical, Religious and Institutional Context: Prehistory to AD 1450* (Chicago: University of Chicago Press, 2010): p. 86 ff.

[64] For a more detailed examination of these causes, see: Michael Boylan "The Galenic and Hippocratic Challenges to Aristotle's Conception Theory" *Journal of the History of Biology* 17.1 (1984): 83–112.

[65] Cf. *G.A.* 723a 30; 772 a 17; 777b 25.

[66] Διέστηκε μὲν οὖν ὅλως πρὸς ἄλληλα τό τε θῆλυ καὶ τὸ ἄρρεν πρὸς τὴν ἀρρενογονίαν καὶ θηλυγονίαν διὰ τὰς εἰρημένας αἰτάς, οὐ μὴν ἀλλὰ καὶ δεῖ συμμετρίας πρὸς ἄλληλα πάντα γὰρ τὰ γινόμενα κατὰ τέχνην ἢ φύσιν λόγῳ τινί ἐστιν. *G.A.* 779b 25–27. (Text follows A.L. Peck, (Loeb).).

Introduction

difference (*summetrias*) since both art (*techne*) and Nature (*phusis₁*) must follow a logical structuring in the formal cause (*eidos*) to bring about the purpose (*telos*).

Another example from Aristotle on how vision is best achieved via the eye G.A. 779b 25-27.[67]

> ... We should take it that the following is the cause of the phenomena [differential sight under various circumstances—like darkness]. Some eyes contain too much fluid, some too little to suit the right movement others contain just the right amount (*summetron*). [my tr. after Peck][68]

The point here is that many biological expressions occur on a continuum of "the more" and "the less." Whenever one is at the extreme on this line segment, they are generally wrong. Balance (here understood as symmetry) is a more reliable principle than the extremes (see Fig. 3).[69]

Figure 3 illustrates the role of the more and the less as a continuum. The ideal is at the center. As one goes to the extremes, one is less healthy. This has some analogues that have stayed with us in medicine up until the present day.[70]

In another example along these lines Aristotle talks about why people might get migraine headaches. He believes it is all about the blood being thin and clear rather than thick and muddy because of an imbalance of *pneuma* (a power that exists in the blood that can explain neurological functions and other essential powers necessary for life).[71]

> This explains why fluxing [pain] exists in the head. They occur in those parts of the brain that are colder than a *symmetrically proportioned* blend that should be. [My tr. after Peck][72]

So, a second prominent sense of *symmetry* is proper proportion. In Aristotle's biology and ethics this is the most common usage.[73] However, Aristotle also engages in mirror image symmetry. For example, when discussing body parts that occur on the

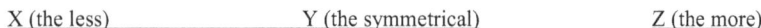

Fig. 3 The more, the less, and the symmetrical

---

[67] Aristotle is arguing against Empedocles here.

[68] ταύτην αἰτἀν ὑποληπτέον εἶναι τῶν εἰρημένων. οἱ μὲν γὰρ ἔχουσι τῶν ὀφθαλμῶν πλέον ὑγρον, οἱ δ' ἔλαττον τῆς συμμέτρου κινήσεως, οἱ δὲ σύμμετρον. (Text follows A.L. Peck (Loeb)).

[69] Note that this is also Aristotle's methodology in the *Nicomachean Ethics*: adopting the mean between extremes, cf. *NE* 2.5: 1105b 20-1107a 9.

[70] I evaluate good and bad uses of this paradigm in the context of present-day medicine in: Michael Boylan, "Health as Self-Fulfillment" in Michael Boylan, ed. *Medical Ethics* 2nd ed. (Oxford: Wiley-Blackwell, 2014): 44–57—compare to the Appendix which is based upon my earlier essay.

[71] See Michael Boylan, *The Origins of Ancient Greek Science* (New York and London: Routledge, 2015): 81, 87, 112, 118, 124.

[72] διὸ καὶ τὰ ῥεύματα τοῖς σώμασιν ἐκ τῆς κεφαλῆς ἐστι τὴν ἀρχήν, ὅσοις ἂν ᾖ τὰ περὶ τὸν ἐγκέφαλον ψυχρότερα τῆς συμμέτρου κράσεως. P.A. 652b 35-36 (Text follows A.L. Peck (Loeb)).

[73] Michael Boylan, *Method and Practice in Aristotle's Biology* (New York and Oxford: UPA/ Rowman and Littlefield, 1983): ch. 4.

right side of the body of an animal and also the left side, the explanation of this is given as mirror image symmetry.[74]

Another example of symmetry in the world of medicine is the humor theory coming from the Hippocratic writers and developed by Galen.

The Hippocratic Corpus writers used the concept of symmetry as proper proportion both externally and internally. In *Airs, Waters, and Places, Regimen,* and *Humours* there is an emphasis for the physician to take into account how the patient's extreme environment (see Fig. 3) might be a cause for disease.[75]

Internally, the four humors are: bood, phlegm, yellow bile, and black bile. Diocles believed more in the internal causes for diseases against those who were environmentally oriented. The balancing agent was *pneuma* (energized, vital air transformed within the body).[76]

The author of *Ancient Medicine* discusses balancing with various internal inputs such as the contraries: hot, cold, wet, and dry.

> For if there be such a thing as hot or cold or dry or wet which hurts a man, it necessarily follows that the scientific healer will counteract cold with hot, hot with cold, wet with dry and dry with wet. [my tr. after Peck][77]

Galen also developed his humor theory based upon internal balance as a critical feature for human health.[78] For these ancient practitioners of medicine, health (the object of the entire exercise) is about balance. Balance is one of the exhibiting characteristics of symmetry.

Completeness and Coherence

These two criteria for theory evaluation were not generally recognized, as such, outside the field of mathematics. Ian Mueller sets out how ancient axiomatic theory in geometry took on the criteria of completeness and coherence—particularly with regard to Euclid's *Elements*.[79]

---

[74] For example, see Aristotle's discussion of claws in crustaceans, *PA* 662b 1, 683b 30, 684a 15, 30.

[75] A. Thivel "*Cnide et Cos? Essai sur les Doctrines médicales dans la Collection hippocratique* (Paris: Les Belles Lettres, 1981): 279–383, provides numerous examples of this citing these texts.

[76] See 95, 98, 102 in P.J. van der Eijk, "Diocles and the Hippocratic Writings on the Method of Dietetics and the Limits of Causal Explanation" in R. Wittern and P.Pellegrin, eds. *Hippokratische Medzin und antike Philosophie* (Hildesheim, Zurich, and New York: Olms Weidmann, 1966).

[77] εἰ γάρ τί ἐστιν θερμὸν ἢ ψυχρὸν ἢ ξηρὸν ἢ ὑγρὸν τὸ λυμαινόμενον τὸν ἄνθρωπον, καὶ δεῖ τὸν ὀρθῶς ἰητρεύοντα βοηθεῖν τῷ μὲν θερμῷ ἐπὶ τὸ ψυχρόν, τῷ δὲ ψυχρῷ ἐπὶ τὸ θερμόν, τῷ δὲ ξηρῷ ἐπὶ τὸ ὑγρου, τῷ δὲ ὑγρῷ ἐπὶ τὸ ξηρόν. *Ancient Medicine* XIII, 2–8 (text after Peck (Loeb)).

[78] Galen, *On the Natural Faculties,* Kühn (vol. II) as edited by Georg Helmreich; (Leipzig:Teubner, 1893): 2.8.107; 2.8.117; 2.9.134.

[79] Ian Mueller, *Philosophy of Mathematics and Deductive Structure in Euclid's* Elements" (Cambridge, M.A.: Massachusetts Institute of Technology Press, 1981).

However, in the modern era (the seventeenth century forward—Part Two in this book) these two criteria become more important—also because the mathematics has developed and these two criteria are often featured as part of the explanans of artificial, mathematical systems.

The Maker/Artifact Dynamic

The origin of this dynamic is cosmological. What I mean by "cosmology" is as one of the two sub-categories of *metaphysics*. The first sub-category is *ontology*, the study of what things *are* (the *ti esti*). Once we have the things that exist on the table, we can *rank* them as to their relative power/sophistication (*ei esti*)—(cosmology). This creates what is known as *the great chain of being*.[80] I suggest that there are two important strains to this great chain of being: (1) that factually there are groups of beings that vary in power and sophistication so that the more powerful are thought to be "higher" than other living entities on the list (living entities are automatically set to be higher than non-living entities). This dynamic is one of the *inferior* (in power dynamics) viewing a *superior* in power; (2) that from the vantage point of the top → down, that all are the result of a creative force, Nature$_2$, so that Nature$_2$ is honored for being the creator of all. This dynamic is rather one of thanksgiving for one's existence via the agency of another.

Behind these distinctions is the notion of Nature, as an existent ontological entity, being contrasted with things that are created "the artifact." The making in the human realm was connected to a process described as an art or *techne*. Nature was set out as primary to human making because humans do not make themselves.[81] This assessment relies upon a principle about *making*: "There is a priority in *making* that gives to the maker priority over the artifact created (that which is made)."[82] This relationship does not depend upon *knowing* the identity of the maker. For example, Nature (as an all-encompassing material system affecting of all living things and that which supports them) exists and, within this systemic structure, brings about the conditions whereby life and its supporting components: air, land, and water, come-to-be, and are sustained. Humans are a part of this grand system. But Humans did not create it. It was created by some systemic material agency—since by the Principle of Sufficient Reason (a seventeenth century construct—see Part Two), everything that *is* must owe its existence to some material, causal explanation.[83] Under this account the material system, Nature$_1$ (perhaps via the agency of Nature$_2$)

---

[80] See especially Arthur O. Lovejoy, *The Great Chain of Being: A Study of the History of an Idea* (Cambridge, MA: Harvard University Press, 1971 [1936]).

[81] This could apply to Nature$_1$ as well, but the ontological attribution is more diffuse.

[82] I put these two relations as the parts of metaphysics as I characterize them: ontology (the things that *are*) and cosmology (the priority relation between the things that *are*)—see Michael Boylan, *The Good, The True, and the Beautiful* (London: Bloomsbury, 2009): ch. 5.

[83] See: Gottfried Wilhelm Freiheer von Leibniz, *Discourse on Metaphysics*, ed. Albert R. Chandler, trans. George Montgomery (Lasalle, IL: Open Court, 1924, rpt. 1902).

is understood to be the named entity that is responsible for our very human existence. As such, under the hierarchy of the making and the output of the making (things made), we owe a level of respect and precaution when dealing with the operation of Nature's systems when we go about *our* making of artificial devices (*teche* → technology). In a situation in which there was a conflict between human making and the ahistorical, Argument 1:[84]

Argument 1 The Priority of the Maker over the Maker's Output

1. There is a priority in the process of making that puts the *maker* above the output of the *maker: things made*—A
2. The status of priority in the process of making means that the operation of and the interests in *the maker* always trump the operation of and the interests in *the things made*—A
3. Nature's relation to Humans is as *maker* to output *made*—F
4. The operation of and the interests in Nature's systemic operation always trumps the operation of and the interests of Human's own *making*—1-3
5. The operation of and the interests in Human artifacts (technology) are always subservient to the ultimate interests of Humans—1, 4

---

6. The operation of and interests in Human artifacts are always subservient to the operation of and interests in Nature's systemic operation—4,5

Argument 1 has some interesting corollaries that transcend the centuries. Among these are that if we rank the priority normatively, from the vantage point of the thing made, the maker is (in an important sense) normatively good. This would make Nature (from the point of view of Humans and all other living things) *good*. In turn, if Nature is good and if any activity created by Humans harms the operation of Nature, then that human activity is *bad* and ought *not* be adopted. This gradation of the evil is a function of how much humans harm Nature.[85]

A few examples of these two dynamics will be instructive in making this clear in this Introduction to the book. These examples will require us to return to the ancient Greek world to express a foundational beginning to this way of thinking

The Maker/Artifact via Power and Sophistication

Aristotle argues that the power (*dunamis*) that certain animals have through their body parts is produced by general structures that are logically and chronologically prior by natural$_1$ design to their presentation within a given individual situated in a species rather than mere accidental presentations that attach to individuals, as such.

---

[84] For the purposes of this argument, let's think of Nature (*phusis*) as proximately Nature$_1$ with possible ultimate control by Nature$_2$ (*phusis$_2$*).

[85] This is obviously more cogent to humankind in the modern age. See my edited volume: Michael Boylan, ed. *Environmental Ethics,* 3rd ed. (Oxford: Wiley-Blackwell, 2020).

Introduction   xxxiii

> Empedocles was wrong when he said that many of the characteristics which animals have are due to some accident in the process of formation, as when he accounts for the vertebrae of the backbone by saying "the fetus gets twisted and so the backbone is broken into pieces."[86] He was unaware that: (a) the seed which gives rise to the animal must, to begin with, have the appropriate *specific* character, and (b) the producing agent was pre-existent—both logically and chronologically. (Aristotle, *Parts of Animals,* 640a20-25, tr. mine after Peck).[87]

This quotation supports a position that animal parts come about from a logically based, material process (*phusis*₁) that controls development in the logical and genetic order. In the case of the spine, a non-segmented spine would be very limited in power to perform certain tasks.[88] Thus, animals with segmented spines would be more powerful than those without segmented spines—particularly the invertebrates—especially those with cords in the back instead of spines. These processes work on species (*eidos*) of animals such that their characters are set by design and not accidental (as Empedocles contended).

Some of these distinctions are larger than others. These larger groupings Aristotle terms as *genos*. For example, among all animals, Aristotle distinguishes between the "blooded" and the "non-blooded."[89] This distinction roughly follows the vertebrate/invertebrate distinction set out in Linnaeus' system. And like Linnaeus, the systematic structure favors humans at the top of the power structure.

The way that we know that humans are more powerful is that they are able to capture and kill all other animals so that (on the basis of power) humans are at the top of the pyramid. Other vertebrates, too, generally outdo the invertebrates, so that they as a group are more powerful and thus *supervene* those species below them.

Since all humans, animals, and plants are within the realm Nature (and are constrained by the teleological and material rules that apply to them),[90] a recognition of the *logos* of the *scala naturae* would give a place of distinction to the force that created this structure. Thus, one form of the maker/artifact distinction would view Nature (in one of its three forms) as being that controlling logical force that sets out the logical/genetic order as it is. Aristotle's *phusis*₁ account of this is due to the material cause of heat.

---

[86] DK 31 B97.
[87] διόπερ Ἐμπεδοκλῆς οὐκ ὀρθῶς εἴρηκε λέγων ὑπάρχειν πολλὰ τοῖς ζῴοις διὰ τὸ συμβῆναι οὕτως ἐν τῇ γενέσει, οἷον καὶ τὴν ῥάχιν τοιαύτην ἔχειν ὅτι στραφέντος καταχθῆναι συνέβη, ἀγνοῶν πρῶτον μὲν ὅτι δεῖ τὸ σπέρμα τὸ συνιστὰν ὑπάρχειν τοιαύτην ἔχον δύναμιν, εἶθ' ὅτι τὸ ποιῆσαν πρότερον ὑπῆρχεν οὐ μόνον τῷ λόγῳ ἀλλὰ καὶ τῷ χρόνῳ. Aristotle, *Parts of Animals,* 640a20-25.
[88] Boylan (1983): 225–229.
[89] For blooded animals, see *P.A.* 642b10, 650b25, 665a25, 676b 10f. 678a5, 685b35ff. For non-blooded animals see: *PA* 642b10, 650b25, 673a30, 678a25ff.
[90] For the argument for this, see Boylan (1983): 87–127; Boylan (2016): 90–112.

> ... Nature₂'s rule is that the perfect offspring shall be produced by the more perfect sort of parent. Those animals which are hotter (based on their having a lung) [are better].[91] (Aristotle, *Generation of Animals* 732a 3-4, translation mine after Peck)

The non-blooded animals have no lungs and so are more "cold."[92] Aristotle thought that the material account behind the "higher" animals being better was *heat*. Even among homo sapiens, Aristotle thought that men were hotter than women and therefore are more sophisticated, i.e., *better*.[93]

Aristotle describes this force as it relates to making (*poein*) in the first chapter of the *Parts of Animals*: "The origin [of a thing formed] is the *logos* in the production of things in both art and nature₂" (*P.A.* 639b16-17; my tr.).[94] Thus, like the previous passage nature's productive process puts some species above others by being more perfect in the production (meaning they can do more and are more powerful).

Another passage that makes this point comes from Aristotle's *peri psuche* (*de anima*):

> Now some of the faculties of the soul which we have mentioned, some living things, as we have said, have all, others only some, and others again only one. Those which we have mentioned are the faculties for nourishment, for appetite, for sensation, for movement in space, and for thought. (Aristotle, *peri psuche* 414a 29-32, translation Peck)[95]

This passage gives an account of *why* humans are at the top of the *scala naturae*: they have all the powers of the soul while plants only have one and animals have two (of different degrees according to sensation and locomotion). Humans have more powers of the soul. This makes them more sophisticated and thus they are better.

The flip side to this comes from the concept of potentiality/actuality. To have potentiality only is to be less powerful than another entity that possesses more levels of potentiality that actualize. A potentiality that is (or cannot be) actualized can also be thought of as a privation. Thus, if entity A has more unactualized potentiality than entity B, then A exhibits more privation than B and is inferior to B.

This sort of argument occurs in Aristotle's *Metaphysics*:

> We speak of "privation": (a) In one sense if a thing does not possess an attribute which is a natural possession, even if the thing itself would not naturally possess it; e.g., we say that a vegetable is "deprived" of eyes. (b) If a thing does not possess an attribute which it or its genus would naturally possess, e.g., a blind man is not "deprived" of sight in the same sense that a mole is; the latter is "deprived" in virtue of its genus, but former in virtue of himself. (c) If a thing has not an attribute which it would naturally possess, and when it would naturally possess it (for blindness is a form of privation; a man is not blind at *any* age, but only

---

[91] ... οὕτως τὸ τέλειον ἐκ τοῦ τελειοτέρου γίνεσθαι πέφυκεν. τὰ δὲ θερμότερα μὲν διὰ τὸ ἔχειν πλεύμονα, ξηρότερα δὲ τὴν φύσιν. Aristotle, *Generation of Animals* 732a 3–4.

[92] One of the contraries: hot, cold, wet, and dry.

[93] Michael Boylan, "The Galenic and Hippocratic Challenges to Aristotle's Conception Theory" *Journal of the History of Biology* 17.1 (1984): 83–112.

[94] ἀρχὴ δ' ὁ λόγος ὁμοίως ἔν τε τοῖς κατὰ τέχνην καὶ ἐντοῖς φύσει συνεστηκόσιν.

[95] Τῶν δὲ δυνάμεων τῆς ψυχῆς αἱ λεχθεῖσαι τοῖς μὲν ὑπάρχουσι πᾶσαι, καθάπερ εἴπομεν, τοῖς δὲ τινὲς αὐτῶν, ἐνίοις δὲ μία μόνη. δυνάμεις δ' εἴπομεν θρεπτικόν, ὀρεκτικόν, αἰσθητικόν, κινητικὸν κατὰ τόπον, διανοητικόν. Aristotle, *peri psuche* 414a 29–32.

if he lacks sight at the age when he would naturally possess it), and similarly if it lacks an attribute in the medium and organ and relation and manner in which it would naturally possess it.... For we call a thing "unequal" because it does not possess equality (though it would naturally do so) ... (Aristotle, *Metaphysics* V, 1022b 22-35, translated by H. Tredennick).[96]

In this passage Aristotle distinguishes between natural$_1$ privation—like a mole not possessing eyes and therefore not possessing sight—from an accidental privation (a human losing their vision). The former case of privation means that some species do not have parts that can give them power (*dunameis*). This makes them inferior in the *scala naturae* due to a privation of potentialities that they cannot actualize. Those species that are not held back by privation are more powerful and sophisticated: better.

In these above examples Aristotle shows how the *scale naturae* can be justified on the basis of power and sophistication.

The Maker/Artifact and the Honoring of Nature

The second aspect of the maker/artifact depicts humans showing a deference to Nature$_2$ due to the power that this form of Nature possesses. This relationship is built upon religion and a recognition of Divinity. In religion, one generally posits an entity that is more powerful than humans. If humans are atop the *scala naturae*, then above that is the cause of nature$_1$ (as high nature or super nature$_2$[97] that acts either as creator or controller: god(s).

The creator role which elicits praise and thanksgiving is set out by Hesiod in *Theogony*:

> Hail, children of Zeus! Grant lovely song and celebrate the holy race of the deathless gods who are forever, those that were born of Earth and starry Heaven and gloomy Night and them that briny Sea did rear. Tell how at first gods and earth came to be, and rivers, and the boundless sea with its raging swell, and the gleaming stars, and the wide heaven above, and the gods who were born of them, giver of good things, and how they divided their wealth, and how they shared their honours amongst them, and also how at first, they took many-folded Olympus. These things declare to me from the beginning, ye Muses who dwell in the

---

[96] Στέρησις λέγεται ἕνα μὲν τρόπον ἂν μὴ ἔχῃ τι τῶν πεφυκότων ἔχεσθαι, κἂν μὴ αὐτὸ ᾖ πεφυκὸς ἔχειν, οἷον φυτὸν ὀμμάτων ἐστερῆσθαι λέγεται, ἕνα σὲ ἂν πεφυκὸς ἔχειν, ἢ αὐτὸ ἢ τὸ γένος, μὴ ἔχῃ, οἷον ἄλλως ἄνθρωπος ὁ τυφλὸς ὄψεως ἐστέρηται καὶ ἀσπάλαξ, τὸ μὲν κατὰ τὸ γένος, τὸ δὲ καθ' αὐτό. ἔτι ἂν πεφυκὸς καὶ ὅτε πέφυκεν ἔχειν μὴ ἔχῃ (ἡ γὰρ τυφλότης στέρησίς τις, τυφλὸς δ' οὐ κατὰ πᾶσαν ἡλικίαν, ἀλλ' ἐν ᾗ πέφυκεν ἔχειν, ἂν ᾗ πεφυκός καὶ καθ' ὃ καὶ πρὸς ὃ καὶ ὡς, ἂν μὴ ἔχῃ πεφυκός... ἄνισον μὲν γὰρ τῷ μὴ ἔχειν ἰσότητα πεφυκὸς λέγεται.... Aristotle, *Metaphysics* V, 1022b 22-35.

[97] "High nature" would be to live *in* nature but at a higher level—such as atop Mount Olympus. To be "super nature" would be to live in a different realm—e.g., either in the super-lunar realm of the stars or even beyond that, cf. Aristotle, *Metaphysics* Book XII.

house of Olympus, and tell me which of them first came to be. (Hesiod, *Theogony* ll. 104-115, tr. H.G. Evelyn-White).[98]

In this passage we see an argument for praise and thanksgiving to the gods (Zeus and the other gods) because of their creative acts from which all humans benefit. The gods shared their power with the humans (which they didn't have to do), therefore; they were exalted (sophisticated) and from this position wielded power in the interests of humans. For this altruistic act, the gods gave a gift to humans and deserve something back: praise and thanksgiving.

When we turn from the general attitude as depicted by the poet to the general needs of the populace via the confrontation of evil and disease, there arises the invocation of help from the *maker*, Zeus.

> But countless other miseries roam among mankind; for the earth is full of evils, and the sea is full; and some sicknesses come upon men by day, and others by night, of their own accord, bearing evils to mortals in silence, since the counsellor Zeus took their voice away. Thus, it is not possible in any way to evade the mind of Zeus. (Hesiod, *Works and Days* ll. 100-104, translated by Glenn W. Most)[99]

There are miseries and diseases among humans. The sicknesses come of their *own accord* (*phusis$_1$*) but Zeus is in the background and can act if confronted in the right way. This is a text that sets out disease via *phusis$_1$* but reminds individuals that recognition of *phusis$_2$* might make things right again. This sort of recognition is from the "made" (humans) to the "maker" (Zeus).

Zeus needed to be recognized. This was *proper* and could avoid further difficulties (disease or other troubles).[100] If proper attention to the disposition of the gods was not made, then problems might show up. This reinforces the necessity of recognizing the "maker/made distinction."

Again, in the *Iliad* the poet (who had medical interests—possibly a physician)[101] sets out several situations in which disease (even if it has a *phusis$_1$* origin) must include *phusis$_2$* for its resolution. The role of Apollo that opens *Iliad* I is complex and sometimes *does* involve negotiation with the gods in such a way that the inferior

---

[98] Χαίρετε, τέκνα Διός, δ ότε δ'ἱμερόεσσαν ἀοιδήν. κλείετε δ'δ' ἀθανάτων ἱερὸν γένος αἰὲν ἐόντων, οἳ Γῆς τ' ἐξεγένοντο καὶ Οὐρανοῦ ἀστερόεντος, Νυκτός τε δνοφερῆς, οὕς θ' ἁλμυρὸς ἔτρεφε Πόντος. εἴπατε δ', ὡς τὰ πρῶτα θεοὶ καὶ γαῖα γένοντο καὶ ποταμοὶ καὶ πόντος ἀπείριτος, οἴδματι θυίων, ἄστρα τε λαμπετόωντα καὶ οὐρανὸς εὐρὺς ὕπερθεν [οἵ τ' ἐκ τῶν ἐγένοντο θεοί, δωτῆρες ἐάων] ὥς τ' ἄφευος δάσσαντο καὶ ὡς τιμὰς διέλοντο ἠδὲ καὶ ὡς τὰ πρῶτα πολύπτυχον ἔσχον Ὄλυμπον. ταῦτά μοι ἔσπετε Μοῦσαι, Ὀλύμπια δώματ' ἔχουσαι ἐξ ἀρχῆς, καὶ εἴπαθ', ὅ τι πρῶτον γένετ' αὐτῶν. Hesiod, *Theogony* ll. 104–115.

[99] ἀλλα δὲ μυρία λυγρὰ κατ' ἀνθρώπους ἀλάληται, πλείη μὲν γὰρ γαῖα κακῶν, πλείη δὲ θάλασσα, νοῦσοι δ' ἀνθρώποισιν ἐφ' ἡμέρῃ αἱ δ' ἐπὶ νυκτὶ αὐτόμαται φοιτῶσι κακὰ θνητοῖσι φέπουσαι σιγῇ, ἐπεὶ φωνὴν ἐξείλετο μητίετα Ζεύς. Hesiod *Works and Days* ll. 100–104.

[100] The possible anger of Zeus and/or other gods is critical both in sickness and in coming to terms with death—especially in the critical nine days after death in which a seer or priest might be consulted on whether the anger of the gods was involved. See: Brooke Holmes, *The Symptom and the Subject: The Emergence of the Physical Body in Ancient Greece* (Princeton, N.J.: Princeton University Press, 2010): 41–48; 79–83.

[101] H. Frölich, *Die Militärmedicin Homers* (Stuttgart: F. Enke, 1879).

status of humans is acknowledged and the "maker/made" paradigm (via honoring Nature$_2$ is reinforced).[102]

It will be the position of this text that both senses of maker/artifact dynamic (The Maker/Artifact via power and sophistication and The Maker/Artifact and the Honoring of Nature) present themselves as one examines historical texts of biomedicine.

Conclusion

This introduction is intended to ground the foundational interpretative principles that will be used to understand and evaluate this select history of Western medicine starting at its beginnings in the ancient Greek and Roman worlds. It will further be the position of this book to contend that some version of these central principles: (a) the tripartite understanding of n/Nature; (b) the symmetry/simplicity/ elegance criteria; (c) the completeness/coherence criteria; and (d) the maker/artifact dynamic will show themselves to be central in understanding and evaluating the historical development of general science in the Western tradition and its application in medical theory and practice.

---

[102] Homer, *Iliad* I, 8-474, cf. Plato, *Laws* 642d, and Thudydides, *History* 2, 47, 4. See also Vivian Nutton, *Ancient Medicine,* 2nd ed. (London: Routledge, 2013): chapter 3.

Part I
The Ancient Western World

Chapter 1
The Beginnings

Abstract This chapter seeks to explore key concepts in ancient Greek Medicine from the beginning of recorded history. Some key figures in general science are put forward as well as applications to biomedicine. One important distinction that is set out are the three senses of how nature is understood "to be." These are defined in this historical context, but it is suggested that these categories will be applicable as we progress in examining both the history of general science, but of medicine, in particular.

Keywords Nature · Hippocratic Writers · Early physicians · Material causes · Divine causes · Skeptical reaction to empirical evidence · Pre-Socratics · Thales · Empedocles · Aristotle · Homer · Theophrastus · Heraclitus · Pythagoras · Parmenides · Zeno · Anti-realism

This chapter will seek to set the formation of a shared community worldview concept of medicine.[1] To do this we will examine some of the earliest writings on medicine and natural philosophy and how a concept of Nature$_1$ developed from Nature$_2$ (common belief led by those of rank in within villages and cities) and in contest with Nature$_3$ (the realm of some philosophers who think that *skepsis* requires them to have very high standards for intellectual closure on issues of medical diagnosis and treatment). From this clash of ideas (as sketched out in the Introduction) an emerging sense of the medical canon comes-to-be in the persona of the Hippocratic Writers (Chap. 2). This canon has been influential in the development of medical knowledge and practice.

---

[1] By "shared community worldview" I what a well-defined social group takes to be accepted facts and values about the world. For a discussion of this concept see Michael Boylan, *Natural Human Rights: A Theory* (New York and Cambridge: Cambridge University Press, 2014): 107–109.

© The Author(s), under exclusive license to Springer Nature
Switzerland AG 2025
M. Boylan, *A Philosophical History of Western Medicine*,
https://doi.org/10.1007/978-3-031-97806-7_1

Homeric Poetry and Nature

In this section we will briefly examine a baseline account of sickness and injury that is, for the most part, biased toward Nature$_2$.

Let's begin at the beginning of the *Iliad*. The poem opens with problems between the gods and the Greeks.[2] The narrative starts 9 years after the start of the war. The Greeks have sacked a Trojan-allied town and captured two comely maidens: Chryseis and Briseis. Agamemnon claims Chryseis as his prize. When Agamemnon has to give her back to her father after a plague sent by Apollo, he turns his attention to Briseis (who is claimed by Achilles, the top warrior for the Greeks). This causes a conflict.

Skipping back to when Agamemnon claimed Chryseis, her father, Chryses, who was a priest of the god Apollo, begs Agamemnon to return his daughter and offers to pay an enormous ransom. When Agamemnon refuses, Chryses prays to Apollo for help.

Apollo sends a plague upon the Greek camp, causing the death of many soldiers. After 10 days of suffering, Achilles calls an assembly of the Achaean army and requests a soothsayer to reveal the cause of the plague. Calchas, a powerful religious seer, stands up and offers his services. Though he fears the self-interested retribution from Agamemnon, Calchas declares the plague to be a vengeful and strategic move by Chryses and Apollo. Agamemnon flies into a rage and says that he will return Chryseis only if Achilles gives him Briseis as compensation.

Agamemnon's demand humiliates and infuriates the proud Achilles. The men argue, and Achilles threatens to withdraw from battle and take his people, the Myrmidons, back home to Pythia. Agamemnon threatens to go to Achilles' tent in the army's camp and take Briseis himself. Achilles stands poised to draw his sword and kill the Achaean commander when the goddess Athena, sent by Hera, the queen of the gods, appears to him and checks his anger. Athena's guidance, along with a speech by the wise advisor Nestor, finally succeeds in preventing the duel.

That night, Agamemnon puts Chryseis on a ship back to her father and sends heralds to have Briseis escorted from Achilles' tent. Achilles prays to his mother, the sea-nymph Thetis, to ask Zeus to punish the Achaeans. He relates to her the tale of his quarrel with Agamemnon, and she promises to take the matter up with Zeus—who owes her a favor—as soon as he returns from a 13-day period of feasting with the Aethiopians. Meanwhile, the Achaean commander Odysseus is navigating the ship that Chryseis has boarded. When he lands, he returns the maiden and makes sacrifices to Apollo. Chryses' father, overjoyed to see his daughter, prays to the god

---

[2] For some brief discussions on some of these dynamics that are relevant here see: Wolfgang Kullmann, "Gods and Men in the *Iliad* and *Odyssey*" *Harvard Studies in Classical Philology* 89 (1985): 1–23; Daniel Turkeiltaub, "Perceiving Iliadic Gods" *Harvard Studies in Classical Philology* 103 (2007): 51–81; and Robert J. Rabel, "Apollo as a Model for Achilles in the *Iliad*" *American Journal of Philology* 111.4 (1990): 429–440.

to lift the plague from the Achaean camp. Apollo acknowledges his prayer, and Odysseus returns to his comrades. This is an obvious case of *phusis*$_2$.

But the end of the plague on the Achaeans only marks the beginning of worse suffering. Ever since his quarrel with Agamemnon, Achilles has refused to participate in battle, and, after 12 days, Thetis makes her appeal to Zeus, as promised. Zeus is reluctant to help the Trojans, for his wife, Hera, favors the Greeks, but he finally agrees. Hera becomes livid when she discovers that Zeus is helping the Trojans, but her son Hephaestus persuades her not to plunge the gods into conflict over the mortals.

This sort of dynamic (if it depicts any actual social attitudes of the times)[3] illustrates a *phusis*$_2$ orientation both in the depiction of the cause of a plague or in the conditions of warfare (causing war wounds) and in its solution.

To balance this causal inclination toward the gods, we can note in this discussion *Iliad* XI, 514–515 which describes Machaon as a healer, who can protect soldiers in war:

> A healer is a man worth many men in his knowledge
> Of cutting out arrows and putting kindly medicines on wounds.[4] (tr. Lattimore)

In this passage the general worth of the physician (in the context of war) is affirmed for a *techne* that requires knowledge and training. The authors of the *Iliad* also demonstrate some familiarity with the details of critical wounds.[5]

There are also passages that mix the talk of wounds with other sources of pain in war—such as infectious disease (plague).[6] This rounds out medicine from causes that are unique to war and violence: (for soldiers) wounds—to those that also afflict the general populace; (for the general populace) infectious disease. In the early stages of the history of medicine in the West, this sets out a two-pronged approach to the province of medicine: (a) correcting the effects of violence to the body

---

[3] For an historical viewpoint on this see Simon Trépanier, *Early Greek Theology: God as Nature and Natural Gods* (Edinburgh: Edinburgh University Press, 2010) and for insight into the normative insights of day-to-day Atheneans see: Michael Boylan, ed. *Living the Good Life:* 'Virtue' and 'Goodness' in the Social and Political Philosophy of Ancient Greece: The Philosophy of A.W.H. Adkins (Cambridge Scholars Press, forthcoming, 2022).

[4] ἰητρὸς γὰρ ἀνὴρ πολλῶν ἀντάξιος ἄλλων/ ἰούς τ' ἐκτάμνειν ἐπί τ' ἤπια φάρμακα πάσσειν. *Iliad* XI, 514–515

Some latter Greek commentators have questioned the second line as not being "Homeric" because it is not sufficiently laudatory of physicians (mentioning only Machaon). For a presentation of these arguments see: H. Erbse, *Scholia Graeca in Homeri Iliadem,,* vol 3 (Berlin: De Gruyter, 1974): 222–223.

[5] H. Frölich, *Die Militärmedicin Homers* (Stuttgart: F. Enke, 1879): 58–59; K.B. Saunders, "The Wounds in *Iliad* 13–16" *Classical Quarterly* 49 (1999): 345–363; and C.F. Salazer, *The Treatment of War Wounds in Graeco-Roman Antiquity* (Leiden: Brill, 2000): 126–158.

[6] For the general mixture emphasizing pain see: Brooke Holmes, "The *Iliad's* Economy of Pain" *Transactions of the American Philological Association* 137.1 (2007): 45–84. On the plague highlighted see: Daniel R. Blickman, "The Role of Plague in the 'Iliad'" *Classical Antiquity* 6.1 (1987): 1–10; Robert J. Rabel, "Apollo as a Model for Achilles in the *Iliad*" *American Journal of Philology* 14.4 (1990): 429–440.

(whether from war, from other acts of violence, or from accidents—including, for the most part, setting fractures and cauterizing wounds) or (b) correcting the effects of bodily disease (trial and error until the advent of the humor theory that connected the presentation of symptoms with a diagnosis and treatment). For ease in reference, let's call this *the early paradigm of medicine.*

Obviously, since the province of war is *violence* the *techne* of wound healing will take precedence over healing from disease.[7]

Another aspect of the difference between these two provinces of medicine can be summed up as the *external cause* versus the *internal cause.* In the case of war, virtually all the instances of injury are due to external causes. The most prominent of these would be the enemy's sword or spear. But they might also include interference by the gods.[8] Thus the intervention of the gods could affect the presence of war wounds or disease. This gives us a good contrast between a natural event that is fundamentally *caused* by humans: war wounds ($phusis_1$) or one fundamentally caused by the gods ($phusis_2$). The response in either event is the healer (representing $phusis_1$ or a soothsayer $phusis_2$).[9] Thus, the soothsayer might be brought forward—especially when the material healer cannot see his way to curing the patient before him.

Disease can be seen either as caused by external causes[10] or from internal causes. It is the job of the healer (one art among many)[11] to solve this. In this bifurcation it is suggested that we have both $phusis_1$ and $phusis_2$ healers. In some ways, this is not too different from our contemporary world when conventional medicine has struggled for a proven treatment in some infectious diseases (as in the AIDS pandemic, the Ebola epidemic, and the COVID-19 pandemic)[12] people turn to religion for hope and miracles ($phusis_2$). This is because humans seek *explanations* for everything.[13]

---

[7] For more on the dominance of violence in the context of the *Iliad* see: Simone Weil, Rachel Bespaloff, and Hermann Broch, *War and the Iliad*, tr. Mary McCarthy (New York: New York Review of Books, rpt. 2005) and James M. Redfield, *Nature and Culture in the Iliad: The Tragedy of Hector* (Durham, N.C.: Duke University Press, 1994).

[8] Walter Petersen, "Divinities and Divine Intervention in the *Iliad*" *The Classical Journal* 35.1 (1939): 2–16. Of course, Apollo plays a key role in some of the most important scenes in The *Iliad*. In Book 1, he brings a plague to the Acheaens for disrespecting his priest, Chryses, by kidnapping his daughter. This plague launches the conflict between Agamemnon and Achilles.

[9] For a discussion of this see: Louis Rawlings, "War and Religion" in *The Ancient Greeks at War* (Manchester, UK: Manchester University Press, 2007): ch. 9.

[10] See the Hippocratic work, *Airs, Waters and Places* for examples of this.

[11] Vivian Nutton, 1992: "Healers in the market place: towards a social history of Graeco-Roman medicine" in A. Wear (ed.), *Medicine in Society* (Cambridge: Cambridge University Press, 1992): 15–58 and Robert Arnott, "Healers and Medicines in the Mycenaean Greek Texts" in *Medicine and Healing in the Ancient Mediterranean* (Oxford: Oxbow Books, 2014): 44–53.

[12] Some of my thoughts on this can be found in my edited book: Michael Boylan, ed. *Ethical Public Health Policy Within Pandemics* (Cham, Switzerland: Springer, 2022): 13–42.

[13] In philosophy this is called "the principle of sufficient reason" that was elaborated by Gottfried Leibniz, *New Essays on the Human Understanding and Other Philosophical Writings,* ed. Robert Latta (Oxford: Oxford University Press, 1948): 357–85.

This epistemological principle is very important to affirm. It is this writer's judgment that it's better to present even an empty placeholder as an explanation than it is to assert that no possible explanation can be given—viz., it just spontaneously occurred.[14]

So, we leave the era of the seventh to eighth centuries BCE with a mixture of *phusis*$_1$ and *phusis*$_2$ accounts, that because of the limited development of treatment in the shared community understanding of medicine as disease, accidents, and wounds due to violence is still forced to view the soothsayer as the primary resource for some medical situations in *the early paradigm of medicine*.

A Shift to Material *Arche*: Thales and Empedocles

In most histories of pre-Socratic philosophy Thales of Miletus is brought forward as offering a different intellectual starting point (*arche*) for the practice of natural philosophy. Part of this has to do with Thales setting out a material entity, water, as the primary component (*arche*) of the world, *phusis*$_1$. The other aspect is Thales' use of mathematics, particularly geometry, to help explain nature—particularly the movements of heavenly bodies.[15]

Because Thales is said to have predicted the solar eclipse of 585 BCE (DK 11A5), it seems reasonable to date him as a sixth century BCE figure. By tradition, he is said to have travelled to Egypt (KRS 67, 68).[16] Egypt, in this period of history was a center of *phusis*$_1$ - based natural philosophy.[17] His written works seem to have

---

[14] Cf. Aristotle's account of some insect generation in *History of Animals* 551a 13.

[15] For a survey of these two attitudes (Thales as setting out a material *arche* and Thales using mathematics—particularly geometry—to explain nature) see: Susan W. Kline, "The First Philosopher of the Western World" *The Classical Journal* 35.2 (1939): 81–85; Bruno Snell, "Die Nachrichten über die lehren des Thales und die Anfänge der griechischen Philosphie und Literaturgeschichte" *Philogus* 96 (1944): 170–182. D.R. Dicks, "Thales," *Classical Quarterly* 9.2 (1959): 294–309; Maria Machela Sassi, "Thales, Father of Philosophy?" in *The Beginnings of Philosophy in Greece* (Princeton, N.J.: Princeton University Press, 2018): 1–3; Eugene Jost and Eli Maor, *Beautiful Geometry* (Princeton, N.J.: Princeton University Press, 2014): 1–3; Gerald Feinberg, "Physics and the Thales Problem" *The Journal of Philosophy* 63.1 (1966): 5–17; and Dirk L. Couprie, "How Thales was able to 'Predict' a Solar Eclipse without the Help of Mesopotamian Wisdom" *Early Science and Medicine* 9.4 (2004): 321–337.

[16] DK refers to Hermann Diels and Walther Kranz, *Die Fragmente der Vorsokratiker*, 6th ed. (Berlin: Weidmann, 1951) and KRS refers to G.S. Kirk, J.E. Raven and M. Schofield, *The Presocratic Philosophers*. 2nd ed. (Cambridge: Cambridge University Press, 1995), cf. also LM (with their internal notation) which refers to André Laks and Glenn W. Most, *Early Greek Philosophy* (London and Cambridge, MA: Loeb, 2016).

[17] For a medical example of this see: Roger Forshaw, "Before Hippocrates. Healing Practices in Ancient Egypt" in *Medicine, Healing, and Performance* (Oxford: Oxbow Books, 2014): 25–41. Regarding Thales, it should be noted that as one of the seven sages of the ancient world, it was often part of their equipage that they made a trip to Egypt, cf. DK 11 b2, 11a18 these astronomical calculations are valid for Egypt but not for Greece. This is another possible piece of evidence of Thales having travelled to Egypt.

been destroyed rather early so that much of his influence is perhaps more oral than written:

> According to some, he [Thales] did not leave behind a written treatise [. . .] Diogenes Laertius (DK 11; LM D1)[18]
>
> . . . we are not able to demonstrate on the basis of a treatise by Thales that he declared that water was the only element, even if this is what everyone believes. Galen, *Commentary on Hippocrates'* On the Nature of Man (LM D2)[19]

Assuming that there is a mixture of some texts that were extant along with oral transmission, we can have some confidence in Aristotle's attribution of Thales' position that the *arche* was water.

> Others say that the earth rests on water. For this is the most ancient account we have received, which they say was given by Thales the Milesian, that it stays in place through floating like a log or some other such thing (for none of these rests by nature on air, but on water)—as though the same argument did not apply to the water supporting the earth as to the earth itself. Aristotle, *de caelo* 294a 28 (tr. KRS)[20]

However, not all [of the earliest philosophers who assert that things come from a substrate] say the same thing regarding the number and kind of a principle of this sort. But Thales [. . .] says it is water (and it is for this reason that he declared that the earth rests upon water) [. . .]. Aristotle, *Metaphysics* 983b 18–22 (tr. LM).[21] Further, the power of this *arche* is set out by Hippolytus:

> [. . .] He said that the beginning of everything and its end is water. For it is out of this that all things are formed, when it solidifies and liquefies in turn, and all things rest upon it, and it is also from this that earthquakes, concentrations of winds, and the motion of the stars come [. . .] Hippolytus, *Refutation of All Heresies* (LM D4 tr. LM).[22]

Let's think a bit on the proposition that there must be an *arche*. This is an important cosmological issue. Since *arche* can be a temporal or a logical beginning, these must be addressed in order. First, is the point of a temporal beginning. Is there a beginning to the universe? There are certainly some natural philosophers who asserted this (Plato, *Timaeus*, 28a). Others, like Aristotle deny an origin of the universe but assert the "steady state" theory that the universe always was, is, and always

---

[18] καὶ κατά τινας μὲν σύγγραμμα κατέλιπεν οὐδέν [. . .]; Diog. Lart. 1.23.

[19] [. . .] ὅτι Θαλῆς ἀπεφήνατο στοιχεῖον εἶναι τὸ ὕδωρ, ἐκ συγγράμματος αὐτοῦ δεικνύναι οὐκ ἕξομεν, ἀλλ' ὅμως ἅπασι καὶ τοῦτο πεπίστευται.; Gal. *In. Hipp. Nat. hom.* 1.27

[20] οἱ δ' ἐφ' ὕδατος κεῖσθαι. τοῦτον γὰρ ἀρχαιότατον παρειλήφαμεν τὸν λόγον, ὅν φασιν εἰπεῖν Θαλῆν τὸν Μιλήσιον, ὡς διὰ τὸ πλωτὴν εἶναι μένουσαν ὥσπερ ξύλον ἤ τι τοιοῦτον ἕτερον (καὶ γὰρ τούτων ἐπ' ἀέρος μὲν οὐθὲν πέφυκε μένειν, ἀλλ' ἐφ' ὕδατοσ), ὥσπερ οὐ τὸν αὐτὸν λόγον ὄντα περὶ τῆς γῆς καὶ τοῦ ὕδατος τοῦ ὀχοῦντος τὴν γῆν; Arist. *De Cael.* 294a 28.

[21] τὸ μέντοι πλῆθος καὶ τὸ εἶδος τῆς τοιαύτης οὐ τὸ αὐτὸ πάντες λέγουσιν, ἀλλὰ Θαλῆς μὲν [. .] ὕδωρ φησὶν εἶναι (διὸ καὶ τὴν γῆν ἐφ' ὕδατος ἀπεφήνατο εἶναι) [. . .]; Ar. *Meta.* 983b 18–22.

[22] [. . .] οὗτος ἔφη ἀρχὴν τοῦ παντὸς εἶναι καὶ τέλος τὸ ὕδωρ. ἐκ γὰρ αὐτοῦ τὰ πάντα συνίστασθαι πηγνυμένου καὶ πάλιν διανιεμένου ἐπιφέρεσθαί τε αὐτῷ τὰ πάντα, ἀφ' οὗ καὶ σεισμοὺς καὶ πνευμάτων συστροφὰς καὶ ἄστρων κινήσεις γίνεσθαι [. . .]; Hippol. *Ref.* 1.1.

will be (Aristotle, *de caelo* 279b 4f. and 279b 17–31).[23] Since the depiction of Thales' positions always comes from later writers, and since the passage I cited earlier on *arche* and water from the *Metaphysics* comes from Aristotle, an advocate of the steady state theory, might have taken an ambiguous Thales and *read him* as advocating the steady state that he, Aristotle, endorsed. Thus, the temporal understanding of *arche* in Thales is, at best, ambiguous.

Second, is there a logical point upon which all things depend? This is the second sense of *arche*. This would refer to an ever-present substrate that acts as the primal cause of everything else (whether or not it ever acted as an original creative force). This logical dependency is similar to the logical dependency of primary posits and rules of inference in axiomatic systems—such as geometry.[24] Under this interpretation water would be viewed as the *most important* of all the material elements in the universe. This could mean several different things. A. Water exists among other elements and is the most important; B. Water exists and is not only the most important, but it is *that out of which* any other material comes-to-be.

One text supporting this "B" reading of Thales comes from the Christian apologist, Minucius Felix.

> Let Thales of Miletus be first of all he who was the first of all to discuss celestial phenomena. This same Thales of Miletus said that water is the beginning of things, but that god is the mind (*mens*) that formed all things out of water. Minucius Felix, *Octavius* (LM R42, tr. LM).[25]

This text suggests that a *phusis*$_2$ force (*deus*) takes the material, water, and forms all things from that material. This would represent a mixed-mode source of genesis using both *phusis*$_1$ and *phusis*$_2$. This is the strongest text that suggests interpretation B.[26]

Another text slides closer to the A interpretation. It comes from the pseudo-Galen text, *Commentary on Hippocrates' On Humors*.

> Although Thales says that all things are constituted out of water, nonetheless he also wants this [i.e., that the elements are transformed into one another]. It is better to cite his own words from Book 2 of *On the Principles*, which are as follows: "Therefore the celebrated four, of which we say that the first is water and posit it as being as it were the only element,

---

[23] For a discussion of Aristotle's position see: Friedrich Solmsen, "Aristotle and Pre-Socratic Cosmology" *Harvard Studies in Classical Philology* 6.3 (1958): 265–283. For an historical treatment of the issue of "steady state" cosmology see: Adolf Grünbaum, "The Pseudo-Problem of Creation in Physical Cosmology" *Philosophy of Science* 56 (1989): 373–394.

[24] For an explanation of how this mechanism works see: Ian Mueller, *Philosophy of Mathematics and Deductive Structure in Euclid's* Elements (Cambridge, MA: M.I.T. Press, 1981).

[25] sit Thales Milesius omnium primus, qui primus omnium de caelestibus disputavit. idem Milesius Thales rerum initium aquam dixit, deum autem eam mentem, quae ex aqua cuncta formaverit; Min. Fel. *Octav.* 19.4.

[26] Charles H. Kahn in *Anaximander and the Origins of Greek Cosmology* (New York: Columbia University Press, 1960): 29–32, argues that using *arche* for material elements (as per the "B" interpretation) begins with Anaximander, who was a little younger than Thales. This is another argument for the "A" view for Thales.

mix with one another for the combination, solidification, and composition of the things in this world. How this happens we have already said in Book 1." (LM, R44, tr. LM).[27]

This text is a bit different from Minucius Felix as it sets out a model in which the root elements (presumably: water, earth, air, and fire) are mixable to form compounds—but that *water* is the most important of these (but not the *only* original element out-of-which all other elements come-to-be as *interpretation B* suggests). The pseudo-Galen text is interesting conceptually for a theory of composition involving mixing of the root elements into all that we observe. However, it is incomplete without some sense of *how* something fluid could account for things that were hard or earthy.

Nonetheless, it seems to this writer that despite these shortcomings, (A) is the most justifiable explanation among these two because of the existing accounts of Thales, there is no suggestion of a *process* whereby water would *become* earth, for example. Certainly, water becomes solid when it freezes, but that does not make it into an earthy substance. Indeed, the whole picture of earth (either flat or as a sphere—most think that Thales believed the earth to be a flat island floating in water, but this is controversial)[28] floating in water suggests that earth, as a substance, is different from water. However, that does not disallow that they can mix: as in rain on the ground increasing crop yield. *Gaeia*, as potentially fertile material, can be actualized to be more fertile by the mixture of water. In fact, without water, earth will be barren (pure potentiality only). Thus, water is a critical, vital factor for crop yield which is essential to maintain life. This primacy, that Thales emphasized, continues today and is especially prevalent in contemporary astrophysics in assessing the possibility of life on other planets.[29]

Now there are some who might assert that Thales was also committed to *phusis*$_2$ as per Aristotle's declaration in *de anima* 411a 7–8 "Some people say that it [i.e., the soul] is mixed in with the whole, which is perhaps also the reason why Thales thought that all things are full of gods."[30] (tr. LM). However, the context here is important. Aristotle also talked about Heraclitus warming himself by a stove in the kitchen, using a similar expression in the *Parts of Animals*, "Come on in and don't

---

[27] Θαλῆς μὲν εἴπερ καὶ ἐκ τοῦ ὕδατός φησι συνεστάναι πάντα, ἀλλ' ὅμως καὶ τοῦτο βούλεται. ἄμεινον δὲ καὶ αὐτοῦ τὴν ῥῆσιν προσθεῖναι ἐκ τοῦ δευτέρου Περὶ τῶν ἀρχῶν ἔχουσαν ὡδέ πως, τὰ μὲν οὖν πολυθρύλητα τέτταρα,ὧν τὸ πρῶτον εἶναι ὕδωρ φαμὲν καὶ ὡσανεὶ μόνον στοιχεῖον τίθεμεν, πρὸς σύγκρισίν τε καὶ πῆγνυσιν καὶ σύστασιν τῶν ἐγκοσμίων πρὸς ἄλληλα συγκεράννυται. πῶς δέ, ἤδη λέλεκται ἡμῖν ἐν τῷ πρώτῳ; Ps.-Gal. *In Hipp. Hum.* 1.1.

[28] For a discussion of these possibilities see Patricia O'Grady's article in the *Internet Encyclopedia of Philosophy*, https://www.iep.utm.edu/thales/ (accessed July 1, 2020).

[29] For a general example of this see: James F. Kasting, Ravikumar Kopparapu, Ramses M. Ramirez, and Chester E. Harman, "Remote Life-Detection Criteria, Habitable Zone Boundaries, and the Frequency of Earth-Like Planets around M and Late K Stars," *Proceedings of the National Academy of Sciences of the United States of America* 111.35 (2014): 12641–12646.

[30] καὶ ἐν τῷ ὅλῳ δέ τινες αὐτὴν μεμεῖχθαί φασιν, ὅθεν ἴσως καὶ Θαλῆς ᾠήθη πάντα πλήρη θεῶν εἶναι; Arist. *De an.* 411a 7–8.

be afraid, there are gods even in here" (*PA* 645a 23, my tr.).[31] It seems to me that there are several overlapping terms that can best be understood via the *phusis*₁ through perspective.

First, and most important is what to make of material events that have no visible causes. In the case of Thales, there is the case of the magnetic loadstone. Key passages on this are:

> Thales too seems, from what is reported, to have thought that the soul is something that moves, for he says that the stone [i.e., the magnet] has a soul, given that it moves iron. Aristotle, *On the Soul* 405a 19, LM D11-a (DK A22), tr. LM.[32]
>
> Aristotle and Hippias say that he attributed a soul to inanimate beings too, judging from the evidence of the magnet and of amber. Diogenes Laertius LM D11-b (DK A1), tr. LM.[33]

The soul (*psuche*), for the most part,[34] is understood by Aristotle via the terms: form (*eidos*) which structures matter (*hule*) both of which would be understood via *phusis*₁ rather than *phusis*₂—even though the forces involved are not visible.[35] Therefore, the fact that Thales attributed the power of the loadstone to *soul* does not mean that he was equating *soul* with *divinity*.

Second, is the more general usage of "gods" which refer to no specific god in Greek mythology. When "god" is equated to "soul" the sense is really "an invisible causal agent is involved." Call it "god" or call it "soul" it really is a *stand-in* for invisible causal principles. Soul definitely has a material meaning as per Aristotle (our primary source for the attribution to Thales), therefore; it is at least plausible that the attribution to Thales of nature being full of gods, can refer to *phusis*₁ principles that are not visible to the naked eye. This will be the interpretation of this author. However, it should be noted that there is some ambiguity here.

Thus, this presentation will put forth "A" as the proper way to understand how Thales viewed water as the *arche* for materials in nature and the cosmic environment of the earth, itself. This is decidedly a *phusis*₁ dominant worldview.

In Empedocles[36] (we have a very influential background thinker (and medical practitioner)[37] to Ancient Greek Medicine from the *phusis*₁ material point of view

---

[31] ἐκέλευε γὰρ αὐτοὺς εἰσιέναι θαρροῦντας εἶναι γὰρ καὶ ἐνταῦθα θεούς; Arist. *PA* 545a 23.

[32] ἔοικε δὲ καὶ Θαλῆς ἐξ ὧν ἀπομνημονεύουσι κινητικόν τι τὴν ψυχὴν ὑπολαβεῖν, εἴπερ τὸν λίθον ψυχὴν ἔχειν ὅτι τὸν σίδηρον κινεῖ; Arist. *De an.* 405a 19.

[33] Ἀριστοτέλης δὲ καὶ Ἱππίας φασὶν αὐτὸν καὶ ἀψύχοις μεταδιδόναι ψυχῆς, τεκμαιρόμενον ἐκ τῆς μαγνήτιδος καὶ τοῦ ἠλέκτρου; Diog. Laert. 1.24.

[34] Aristotle's possible exception in de anima for rational soul since it does not come from a physical organ: *De An.* 429a24–27.

[35] For a discussion of this in the context of Aristotle see: Michael Boylan, *Method and Practice in Aristotle's Biology* (Lanham, MD and London: UPA/Roman and Littlefield, 1983): 50–59; 224–225.

[36] Readers of Empedocles are encouraged to look at Jean-Claude Picot's wonderful bibliography of secondary works: https://sites.google.com/site/empedoclesacragas/bibliography-a-z (accessed September 1st, 2021).

[37] Though Empedocles is reputed to have been a healer, among other things—including a god—his works fit into this work as general scientific background and some general biological speculation.

(even though Empedocles did subscribe to various ideas about spiritual topics—such as reincarnation and purifying powers of vegetarianism).[38] According to Diogenes Laertius (DK 31 A1) Empedocles was probably born between 495 and 485 B.C. E. in Acragas (west coast of Sicily) and later moved to Syracuse (on the east coast of Sicily).

I will examine two groups of Empedoclean texts: first those describing the broad background conditions of science and then some texts that are more focused upon biology.

The first group begins with his famous poem "On Nature."[39]

> Twofold is what I shall say: for at one time they grew to be only one/ Out of many, at another time again they separate to be many out of one, / Fire, Water, Earth, and the immense height of Air; /And baleful Strife is separate from them, equivalent everywhere,/ And Love (*Philotês*) in them, equal in length and in breadth./ Look you upon her with your mind (*noos*)—and do not sit there with astounded eyes. Strassbourg Papyrus; Simplicius, LM D73, ll. 249–255, tr. LM.[40]

And

> Moreover, he was the first to make the material "elements" four. Aristotle, *Metaphysics*. DK 31a 37, tr. KRS.[41]
>
> Of the vortex, and Love comes to be in the center of the whirling, / Under her dominion all things [i.e., the elements] come together to be [only] one. Strassbourg Papyrus; Simplicius, LM D73, ll. 288–292, tr. LM.[42]

And

> This one [i.e., Empedocles] says that the corporeal elements are four, fire, air, water, and earth, which are eternal, in large or small quantity, but which change according to their union and separation; but the principles properly speaking, by which these move, are Love and Strife. For it is necessary that the elements continue to exchange their places recipro-

---

[38] See *The Purifications* (LM D4-D40).

[39] The two most famous works by Empedocles are "On Nature" (*peri phuseos*) and "Purifications" (*katharmoi*) which may be two works or two parts of the same work. For some of the issues on this question from the position that they are two poems see: Friedrich Wilhelm Sturz, *Empedocles Agrigentinus* (Liepzig: Van der Ben, 1805) and Friedrich Solmsen, "Hymn to Apollo" *Phronesis* 25 (1980): 219–227; and David Sider, "Empedocles' *Persika*" *Ancient Philosophy* 2 (1982): 76–78. For arguments that they are a part of one poem see: Catherine Osborne, "Empedocles Recycled" *Classical Quarterly* 37 (ns) (1987): 24–50 and Tom Mackenzie, "The Contents of Empedocles' Poem: A New Argument for the Single Poem Hypothesis" *Zietschrift für Papyrobgie und Epigraphik* Bd. 200 (2016): 25–32.

[40] δίπλ᾽ ἐρέω τοτὲ μὲν γὰρ ἐν ηὐξήθη μόνον εἶναι/ ἐκ πλεόνων, τοτὲ δ᾽ αὖ διέφυ πλέον᾽ ἐξ ἑνὸς εἶναι,/ πῦρ καὶ ὕδωρ καὶ γαῖα καὶ ἠέρος ἄπλετον ὕψος,/ Νεῖκός τ᾽ οὐλόμενον δίχα τῶν, ἀτ ἀλαντον ἀπάντηι,/ καὶ Φιλότης ἐν τοῖσιν, ἴση μῆκός τε πλάτος τε/ τὴν σὺ νόωι δέρκευ, μηδ᾽ ὄμμασιν ἧσο τεθηπώς P. Strasb.gr. Inv. 1665–66, v. 232–308, ed. Primavesi 2008; v. 233–266 (Simp. *In Phys.*, pp. 158.1–159.4, et. al.).

[41] ἔτι δὲ τὰ ὡς ἐν ὕλης εἴδει λεγόμενα στοιχεῖα τέτταρα πρῶτος εἶπεν. Aristotle, *Met.* A4 985a 31–3.

[42] δ[ίνη]ς, ἐν δὲ μέσ[ηι] Φ[ιλ]ότης στροφά [λιγγι] γένηυαι, / ἐν [τῆι] δὴ τάδε πάντα συνέρχεται ἓν μόνον εἶναι. P. Strasb.gr. Inv. 1665–1666, v. 232–308, ed. Primavesi 2008; v. 233–266 (Simp. *In Phys.*, pp. 158.1–159.4, et. al.).

cally, being at one time united by Love, at another time separated by Strife. So that the principles according to him are also six. For in certain passages, he attributes the efficient power to Strife and Love, when he says [D73. 239–240], but sometimes he assigns these too to the four as belonging to the same series, when he says . . . [D73. 248b–51]. Simplicius, *Commentary on Aristotle's Physics*. LM D 80, tr. LM.[43]

And

[. . .] as Empedocles says that the earth is at rest because of the vortex. Aristotle, *On the Heavens*. LM D117-a. tr. LM.[44]

And

[. . .] Others, like Empedocles, [scil. say] that the motion of the sky, moving in a circle and more rapidly than the motion of the earth, prevents the latter, like what happens to water in cups. For it is for the same reason that this latter too, although it often ends up being below the bronze, all the same does not move when the cup is moved in a circle,[45]

Empedocles' Cosmic Cycle

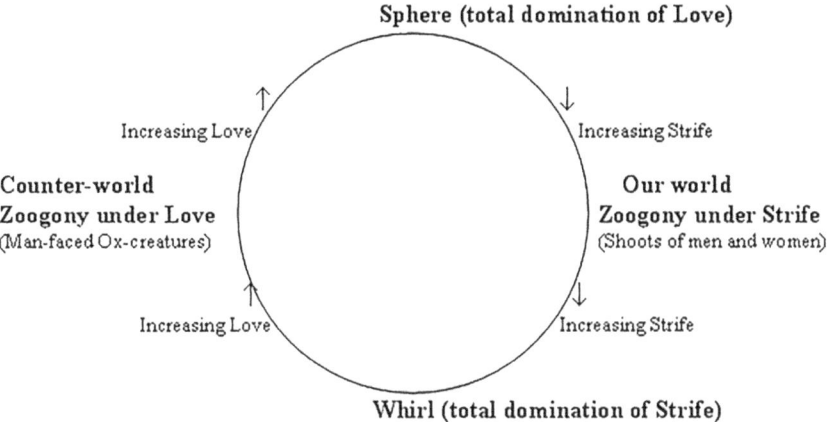

---

[43] οὗτος δὲ τὰ μὲν σωματικὰ στοιχεῖα ποιεῖ τέτταρα, πῦρ καὶ ἀέρα καὶ ὕδωρ καὶ γῆν, ἀΐδια μὲν ὄντα πλήθει καὶ ὀλιγότητι, μεταβάλλοντα δὲ κατὰ τὴν σύγκρισιν καὶ διάκρισιν, τὰς δὲ κυρίως ἀρχάς, ὑφ' ὧν κινεῖται ταῦτα, Φιλίαν καὶ Νεῖκος. δεῖ γὰρ διατελεῖν ἐναλλὰξ κινούμενα τὰ στοιχεῖα, ποτὲ μὲν ὑπὸ τῆς Φιλίας συγκρινόμενα, ποτὲ δὲ ὑπὸ τοῦ Νείκους διακρινόμενα, ὥστε καὶ ἓξ εἶναι κατ' αὐτὸν τὰς ἀρχάς. καὶ γὰρ ὅπου μὲν ποιητικὴν δίδωσι δύναμιν τῷ Νείκει καὶ τῇ Φιλίᾳ ὅταν λέγῃ [. . . = D73.239–40] [. . .] ποτὲ δὲ τοῖς τέτταρσιν ὡς ἰσόστοιχα συντάττει καὶ ταῦτα, ὅταν λέγῃ [. . . = D73.248 (τοτὲ δ') -51. Simpl. *In Phys.*, pp. 25.21–26.4.

[44] [. . .] καθάπερ φησὶν Ἐμπεδοκλῆς τὴν γῆν ὑπὸ τῆς δίνης ἠρεμεῖν. Arist. *Cael.* 300b 2–3.

[45] Gordon Campbell, "Empedocles" in the *Internet Encyclopedia of Philosophy* (accessed November 2, 2020) puts this usefully in the following chart:

even though its nature is to move downward. Aristotle, *On the Heavens*. LM D117-b. tr. LM.[46]

These five texts form a more comprehensive context that mixes clearly material elements (earth, air, fire and water), *phusis*$_1$, with non-material causal agents of synthesis and analysis, *phusis*$_2$ (Love and Strife). However, even here each side is also mixed with its complement. The *phusis*$_1$ material elements get linked with particular gods: Zeus, Hera, Aidoneus, and Nestis, and her tears (Aëtius 1, 3, 20, Fr. 6). If Nestis' tears associate with water, then there has always been some disagreement about which god to associate with the other three. Theophrastus suggested Zeus with fire, Hera with air, and Aidoneus (i.e., Hades) with earth (Aëtius 1,2, 20). In any event, the linking of material *phusis*$_1$ with gods *phusis*$_2$ is consistent with Empedocles' idea of constant mixing and change.[47]

If the cycle of analysis is driven by the Strife through the agency of whirling, then at the point of maximum dissolution, there is a point of calm in the center of the whirling. From this center, the whirling vortex starts to come together again under the agency of Love until maximum unity is achieved—though never perfect unity. Unlike Parmenides, "the one" is an unstable condition that is waiting to fall apart.

The result is a dynamic system of Nature that is neither *one* nor *many*, only. Rather, it is both, but via intentionally unstable systematics. This is a very suggestive thought that is especially cogent to the history of biomedicine, which is the principal concern of this book.

It should also be noted here that this vibrant understanding of Nature gives the verb "to be" (*einai*) an existential connection to its objects (*tauta*) as themselves (*auta*).[48] Because of this vibrancy of the existential understanding of "is" Empedocles can make his synthetic principles of *likeness* (Love) into a principle of epistemology via the process of analogy. Analogy is an important concept in the history of science—particularly in biology via the notion of similarity between different sorts of animals. The principle itself, is a priori (therefore *phusis*$_2$). But its use as a regulatory principle in material analysis will connect Empedocles to later thinkers who are examined in this book—especially in the seventeenth century and after.[49]

---

[46][. . .] οἱ δ' ὥσπερ Ἐμπεδοκλῆς τὴν τοῦ οὐρανοῦ φορὰν κύκλῳ περιθέουσαν καὶ θᾶττον φερομένην ἢ τὴν τῆς γῆς φορὰν κωλύειν, καθάπερ τὸ ἐν τοῖς κυάθοις ὕδωρ, καὶ γὰρ τοῦτο κύκλῳ τοῦ κυάθου φερομένου πολλάκις κάτω τοῦ χαλκοῦ γινόμενον ὅμως οὐ φέρεται, κάτω πεφυκὸς φέρεσθαι, διὰ τὴν αὐτὴν αἰτίαν. Arist. *Cael.* 295a 16–21.

[47]For one account connecting Empedocles' interest in material nature with divine nature see: Spyridon Rangos, "Empedocles on Divine Nature" *Revue de Métaphysique et de Morale* 3 *Les Dieux, Le Sacrifice et la Grâce* (2012): 315–338.

[48]This is set out in more detail by Denis O'Brien, "Empedocles on the Identity of the Elements" *Elenchos* 37.1/2 (2016): 5–28.

[49]The use of "likeness" via Love and the application of abstract analogy to concrete referents are brought out by: Rachana Kamtekar, "Knowing by Likeness in Empedocles" *Phronesis* 54.3 (2009): 215–238 and by Patricia Curd, "Where are Love and Strife? Incorporeality in Empedocles" from *Early Greek Philosophy* (Washington, D.C.: Catholic University Press of America, 2013): 113–138.

In the last text from this section material motion is taken on from one of the four elements, water. Empedocles seems to have contended that when a cup is moved in a circle, then the water within the cup will also create a waving function that mimics the cup. This is both true and false. It is true if the cup, itself, is set to whirling about a central axis. This would make the cup the "universe" in question and it would affect all within.

On the other hand, if the cup itself does not whirl, but is stationary, but (as stationary) is moved into a circulatory pattern by the person holding the cup and moving in a circle, then the water within the cup won't move at all. This is an important insight that either Empedocles or the existent texts fail to recognize. Given that Empedocles is suggesting a dynamic theory of change, such lapses suggest that despite his four material elements, he is really leaning toward $phusis_2$.[50]

The second group of texts concerns biology in particular—of interest to a healer.

> Empedocles [. . .] thought that the heart develops before all the other parts, because it is this on which the life of man depends most of all. Censorinus, *The Birthday.* LM D167, tr. LM.[51]

And

> The others say that it [i.e., the differentiation of the sexes] occurs in the womb, like Empedocles. For he says that what penetrates into a warm uterus becomes male, what into a cold one female, and that the cause of the warmth or coldness is the flow of the menstrual fluids, which is colder or warmer, and older or more recent. Aristotle, *Generation of Animals*, LM D173, tr. LM.[52]

And

> Empedocles: similarities [scil. With parents and ancestors] come about as a result of the preponderance of the spermatic seeds, dissimilarities when the heat in the sperm evaporates. Aëtius, LM D181, tr. LM.[53]

And

> Empedocles: [scil. The reason why some children are born resembling other people and not their parents is that] the shape of the embryos is determined by the imagination of the

---

[50] This inclination to change is sometimes described as indicating an early advocate of a sort of evolutionary theory. Sometimes this is also included Anaximander. For an overview of some of Anaximander's positions, see: Andrew Gregory, "Anaximender and the Ionian Contribution: Brilliant but No Miricle" from Hahn R. Herda, ed., *Ex Ionia Scientia: Knowledge in Archaic Greece* (Cambridge, MA: Harvard University Press, 2021): 315–322. Compare to the discussion on evolution in nineteenth century Europe in Chap. 7.

[51] Empedocles [. . .] ante omnia cor iudicavit increscere, quod hominis vitam maxime contineat.. Cens. *Die. Nat.* 6.1.

[52] οἱ δ' ἐν τῇ μήτρα, καθάπερ Ἐμπεδοκλῆς, τὰ μὲν γὰρ εἰς θερμὴν ἐλθόντα τὴν ὑστέραν ἄρρενα γίνεσθαί φησι, τὰ δ' εἰς ψυχρὰν θήλεα, τῆς δὲ θερμότητος καὶ τῆς ψυχρότητος τὴν τῶν καταμηνίων αἰτίαν εἶναι φησιν, ἢ ψυχροτέραν οὖσαν ἢ θερμοτέραν, καὶ ἢ παλαιοτέραν ἢ προσφατωτέραν. GA 4.1 764a 1–6.

[53] Ἐμπεδοκλῆς ὁμοιότητας γίνεσθαι κατ' ἐπικράτειαν τῶν σπερματικῶν γόνων, ἀνομοιότητας δὲ τῆς ἐν τῷ σπέρματι θερμασίας ἐξατμισθείσης. Aët. 5.11.1 (Ps.-Plut.).

woman at the moment of conception. For women have often fallen love with statues and portraits and have given birth to children resembling them. Aëtius, LM D 182, tr. LM.[54]

And

It is in this way that all [scil. Probably: living beings] inhale and exhale: for all, channels of flesh, which the blood leaves, extend to the surface of their bodies; and at the openings, the furthest limits of their skin (*rhinon*) are perforated through and through with dense furrows, so that the blood lies hidden, while easy access is cut by these passages for the air (*aither*). When from here the delicate blood then rushes backward, the air, boiling, rushes after it in a raging surge, but when it [i.e., the blood] leaps back, the other is exhaled again—just as when a child plays with a clepsydra of handy copper: when she places the opening of the pipe against her well-formed hand and dips it into the delicate body of silvery water, liquid no longer enters into the container, but is prevented from doing so by the mass of air falling from inside the dense holes, as long as she restrains the thick flow [scil. of air]; but then, when the breath is lacking, water enters in the predetermined amount. In the same way, when she keeps the water in the depths of the copper vessel, blocking with her mortal skin the opening and the passage, the air outside, desiring to come inside, repels the liquid around the gates of the dull-sounding sieve, dominating the surface, until she lets go with her hand; then again inversely, in reverse of earlier, the breath now falling into it, the water runs out in the destined amount. In the same way the delicate blood, trembling through the limbs, when turning back, it leaps toward the nooks inside, at once the flow of air (*aither*) pursues it, rushing in its surge, but when it leaps back, it exhales again, in the same amount, backward. Aristotle, *On Respiration*. LM D 201, tr. LM.[55]

And

But they [i.e., the organs of sight] are not constituted in the same way [but the ones come more from what is similar], the others from the opposites; and for some the fire is in the center, for others outside. This is also why some animals have sharper sight by day, others at night: those that have less fire, by day, for their inner light is compensated by the external one; those that have less of the opposite element, at night, for their lack too is compensated by an addition. And both of these processes depend upon contraries [in a contrary manner]. For those that have an excess of fire see also more dimly, since, increased further by day, it spreads out over the passages of water and covers them; while for those with an excess of water, the same thing happens at night (for the fire is covered by water). [And this happens]

---

[54] [πῶς ἄλλοις ὅμοιοι γίγονται οἱ γεννώμενοι καὶ οὐ τοῖς γονεῦσιν] Ἐμπεδοκλῆς τῇ κατὰ τὴν σύλληψιν φαντασίᾳ τῆς γυναικὸς μορφοῦσθαι τὰ βρέφη, πολλάκις γὰρ ἀνδριάντων καὶ εἰκόνων ἠράσθησαν γυναῖκες, καὶ ὅμοια τούτοις ἀπέτεκον. Aët. 5.12.2 (Ps.-Plut.).

[55] ὧδε δ᾽ ἀναπνεῖ πάντα καὶ ἐκπνεῖ, πᾶσι λίφαιμοι σαρκῶν σύριγγες πύματον κατὰ σῶμα τέτανται, καί σφιν ἐπὶ στομίοις πυκναῖς τέτρηνται ἄλοξιν ῥινῶν ἔσχατα τέρθρα διαμπερές, ὥστε φόνον μὲν κεύθειν, αἰθέρι δ᾽ εὐπορίην διόδοισι τετμῆσθαι. ἔνθεν ἔπειθ᾽ ὁπόταν μὲν ἀπαίξῃ τέρεν αἷμα, αἰθὴρ παφλάζων καταΐσσεται οἴδματι μάργῳ, εὖτε δ᾽ ἀναθρῴσκῃ, πάλιν ἐκπνέει, ὥσπερ ὅταν παῖς κλεψύδρῃ παίζῃσι δι᾽ εὐπετέος χαλκοῖο, εὖτε μὲν αὐλοῦ πορθμὸν ἐπ᾽ εὐειδεῖ χερὶ θεῖσα εἰς ὕδατος βάπτῃσι τέρεν δέμας ἀργυφέοιο, οὐκέτ᾽ ἐς ἄγγοσδ᾽ ὄμβρος ἐσέρχεται, ἀλλά μιν εἴργει ἀέρος ὄγκος ἔσωθε πεσὼν ἐπὶ τρήματα πυκνά, εἰσόκ᾽ ἀποστεγάσῃ πυκινὸν ῥόον, αὐτὰρ ἔπειτα πνεύματος ἐλλείποντος ἐσέρχεται αἴσιμον ὕδωρ. ὣς δ᾽ αὔτως ὅθ᾽ ὕδωρ μὲν ἔχῃ κατὰ βένθεα χαλκοῦ πορθμοῦ χωθέντος βροτέῳ χροῒ ἠδὲ πόροιο, αἰθὴρ δ᾽ ἐκτὸς ἔσω λελιημένος ὄμβρον ἐρύκει ἀμφὶ πύλας ἠθμοῖο δυσηχέος, ἄκρα κρατύνων, εἰσόκε χειρὶ μεθῇ, τότε δ᾽ αὖ πάλιν, ἔμπαλιν ἢ πρίν, πνεύματος ἐμπίπτοντος ὑπεκθέει αἴσιμον ὕδωρ. ὣς δ᾽ αὔτως τέρεν αἷμα κλαδασσόμενον διὰ γυίων, ὁππότε μὲν παλίνορσον ἐπαΐξειε μυχόνδε, αἰθέρος εὐθὺς ῥεῦμα κατέρχεται οἴδματι θῦον, εὖτε δ᾽ ἀναθρῴσκῃ πάλιν ἐκπνέει ἶσον ὀπίσσω. Arist. *Resp.* 7 473b 9–474a6.

until for the ones the water is removed by the external light and for the others the fire is removed by the air. For the remedy comes for each of them from the opposite. The one [scil. organ of sight] that is composed of both elements in the same quantity is the one that is best mixed and is the best one. Theophrastus, *On Sensations*. LM D218, tr. LM.[56]

And

The sensation of pleasure is caused by what is similar, in terms of parts and their mixture, that of pain by contraries [. . .]. Theophrastus, *On Sensations*. LM D235, tr. LM.[57]

And

Empedocles: it [scil. the directing organ of the soul is located] in the composition of the blood. Aëtius. LM D238a.

And

[. . .] Empedocles [. . .]: the intellect and the soul are the same thing [. . .]. Aëtius. LM D238b.[58]

As per the methodology of this book (a select history of medicine), comments on background science are followed by insights into zoology, human biology, and diagnosis and treatment of human ailments, respectively. The above texts fit into the human biology bin. Let's look at them in order. First, in D167 there is an interesting conjecture on the role of the heart in the human body. This is a critical issue on two fronts. In the first part some organ may be crucial (Greek *arche*, Latin, *principia*) if it is first in time (developmental biology) or first in importance (where such organ fits into the triage of the organism). Aristotle thought the heart was both. For Aristotle, for example, assumed that humans developed in the same manner as chickens (whose eggs he observed at regular intervals) and put the heart first *in time*.[59] He also assumed that it was primary to the existence of the organism so that the loss of the heart's functioning would be the point of death (a judgment that was universally held by most of the world until the last proponent, Japan, turned to brain activity as the signifier of death in 1985).[60] The argument for the heart being the key

---

[56] συγκεῖσθαι δ' οὐχ ὁμοίως, [ἀλλὰ τὰς μὲν μᾶλλον ἐκ τῶν ὁμοίων] τὰς δ' ἐκ τῶν ἀντικειμένων, καὶ ταῖς μὲν ἐν μέσῳ ταῖς δ' ἐκτὸς εἶναι τὸ πῦρ, διὸ καὶ τῶν ζῴων τὰ μὲν ἐν ἡμέρᾳ τὰ δὲ νύκτωρ μᾶλλον ὀξυωπεῖν, ὅσα μὲν πυρὸς ἔλαττον ἔχει μεθ' ἡμέραν, ἐπανισοῦσθαι γὰρ αὐτοῖς τὸ ἐντὸς φῶς ὑπὸ τοῦ ἐκτός, ὅσα δὲ τοῦ ἐναντίου νύκτωρ, ἐπαναπληροῦσθαι γὰρ καὶ τούτοις τὸ ἐνδεές. ἐν δὲ τοῖς ἐναντίοις ἑκάτερον, ἀμβλυωπεῖν μὲν γὰρ καὶ οἷς ὑπερέχει τὸ πῦρ, ἐπεὶ αὐξηθὲν ἔτι μεθ' ἡμέραν ἐπιπλάττειν καὶ καταλαμβάνειν τοὺς τοῦ ὕδατος πόρους, οἷς δὲ τὸ ὕδωρ, ταὐτὸ τοῦτο γίνεσθαι νύκτωρ, καταλαμβάνεσθαι γὰρ τὸ πῦρ ὑπὸ τοῦ ὕδατος. [γίγνεσθαι δὲ ταῦτα] ἕως ἂν τοῖς μὲν ὑπὸ τοῦ ἔξωθεν φωτὸς ἀποκριθῇ τὸ ὕδωρ, τοῖς δ' ὑπὸ τοῦ ἀέρος τὸ πῦρ, ἑκατέρων γὰρ ἴασιν εἶναι τὸ ἐναντίον. ἄριστα δὲ κεκρᾶσθαι καὶ βελτίστην εἶναι τὴν ἐξ ἀμφοῖν ἴσων συγκειμένην. Theophr. *Sens*. 7–8.

[57] [. . .] ἥδεσθαι δὲ τοῖς ὁμοίοις κατὰ [τά] τε μόρια καὶ τὴν κρᾶσιν, λυπεῖσθαι δὲ τοῖς ἐναντίοις [. . .]. Theophr. *Sens*. 9.

[58] A= Ἐμπεδοκλῆς ἐν τῇ τοῦ αἵματος συστάσει. And B= [. . .] Ἐμπεδοκλῆς [ ] ταὐτὸν νοῦν καὶ ψυχήν [. . .]. Aët. 4.5.8 et 12 (Ps.-Plut., Stob.).

[59] See Boylan (1983: 170ff).

[60] Mashiro Morioka, "Reconsidering Brain Death" *Hastings Center Report* 31.4 (2001): 41–46.

sign showing when life begins and when it ends is enhanced by a *symmetry* between the point at which life begins (particularly human life) and the point at which it ends. The Greek word *arche* fits the bill here as the principle of life.

According to Empedocles and later Aristotle, this principle is the heart.[61]

In the second group of passages (LM D173; LM D 181; LM D182), the move is toward conception theory and inheritance.[62] There are two issues here: sex/gender presentation and other morphology (perhaps behavioral?). Empedocles, like Aristotle later, seems to associate the contrary *cold* with the feminine and *hot* with the masculine. This is put into an analogy to cooking which I have linked between Aristotle and Empedocles.[63] The "hot" in cooking makes for more palatable (better) meat, so also does an embryo that is hotter yield a more developed and better: i.e., male. This blatant sexism is consistent with the times.[64]

I have argued in the past that conception theory gives philosophical historians a good insight on how theory and limited observation link together. One has to utilize general principles acquired otherwise in one's *phusis*$_1$ design to engage with novel problems. This is what it means to have a systematic rather than an aporetic approach.

Another key issue in the LM D 181 passage is the slight lean toward pangenesis. As footnoted above, pangenesis (the theory in which all parts of the body send material to the gonads of the male and female so that the entire body is represented in the *sperma* and the *katamenia*)[65] is one of the two competing conception theories in the ancient world—epigenesis (the theory that development occurs via stages of development instead of an ever-developing homunculus). Though this passage is rather contracted, other pangenesis writers (from the Hippocratic canon) who I cited in my two articles on the subject. When these two principles meet up in the uterus, they have a battle in which either the male or the female predominate.

The LM D182 is absolutely bizarre. It is asserted that if the woman is imagining the image of a statue (such as an idealized male—like Achilles), then sometimes it is the case that these imaginative thoughts might have an influence on the *morphe* (shape) of the offspring. What could possibly be the mechanism for this? It is one thing to imply pangenesis and the effects of the contraries—especially hot and cold, because these are material events that could, in principle, affect material. But *thinking* or *imagining* is not a material event. How could this possibly affect the developing offspring? We have no extant texts to help us here.

---

[61] Boylan (1983: 195–199; 201–202).

[62] For more detail on the dynamics of this see: Michael Boylan, "The Galenic and Hippocratic Challenges to Aristotle's Conception Theory" *Journal of the History of Biology* 17.1 (1984): 83–112, and "Galen's Conception Theory" Journal *of the History of Biology* 19.1 (1986): 44–77.

[63] Boylan (1983: 152, cf. 151, 185, 200).

[64] For some background on the role of women in the society of ancient Greece, see: Bonnie MacLachlan, *Women in Ancient Greece: A Sourcebook* (London: Continuum, 2012).

[65] The competing theory of conception is *epigenesis* in which a potential whole is created that actualizes in stages. See my conception theory articles, *loc. cit.*

The next text (LM D 201) I have chosen to bring forward concerns *skin breathing*. This is an important concept in ancient medicine. The act of breathing clearly is important to life. Lung breathing is clearly an important source of air which is then converted into *pneuma* an active, animated form of air that provides various functions—many of which are similar to the neurological system as understood today. However, there are two key factors that make this a more difficult problem for proponents of ancient medicine. The first is that there was a debate on where thinking took place. The two principal candidates are the brain (an organ) and the region of the heart (non-organ centered). One of the problems of the brain theory is that the brain is located in an extremity of the body. With no transmission mechanism, this could cause a problem to account for a human wanting to lift her right arm and then lifting it (which is in the province of *pneuma*, air that has been transposed into having extra-dynamic properties). If, however, air came in from all over the body and if it were integrated into the blood and became *pneuma*, and if all *pneuma* could, by nature of its $phusis_2$ properties, connect to other somatic *pneuma*, then the issue of *transmission* could be solved. However, there is one small problem: we move from an organ-oriented account of consciousness (the brain) which is clearly $phusis_1$ to a communications mechanism that is clearly $phusis_2$. How do these different metaphysical *types* interact? They operate on different planes.[66]

The non-brain (as the seat of consciousness) theory would likewise require another conduit. Since the area of the heart (à la Aristotle) is connected with blood, this would seem to be an interesting option for this group. Diocles would be a leading candidate here.[67]

At any rate, Empedocles felt committed to skin breathing for perhaps these very reasons. The experiment itself with the *klepsudra* is meant to prove this. The copper vessel is used in this instance to prevent nasal breathing (on the one hand) and to show, mechanically, how air can make water flow in or out of the hole in the vessel (on the other hand). In the first case, the subject would die were there not skin breathing. In the second case the analogy is merely *how* the material mechanics of skin breathing might occur—just like the vessel submerged in water allows water to rush in and when raised again allows water to exit.[68] This passage would be broadly influential among biomedical writes up to, and including, Galen.

In DM D 218, Empedocles speculates on the power of vision. This passage plays upon two points: (a) it seeks to explain sight using the material *fire*. Fire gives forth light. (b) we all see better in the light than in the darkness. These two factors play together to give the explanation a quasi-plausibility from the $phusis_1$ standpoint. However, the principal problem with explaining vision is where the causal nexus

---

[66] In the Seventeenth Century, this sort of problem became critical to René Descartes who tried to solve the problem by creating a gland in the brain that had both properties and could mix them: the pineal gland. See: Descartes, *The Passions of the Soul,* tr. Elizabeth Haldane and G.R.T. Ross (Cambridge and London: Cambridge University Press, 1911 [1649]).

[67] See my discussion in Boylan (2015: 77–80).

[68] See my discussion in relation to Galen in Boylan (2015: 122–125).

begins. Plato firmly sets his theory based upon a power withing the eye (it could be related to fire) that sends out a beam that is reflected by the object back into the eye. This can be called the *emission* or *extra-mission theory*.[69] Since Empedocles does not fill in the gaps, it seems to this author that this Platonic reading of Empedocles' earlier position makes the most sense. If it is correct, then Empedocles is seen to influence Plato's theory of sight.

In DM D235, the concept of *mixture* is brought to the fore. As the contraries exist within the human body, when they are balanced, then a pleasant outcome is achieved. This is a very important concept that will develop later into the humor theory that is also based upon the existence of contraries withing the body and that if any single agent is allowed to predominate, it will bring about disease. A balanced mixture is the ideal. Such an ideal, once again, engages *symmetry* (one of the key concepts in the development of medicine—which itself is a non-empirical/non-physical category ($phusis_2$)).[70]

In DM D238a & b we return to the blood as the conduit of the soul and the intellect. To connect *nous* and *psuche* in the context of the blood is a big step. If *psuche* is both a vital principle and one of particular function, then it becomes an essential concept of human life.[71] From the perspective of the medical practitioner, if *psuche* was carried about the body materially in the blood—perhaps by the physical agency of *pneuma*—then when something went wrong with the patient, manipulation with blood (either by control of diet (that caused the creation of blood) or by bloodletting) would provide the physician with critical tools by which to plan a treatment that might bring health back again. At this point, everything is on the horizon, but Empedocles' biological writings set the foundation for later work in this direction.

Empedocles offers more than his four roots: earth, air, fire, and water. He was a very speculative figure whose own work in medicine allowed him to make applications to human biology that would prove to be prescient for future medical writers in the ancient world.

Dueling Principles: Heraclitus

Our next stage in the process of the development of ancient Greek science that serves as a foundation for medicine, will turn to Heraclitus. Heraclitus is interesting in this historical journey because he engages in the identity of the *phusis* journey with a lens of theory of knowledge (epistemology) and of a study of the entities that

---

[69] Plato, *Timaeus* 45b and 46b. For commentaries on this see: Arnold Reymond, *History of the Sciences in Greco-Roman Antiquity* (London: Methuen, 1927): 182 and Stanley Finger, *The Origins of Neuroscience* (Oxford: Oxford University Press, 1994): 67–69.

[70] This passage does not do so, but it would be consistent with Empedocles' concepts of Love and Strife, here to link Love to the positive mixture that brings about pleasure.

[71] Later, of course, Aristotle would declare it to be the key concept to understand all life on earth: plants, animals, and humans.

are and *how they can be ranked*, metaphysics.[72] The lens of epistemology examines the way we come-to-know facts about the physical world. This information is useful in the exposition of and ranking of those facts.

Heraclitus's life is put by Diogenes Laertius around the 69th Olympiad (504/500 BCE).[73] He comes after Thales and Anaximenes by 40 or so years. He is just a little older than Empedocles. Heraclitus was aware of the work of Pythagoras (KRS 190), Xenophanes, and Parmenides (KRS 293). More analysis of the anti-realists will follow in the next section.

Regarding epistemology and metaphysics, we have the following texts.

The Role of Logos

> And of this account (*logos*) that is—always—humans are uncomprehending, both before they hear it, and once they have first heard it. For, although all things come about according to this account (*logos*), they resemble people without experience of them when they have experience of them, when they have experience both of words and of things of the sort that I explain when I analyze each [of them] in conformity with its nature and indicate how it is. But other men are unaware of all they do when they are awake, just as they forget all they do while they are asleep. Sextus Empiricus, *Against the Logicians,* LM D1, tr. LM[74]

The Role of the Senses

> All things of which sight and hearing are knowledge (*mathesis*) I honor most. (Ps?) Hippolytus, *Refutation of all Heresies,* LM D31, tr. LM[75]

And

> The eyes are more accurate witnesses than the ears, Polybus, *Histories,* LM D32, tr. LM[76]

And

---

[72] Strictly speaking metaphysics is the category that comprises the sub-categories of *ontology* (the study of what entities exist) and *cosmology* (the ranking of the entities that exist). For a brief discussion see: Michael Boylan, *The Good, the True, and the Beautiful* (London: Bloomsbury, 2008): 82–84.

[73] Diogenes Laertius IX 1 (DK 22 A1).

[74] τοῦ δὲ λόγου τοῦδ' ἐόντος ἀεὶ ἀξύνετοι γίγνονται ἄνθρωποι, καὶ πρόσθεν ἢ ἀκοῦσαι, καὶ ἀκούσαντες τὸ πρῶτον, γινομένων γὰρ πάντων κατὰ τὸν λόγον τόνδε ἀπείροισιν ἐοίκασι, πειρώμενοι καὶ ἐπέων καὶ ἔργων τοιούτων, ὁκοίων ἐγὼ διηγεῦμαι κατὰ φύσιν διαιρέων ἕκαστον καὶ φράζων ὅκως ἔχει. τοὺς δὲ ἄλλους ἀνθρώπους λανθάνει ὁκόσα ἐγερθέντες ποιοῦσιν, ὅκωσπερ ὁκόσα εὕδοντες ἐπιλανθάνονται. Sext. Emp. *Adv. Math.* 7.132 (et al.)

[75] ὅσων ὄψις ἀκοὴ μάθησις, ταῦτα ἐγὼ προτιμέω. Hippol. *Ref.* 9.9.5.

[76] ὀφθαλμοὶ τῶν ὤτων ἀκριβέστεροι μάρτυρες. Polyb. 12.27.

Bad witnesses for humans are the eyes and ears of those who possess barbarian souls. Sextus Empiricus, *Against the Logicians.* LM D33, tr LM[77]

And

If all the things that exist became smoke, the nostrils would be able to identify them. Aristotle, *On Sensation,* LM D34, tr. LM[78]

The Metaphysical Characteristics of Nature

Nature tends to hide. Themistius, *Oration.* LM D35, tr. LM[79]

And

We step and we do not step into the same rivers, we are and we are not. *Homeric Allegories,* LM D65a, tr. LM[80]

And

It is always different waters that flow toward those who step into the same rivers. Cleanthes in Arius Didymus in Eusebius, *Evangelical Preparation,* LM D 65b, tr. LM[81]

And

[Socrates:] Heraclitus says something like this: that all things flow and nothing remains; and comparing the things that are to the flowing of a river, he says that you could not step twice into the same river. Plato, *Cratylus,* LM D 65c, tr. LM[82]

And

[. . .] the Heraclitean doctrines according to which all perceptible things are constantly flowing and there is no knowledge about them [. . .] Aristotle, *Metaphysics* LM D66, tr. LM[83]

Let's address these in order. In section "The Role of Logos," the role of *logos* we have a meta-principle set forth. Since *logos* has a wide denotation from "word," to "account," to "supernatural pattern" (akin to Plato's form),[84] it is difficult to fathom

---

[77] κακοὶ μάρτυρες ἀνθρώποισιν ὀφθαλμοὶ καὶ ὦτα βαρβάρους ψυχὰς ἐχόντων. Sext. Emp. *Adv. Math.* 7.126.

[78] εἰ πάντα τὰ ὄντα καπνὸς γένοιτο, ῥῖνες ἂν διαγνοῖεν. Aristotle, *Sens.* 5 443a 23.

[79] φύσις κρύπτεσθαι φιλεῖ. Them. *Orat.* 5, p. 69b; cf. 12, p. 159b.

[80] ποταμοῖς τοῖς αὐτοῖς ἐυβαίνομέν τε καὶ οὐκ ἐμβαίνομεν, εἶμέν τε καὶ οὐκ εἶμεν. Hercl. *Alleg.* 24.4.

[81] ποταμοῖσι τοῖσιν αὐτοῖσιν ἐμβαίνουσιν ἕτερα καὶ ἕτερα ὕδατα ἐπιρρεῖ. Cleanthes apud Arius Didymus. In Eus. *PE* 15.20.2.

[82] [ΣΩ] λέγει που Ἡράκλειτος ὅτι πάντα χωρεῖ καὶ οὐδὲν μένει καὶ ποταμοῦ ῥοῇ ἀπεικάζων τὰ ὄντα λέγει ὡς δὶς ἐς τὸν αὐτὸν ποταμὸν οὐκ ἂν ἐμβαίης. Plato, *Crat.* 402a.

[83] [. . .] ταῖς Ἡρακλειτέιοις δόξαις, ὡς ἁπάντων τῶν αἰσθητῶν ἀεὶ ῥεόντων καὶ ἐπιστήμης περὶ αὐτῶν οὐκ οὔσης [. . .]. Aristotle, *Metaph.* 987a 32.

[84] Cf. Ἐν ἀρχῇ ἦν ὁ λόγος, καὶ ὁ λόγος ἦν πρὸς τὸν θεον, καὶ θεὸς ἦν ὁ λόγος. Gospel of John, Christian Bible. 1:1.

exactly what Heraclitus was getting at. But that may be part of the point (cf. his doctrine of conflicting explanations (below). *Logos* is a part of the expression of how humans experience the world. This would fix it as *phusis*$_1$ or *phusis*$_3$. But as a "real" account (that might exist separately from the physical world in which we live), it is more aligned to *phusis*$_2$. It seems to this author that because of the doctrine of conflicting explanations, that Heraclitus would probably intend both—but with some sort of priority attached to the *logos* as real. This would set it up as a referee when empirical data seem to be contradictory or changeable. It is the principal verité for natural philosophy. In Heraclitus' scheme the *logos* provides a cornerstone to support epistemological and metaphysical realism. It is epistemological because the natural philosopher can know *logos*, which is unchanging and true. It is metaphysical because it exists in its own realm: unchangeable and true.

This strategy is important structurally in setting up an account of nature. In the history of science, the positing of a limited number of "super principles" (such as evolution in contemporary biology) is thought to be, structurally, a good strategy in setting up the architecture of a comprehensive account of nature.

In section "The Role of the Senses," we have a more variable universe. First, the senses create the possibility of learning (that can lead to knowledge—LM D 31). The mechanics of how this happens are left out.[85]

In LM D32, Heraclitus asserts the primacy of vision over other options (in this case *hearing* gets second place).[86] What is important here is distinguishing various forms of sensory input and ranking them in some way. Most people have experienced instances of having sensed x, and then finding out that either the sensory impression was false or the judgment about the sensory impression was wrong. Error occurs in sensory events and it must be accounted for somehow.[87]

For Heraclitus, the way to respond to this epistemological problem is to posit dual explanations that can either make him an epistemological anti-realist or a realist under some kind of "process-explanation."

In section "The Metaphysical Characteristics of Nature," we have a presentation of the changeability of the empirical world. In LM D35 there is a depiction of nature as something that hides itself. This suggests an entity that either has the power to hide, *phusis*$_2$ or it is in its character an elusive entity. The latter seems more plausible to this author.

Then there is the famous pronouncement about river stepping (LM D65 a–c) which refers to a conflict between *particularity* and *law-like nomic structure*.[88] In the first place, particularity refers to how *this bit of material* may change all the

---

[85] This is generally the case prior to Aristotle in *APo.* II.19.

[86] It is Aristotle in *De Anima* II. 7–11 who sets out the number of 5 senses.

[87] The key positions on this dispute—over the ages—are represented by Augustine and Descartes. For a discussion of these issues see: Michael Boylan, *The Good, The True, and The Beautiful* (London: Bloomsbury, 2008); ch; 4.

[88] By "lawlike nomic" structure I am referring to a situation in which some statement about nature could act like a proposition set out in later thinkers like Aristotle in the *Posterior Analytics,* or Euclid in *Elements.* Some think that that there are interesting connections here: see Ian Mueller, *Philosophy of Mathematics and Deductive Structure in Euclid's Elements* (Cambridge, MA: M.I.T. Press, 1981).

time. In this instance it is water. Because of its permeable nature (like air) water (in its particularity) is always changing. If we were to assign a "name" as per contemporary philosophy in its protocol of the rigid designation of identity, then that name would always be changing.[89] One could, by extension, claim that if there is nothing rigid to designate individuals within a given sample space, then there could be no coherent understanding of group names that depend upon some actual-non-changing individual natures. This could put the entire process of creating group concepts (*eidos* and *genos*) into question.

Dual Explanations

A way around some of these conundrums is by the use of redundant explanations. This is a strategy in the ancient Greek medical/biological world that I have described before, beginning with Aristotle's use of the four causes and then his grouping them into two larger categories (material accounts and teleological accounts).[90] In this case Heraclitus, is doing something a little different: he's showing how contradictions occur whenever one tries to create a coherent account of nature. Let's view a few of these theoretical views.

> Conjoinings: wholes and not wholes, converging and diverging, harmonious dissonant; and out of all things one, and out of one all things. Pseudo Aristotle, *On the World*, LM D47, tr. LM[91]

And

> They do not comprehend how, diverging, it accords with itself: a backward-turning, fitting-together (*harmonie*), as of a bow and lyre. Hippolytus, *Refutation of All Heresies* LM D49, tr. LM[92]

And

---

[89] You could imagine taking a bucket of some particular collection of water at t_1 and comparing it to another bucket collected from the same stream at t_2 and if each were *named,* they would possess different names. For a discussion of this in the context of philosophy of language see Saul Kripke, *Naming and Necessity* 4th ed. (Cambridge, MA: Harvard University Press, 1980). For some critics on this groundbreaking study see: Scott Soames, *The Unfinished Semantic Agenda of Naming and Necessity* (Oxford: Oxford University Press, 2001); Osamu Kiritani, "Naming and Necessity from a Functional Point of View" *Croatian Journal of Philosophy* 13.37 (2013): 93–98; and Andrea Strollo, "If I were Kripke . . . Attributable Names and the Necessary *A Posteriori*" *Philosophical Forum* 50.1 (2019): 116–134.

[90] See Michael Boylan, *Method and Practice in Aristotle's Biology* (New York and Oxford: UPA/Rowman and Littlefield, 1983): 88, 119–120, 227; and Michael Boylan, *The Origins of Ancient Greek Science: Blood—A Philosophical Study* (London and New York: Routledge, 2015): 28–35.

[91] συνάψιες ὅλα καὶ οὐχ ὅλα, συμφερόμενον καὶ διαφερόμενον, συνᾷδον διᾷδον, καὶ ἐκ πάντων ἓν καὶ ἐξ ἑνὸς πάντα. Ps-Arist. *Mund.* 5 396b 20–22.

[92] The bow and the lyre are the associated with Apollo. οὐ ξυνιᾶσιν ὅκως διαφερόμενον ἑωυτῷ ὁμολογέει, παλίντροπος ἁρμονίη ὅκωσπερ τόξου καὶ λύρης. Hippol. *Ref.* 9.9.9, et al.

Invisible fitting-together (*harmonie*) [is] stronger than a visible one. Plutarch, *On the Generation of the Soul in* Plato's *Timaeus.* LM D50, tr. LM[93]

And

The way upward and downward: one and the same. Hippolytus, *Refutation of All Heresies.* LM D51, tr. LM[94]

And

The name of the bow is life (*bíos*), but its work is death. Etymologicum Magnum. LM D53, tr. LM[95]

And

Doctors, Heraclitus says cutting, cauterizing, badly mistreating their patients in every way, complain that they do not receive an adequate payment from their patients—and are producing the same effects [benefits and diseases]. Hippolytus, *Refutation of all Heresies.* LM D57, tr. LM[96]

And

Changing, it remains at rest. Plotinus, *Enneads* LM D58. Tr. LM[97]

In the D47 account Heraclitus attacks the common problem of the one and the many.[98] It is a thorny problem but Heraclitus' own position of the unifying *logos* demands a solution much like Plato's. Otherwise, coherent discourse would become cacophony.

In LM D49 the bow is a symbol of Apollo's forcefulness that can cause bodily distress or death when delivering an arrow, while the lyre is a musical instrument that is meant to create harmony and balance which bring about health. Thus, the results of each are: (1) bodily harm and destruction and (2) Harmony and health. These are contradictory outcomes, but they come from one single entity, the

---

[93] ἁρμονίη ἀφανὴς φανερῆς κρείττων. Plut. *An. Proc.* 27 1026 C.

[94] ὁδὸς ἄνω κάτω μία καὶ ὠυτή. Hippol. *Ref.* 9.10.4.

[95] τῷ οὖν τόξῳ ὄνομα βίος, ἔργον δὲ θάνατος. *Etym. Mag.* S.v. βίος, p. 198.26.

[96] οἱ γοῦν ἰατροί, φησὶν ὁ Ἡράκλειτος, τέμνοντες, καίοντες, πάντῃ βασανίζοντες κακῶς τοὺς ἀρρωστοῦντας, ἐπαιτιῶνται μηδέν᾽ ἄξιον μισθὸν λαμβάνειν παρὰ τῶν ἀρρωστούντων, ταὐτὰ ἐργαζόμενοι, [τὰ ἀγαθὰ καὶ τὰς νόσους]. Hippol. *Ref.* 9.10.3.

[97] μεταβάλλον ἀναπαύεται. Plot. 4.8.1.

[98] The way that the multiplicity of human experience connects with the necessity to create coherent discourse was a problem for many in the ancient Greek world. Plato's theory of forms was one solution. For a discussion of the broader topic in the ancient Greek world see: Edward C. Halper, *One and Many in Aristotle's* Metaphysics: Books Alpha-Delta (Las Vegas: Parmenides, 2009): ch. 4; Arthur Adkins, *From the Many to the One: A Study of Personality and Views of Human Nature in the Context of Ancient Greek Society, Values, and Beliefs* (Ithaca, N.Y.: Cornell University Press, 1970). ch. 1; André Laks, *The Concept of Presocratic Philosophy: Its Origins, Development, and Significance* (Princeton, N.J.: Princeton University Press, 2018): 35–52; and Anthony Paul Smith, *Thinking from the One: Science and the Ancient Philosophical Figure of One* (Edinburgh: Edinburgh University Press, 2012): 19–41.

Olympian god, Apollo. Once again, symmetry (a key ingredient to harmony) plays a foreshadowing role in future ideas about health.

In LM D50, the *aphanes* harmony is said to be stronger than the *phaneres*. Why would this be the case? It seems to this author that the obvious answer is the *logos*. The *logos* is invisible and that is the realm of unification of the many into the one. Since the visible realm is the realm of the many that is discordant, the only harmonies are those that are accidental or illusory—or active and invisible (a case of *phusis$_2$*).

In LM D51, we have the beginnings of Paramedian epistemological antirealism. "Upward and downward" are similar in that both are measured by a background. This similar to some of Zeno's paradoxes of motion that only become paradoxes when we shift the background conditions by which the object is being judged (see below). This theme is repeated in D53 with the depiction of the bow. On the one hand, the bow brings life in hunting for the hunter and his family. But the way that he does this is through killing. Or in battle, the successful warrior saves his life by the use of the bow in the killing of his enemy. This is an instance of setting the judgment of some action x against a particular social backdrop. Change the backdrop and you change the judgment. See note 64 on the methodology of the modern Michel Foucault who often uses the same logic.

In LM D57 we get a medical example of sorts. The idea here is that a physician is only compensated for his intervention and not for his results. If one were to set down the background condition of self-interest, then if the physician gets paid whether or not health is the outcome, then what is the motivation of curing patients? Is this cultural concern a cause for Galen to suggest treating patients for free? (Especially easy for wealthy physicians.)

And finally, in LM D58 we have a self-referential proposition.[99] Now, self-referential propositions are very tricky. In the twentieth century the analysis of such propositions (in the context of justifying mathematics as both complete and coherent) caused Kurt Gödel to come up with his famous proof.[100] Here the idea of "changing" means that a non-changing state comes out of constant change: "rest." Thus change (self-referentially) brings about non-change—an apparent contradiction. But like the rest of these dualizing accounts, *contradiction* is being used constructively.

It seems that Heraclitus' explanation for these anomalies of identification is to set out a metaphysics in which the contraries exist in one way, but are for the purpose of some sort of synthesis in another. This is reminiscent of Hegel's

---

[99] One popular example of a self-referential proposition is "Epimenides, the Cretan, says that 'All Cretans are liars.'" If Epimenides' speech act, A, and the content of his speech act B are put side-to-side, then if A is true, then B is false and vice versa. There is an unending repetition here. This is *not* just a silly word game.

[100] For an excellent gloss on this proof see, Ernest Nagel and James R. Newman, *Gödel's Proof* (New York: New York University Press, 1958).

methodology and some post-modern philosophers.[101] The idea here is that there is some embedded truth (*logos*) that can be known, but not directly. It is not like mathematical problems of figuring out how much money one owes the fisherman selling his wares in the agora. Rather, it is a structure that is beyond the sensory and the elementary deduction in that it creates a background against which the seeming contradictions that arise from the ever-changing world of the senses present.

Then there are the most famous pronouncements on the changeability of nature such that knowledge of some fact at t_1 will not necessarily yield knowledge about what would *appear to be the same fact* at t_2. This would amount to a metaphysical claim that *phusis* (however it is understood) is changeable to a high degree so that it might be *inscrutable* -rendering attempts to figure it all out to be futile. This might sound like an assertion of anti-realism and the adoption of skepticism. But this is too quick. There is always the super-sensible *logos* to come to the rescue. But unless it is viewed in its own arena, it is unclear how it can integrate with empirical experience except through intuition. For intuition can work in two ways to solve this problem. For intuition$_1$ the practitioner—for example a physician—might use intuition to directly connect to the *logos* in order to determine the diagnosis so that prognosis and treatment might be possible via this intuitive epistemological leap. This is the direction of the epistemological anti-realists.

Another possibility that is developed in the Hippocratic Canon is to consult a handbook that lists symptoms and treatments for those symptoms. When several sets of symptoms seem plausible from the handbook, then the physician uses intuition$_2$ to select which of competing accounts matches the patient in front of you. This becomes the direction from Hippocrates onward until the present day.[102]

The use of empirical cases counts for *phusis*$_1$. The dueling principles, simpliciter, counts for *phusis*$_3$ while the intuitive connection to the best of several empirically-based alternatives is *phusis*$_2$.

---

[101] For the underpinnings of the concept that "dueling accounts" aka Hegelian Dialectic see *The Logic of Hegel*, tr. William Wallace (Oxford: Oxford University Press, 1965). The dialectical process posits opposites in order to move the mind toward a synthesis that is somewhere in-between. Two more contemporary theorists that use opposites productively are Michel Foucault, *Les Mots et les chosed* (Paris: Gllimard, 1966): esp. ch. 1 and Jacques Derrida, *L'Ecriture et la difference* (Paris: Seuil, 1967): esp. ch. 7. In the case of Foucault, his initial examination of Diego Velázquez's painting *Las Meninas* shows that the structure of a judgment is based upon the underpinnings of historical données. And between these paradigms' contradictions can appear which can call into play the veracity of each. In the case of Derrida, one must search for the contradiction by a process of deconstruction that calls for examination of etymologies that may show radical difference. The process of putting it together resolves the contradiction. In all three instances contradiction alone is not enough to stop the investigative process into truth—rather it is just the beginning.

[102] I discuss these two forms of intuition in the context of action theory in Michael Boylan, *Basic Ethics*, 3rd ed. (London and New York: Routledge, 2020): ch. 9.

Numbers: Pythagorean Mathematical Realism v. Parmenides' and Zeno's Anti-Realism

Pythagoras

Our foray into epistemological anti-realism will begin with the way numbers are used by Pythagoras to support a *phusis$_2$* -driven realism versus the way numbers are used by Parmenides and Zeno to arrive at epistemological anti-realism, *phusis$_3$*. The epistemological anti-realists do not think that it is possible to come up with definite, general answers to the question "what is x" where x is an object or process occurring in nature. In the ancient world, this expressed itself in embracing *skepsis* which has as one of its primary meanings "resisting closure on giving a definitive answer on the general character of x (as a type of thing or process or event)." Rather, the emphasis is upon particularity—a nameable thing (see footnote # 52 in Introduction). To make this relationship clearer throughout the book, when referring to the x's nature as a type of thing or process or event, a capital "N" will be used for the English word Nature. When referring to a particular individual's characteristics, etc. a small "n" will be used for the English word nature.

Diogenes Laertius puts the dating of Pythagoras in the range of 510–540 BCE.[103] He travelled some and became steeped in various cultural and religious traditions. It is perhaps because of this experience that Pythagoras is inclined to see mathematics as a link to divine-based accounts (see below).

Plato praised Pythagoras' character as a decent personal guide and educator (*Republic* 600a–b) and his inclination for applied mathematics via astronomy and musical harmony (*Republic* 530d). It is this inclination towards mathematics that inclines many (ancient and modern) toward *phusis$_2$* accounts. This is because mathematics is non-empirical—even if it is taken to be metaphysically *real*.[104] And this step has inspired critical treatments of Pythagoras of the same vein.[105]

---

[103] LM P1, P3, cf. P4 from Eusebius, *Chronicle* (Jerome) as 530 BCE. P7 suggests that he died at the age of 80 (Diogenes Laertius). P14 Porphyry claims he studied not only with Pherecydes and Hermondamus, but also with Anaximander. Porphyry in his *Life of Protagoras* suggests he travelled to Egypt to study, P. 20.b. Related to this is the claim (by Hippolytus, *Refutation of All Heresies*) that he met with Zaratas (i.e., Zoroaster), P21.

[104] An introduction to some of the issues involved with mathematical realism and its possible relation to physical realism can be found in: Ünsal Cimon, "On Saving the Astronomical Phenomena: Physical Realism in Struggle with Mathematical Realism in Francis Bacon, Al-Bitruji and Averroes" *Hopos* 9.1 (2019): 135–151; Helen De Cruz, "Numerical Cognition and Mathematical Realism" *Philosophers' Imprint* 16.16 (2016): e1–e13; Katharina Felka, *Talking about Numbers: Easy Arguments for Mathematical Realism* (Frankfurt: Verlag Vittorio Klostermann, 2015); and Jacob Busch, "Scientific Realism and the Indispensability Argument for Mathematical Realism: A Marriage Made in Hell" *International Studies in the Philosophy of Science* 25.4 (2011): 307–325.

[105] This interpretation of Pythagoras as balancing between mathematical realism and an application to natural realism can be found in the following: F.M. Cornford, "Mysticism and Science in the Pythagorean Tradition" *Classical Quarterly* 17.1 (1923): 1–12; Walter Burkert, *Weisheit und Wissenschaft* (Nürnberg: H. Carl, 1962); Arturo Sangalli, *Pythatoras' Revenge: A Mathematical*

Some key texts for Pythagoras in the context of numbers and nature are as follows:

> Some of them [i.e., the philosophers] left behind treatises, while others did not write anything at all, like [. . .] according to some people Pythagoras. Diogenes Laertius. LM D1 tr. LM.[106]

And

> Pythagoras was the first who tried to speak about virtue, but he did not do so correctly. For in referring the virtues to numbers he did not establish an appropriate way to study the virtues. For justice is not a number equal times equal [i.e., a square number]. Pseudo-Aristotle. *Magna Moralia.* LM D6 tr. LM.[107]

And

> Apollodorus, the mathematician, says that he [i.e., Pythagoras] offered a hecatomb in sacrifice because he had discovered that the side subtending the right angle in a right triangle is equal in power to the sides containing the right angle [i.e., the square of the hypotenuse is equal to the sum of the squares of the sides.]. LM D7a Diogenes Laertius. tr. LM.[108]

And

> ["In right-angled triangles, the square on the side sub-tending the right angle [i.e., the hypotenuse] is equal to the squares on the sides containing the right angle [scil. Added together]" = Euclid, *Elements,* Book 1, Proposition 47]. One can find, if one reads the authors who investigate antiquity, some who attribute this theorem to Pythagoras and say he sacrificed an ox when he discovered it [. . .]. LM D 7b Proclus, *Commentary on Euclid.* tr. LM.[109]

And

> Heraclides writes the following about this in his *Introduction to Music*: "As Xenocrates says, Pythagoras also discovered that the musical intervals are not generated independently

---

Mysteru (Princeton, N.J.: Princeton University Press, 2009): see esp.: 106–114; 158–168; Ian Hacking, "The Lure of Pythagoreans" *The Jerusalem Philosophical Quarterly* 61 (2012): 103–128; Nicholas Rescher, "Pythagoras's Number" in *A Journey Through Philosophy Through 101 Anectotes* (Pittsburgh, PA: University of Pittsburgh Press, 2015):18–19; and Aaron Segel, "Pythagoreanism: A Number of Theories" *Philosopher's Imprint* 19.26 (2019):e1–e19.

[106] καὶ οἱ μὲν αὐτῶν κατέλιπον ὑπομνήματα, οἱ δ' ὅλως οὐ συνέγραψαν, ὥσπερ [. . .] κατά τινας Πυθαγόρας [. . .]. Diog. Laert. 1.16.

[107] πρῶτος μὲν οὖν ἐνεχείρησεν Πυθαγόρας περὶ ἀρετῆς εἰπεῖν, οὐκ ὀρθῶς δέ, τὰς γὰρ ἀρετὰς εἰς τοὺς ἀριθμοὺς ἀνάγων οὐκ οἰκείαν τῶν ἀρετῶν τὴν θεορίαν ἐποιεῖτο, οὐ γάρ ἐστιν ἡ δικαιοσύνη ἀριθμὸς ἰσάκις ἴσος. Ps.-Arist. *MM* 1.1 1182a 11–14.

[108] φησὶ δ' Ἀπολλόδωρος ὁ λογιστικὸς ἑκατόμβην θῦσαι αὐτόν, εὑρόντα ὅτι τοῦ τριγώνου ὀρθογωνίου ἡ ὑποτείνουσα πλευρὰ ἴσον δύναται ταῖς περιεχούσαις. Dior. Laert. 8.12.

[109] "ἐν τοῖς ὀρθογωνίοις Δ τὸ ἀπὸ τῆς τὴν ὀρθὴν γωνίαν ὑποτεινούσης πλευρᾶς τετράγωνον ἴσον ἐστὶ τοῖς ἀπὸ τῶν περὶ τὴν ὀρθὴν γωνίαν πλευρῶν τετραγώνοις." τῶν μὲν ἱστορεῖν τὰ ἀρχαῖα βουλομένων ἀκούοντας τὸ θεώρημα τοῦτο εἰς Πυθαγόραν ἀναπεμπόντων ἐστὶν εὑρεῖν καὶ βουθύτην λεγόντων αὐτὸν ἐπὶ τῇ εὑρέσει [. . .]. Proc. *In Eucl.* 47, Theor. 33 (p. 426.1–9 Friedlein)

of number [. . .]. LM D8 Xenocrates in Porphyry, *Commentary on Ptolemy's* Harmonics. tr. LM.[110]

And

To indicate this [scil. That the criterion of all things is number], the Pythagoreans had the habit of saying sometimes: all things resemble number. LM D10 Sextus Empiricus, *Against the Logicians*. tr. LM.[111]

And

After these [i.e., Thales and Mamercus (?), Pythagoras transformed the kind of philosophy regarding this [i.e. geometry] by giving it the form of a free discipline, considering its principles from the origin and investigating its propositions independently of matter and in accordance with the intelligible—he who also discovered the study of the irrationals and the arrangement of the cosmic figures. LM D11 Eudemus in Proculus, *Commentary on Euclid's Elements*. tr LM.[112]

And

Pythagoras was the first to call what surrounds all things "*kosmos*" (i.e., a beautiful, organized whole) because of the order (*taxis*) that is found there. LM D13. Aëtius. tr. LM.[113]

And

He was the first to say that the morning star and the evening star are one and the same [. . .]. LM D14 Diogenes Laertius. tr. LM.[114]

There are several remarks that should be made on these texts. First, from LM D1, we are not certain whether the various positions we attribute to Pythagoras were *actually* affirmed by Pythagoras if he did not write anything down—or if he *did* write something down, then those texts were lost rather quickly. I don't see this as an interpretative problem (though it may be an historical problem).

Here I am following the team who put *ideas* (existing texts or not) over a clear historical record clearly linking ideas to existent texts. In such a case, all we need is the oral record of various individuals whose texts have come down by historical

---

[110] γράφει δὲ καὶ Ἡρακλείδης περὶ τούτων ἐν τῇ Μουσικῇ εἰσαγωγῇ ταῦτα, "Πυθαγόρας, ὥς φησι Ξενοκράτης εὕρισκε καὶ τὰ ἐν μουσικῇ διαστήματα οὐ χωρὶς ἀριθμοῦ τὴν γένεσιν ἔχοντα [. . .]. Porph. *In Ptol. Harm.* 30.1–3.

[111] καὶ τοῦτο ἐμφαίνοντες οἱ Πυθαγορικοὶ ποτὲ μὲν εἰώθασι λέγειν τὸ ἀριθμῷ δέ τε πάντ' ἐπέοικεν. Sext. Emp. *Adv. Math.* 7.94.

[112] ἐπὶ δὲ τούτοις Πυθαγόρας τὴν περὶ αὐτὴν φιλοσοφίαν εἰς σχῆμα παιδείας ἐλευθέρου μετέστησεν, ἄνωθεν τὰς ἀρχὰς αὐτῆς ἐπισκοπούμενος καὶ ἀΰλως καὶ νοερῶς τὰ θεωρήματα διερευνώμενος, ὃς δὴ καὶ τὴν τῶν ἀλόγων πραγματείαν καὶ τὴν τῶν κοσμικῶν σχημάτων σύστασιν ἀνεῦρεν. Proc. *In Eucl.* Prol. 2 (p. 65.15–21, Friedlein).

[113] Πυθαγόρας πρῶτος ὠνόμασε τὴν τῶν ὅλων περιοχὴν κόσμον ἐκ τῆς ἐν αὐτῷ τάξεως. Aët. 2.1.1 (Plut.) [περὶ κόσμου] For a further discussion of "wholes" and their relation to ancient medicine see: Chiara Thuminger, ed. *Holism in Ancient Medicine* (Leiden: Brill, 2020).

[114] πρῶτόν τε Ἕσπερον καὶ Φωσφόρον τὸν αὐτὸν εἰπεῖν [. . .]. Diog. Laert. 8.14. This is historically an important distinction that makes its way into contemporary philosophy of language via Gottlob Frege's famous essay, "Über Sinn und Bedeutung" *Zietschrift für Philosophie und Philosophische Kritik* 100 (1892): 25–50.

contemporaries reading and quoting a text before them or versions of the copied and re-copied texts themselves.[115]

In LM D6 we have a quotation from pseudo-Aristotle that Pythagoras was so keen on numbers that he wanted to set up an account of virtue (*arete*) that used principles of proportionality to support an idea of justice (*dikaiosune*). The author is critical of Pythagoras, in this regard—even though Aristotle, himself, uses such a system in *EN* V in discussing justice (cf. Plato, *Republic* VII). With respect to distributive justice, quantitative tools seem to be essential.

LM D7a and LM D7b both describe what Pythagoras is best known for in history: the theorem that allows one to calculate the missing side from two known sides of a right triangle. This is an advance in geometry (which, at the time, was generally is primary to arithmetic and gives a physical instantiation of measurement that will give groundwork to the Hippocratic medical writers).[116]

In LM D8 there is a connection between music and numbers via harmony. This introduces the concept of *elegance* (mentioned in the Introduction). Ancient Greek Science, which forms much of the basis for Ancient Greek Medicine, here give a soft connection to elegance by the connection to beauty through number.

LM D10 is a statement about ontology. If all things are essentially number, then *number* (a *phusis*$_2$, super-sensible entity) is what is *real*. This is similar to Heraclitan *logos*—except that the *logos* is a qualitative concept while *number* is a quantitative, measurement entity. The basic dualism of ontology is common to both, but the nature of the supersensible varies. This structure is continued in LM D11 in which a separate cosmological realm is said to be independent of matter; therefore, a dualistic parallel universe of sorts is created. Often, when such parallel realms are posited, the investigator looks *upward*—as does Pythagoras (according to LM D13) as he is attributed to the naming of this geometrically organized space as: *kosmos*.

The last text that I have highlighted for Pythagoras is LM D14. In this text Pythagoras reveals a glimpse of *phusis*$_1$ by identifying a single entity (Venus) as being both the morning and evening star. Though the object is seen at different times of the day and often has slightly different qualitative properties, the position in the sky (probably) was behind Pythagoras' call. As noted above, this sort of activity (rigidly identifying individuals by key enduring properties) has been a key distinction concerning denotation until the present day.

---

[115] For two thoughts on this see Roland Barthes, "The Death of the Author" translated in Seán Burke, ed., *Authorship: From Plato to Post-Modern: A Reader* (Edinburgh: Edinburgh University Press, 1995 [1967]): Ch. 15. Cf. Foucault, *op. cit.*

[116] Concerning the relationship between geometry and arithmetic see: Arpad Szabo, *The Beginnings of Greek Mathematics* (Dordrecht: D. Reidel, 1978): 2.10–2.12, 3.2; Edward A. Maziarz and Thomas Greenwood, *Greek Mathematical Philosophy* (New York: Frederick Unger, 1968): chapts. 2, 3; and José Ferreirós, "Ancient Greek Mathematics: A Role for Diagrams" in *Mathematical Knowledge and the Interplay of Practices* (Princeton: Princeton University Press, 2016): 112–152. With respect to this concrete connection as useful to "measurement minded" Hippocratic writers see: G.E.R. Lloyd, *The Revolutions of Wisdom* (Berkeley, CA: University of California Press, 1987): 257–270.

Obviously, Pythagoras is keen on seeing nature through numbers. How should we understand what sense of nature is operating here? What is crucial in forming our answer (and as the book develops) is the commitment being made when using numbers to describe natural phenomena. First of all, are numbers conventional devices that are used for one of many possible descriptions and/or measurements of nature? Or is there something that is *real* in mathematical constructs so that they coincide with objects and processes that they are describing? And *if* they do, then are they in any way primary in either our understanding and do they represent the things-in-themselves and their interactions? In either of these two options the result is an understanding or representation of nature as *phusis*$_2$ —not as a divine entity, but as a super sensible entity that is responsible for a natural event.[117]

If the mathematical constructs are not merely conventions, then we may describe this state of affairs as mathematical realism. And if there is there a smooth connection between these real measuring devices and the objects they describe, then we can talk about a synonymy between mathematical and material realism. This connects Pythagorean ontology and cosmology more tightly to subsequent Platonic metaphysics.

Parmenides and Zeno

This section of the chapter will treat the shift in the use of numbers from mathematical realism that actually connects to nature (*phusis*$_2$), to the use of numbers to set forth a doctrine of *skepsis* (in the sense of the non-closure of knowledge claims that will amount to epistemological anti-realism, *phusis*$_3$). The two key proponents of this among the pre-Socratics were Parmenides and Zeno of Elea.

Plato says that Parmenides and Zeno came to Athens for the Great Panathenaea. Parmenides was said to be advanced in years—around sixty and Zeno was almost forty. Socrates was very young at the time (Plato, *Parmenides* 127a = DK 29a 11). Parmenides is said to be (according to Theophrastus in his *Epitome* of Anaximander) the pupil of Xenophanes, but he did not follow his tenets of natural philosophy. He was in his maturity during the 69th Olympiad [around 504-500 B.C.E.] (Diogenes Laertius, IX, 21-23, (DK 28a 1)).

Since we know the years of Socrates: 469–399 B.C.E., and if "quite young man" were to be around 20, then the supposed meeting might have been around 449 B.C.E—putting Parmenides' birth around 515 B.C. E. and Zeno around 490 B.C.E.[118] Some of these chronologies also depend upon the dating the foundation of Elea (540 B.C.E.), which was adopted as the *floruit* of Xenophanes.[119]

---

[117] This refers to some of the same issues discussed in footnote #67 in Introduction.

[118] See KRS on the reading of σφόδρα νέον, p. 240.

[119] For the argument on this see: John Burnet, *Early Greek Philosophy* 4th ed. (London: A&C Black, 1930): 170.

Parmenides is best known for the extant proem [*The Journey to the Goddess*] to his poem that begins his book *On Nature*. For our purposes the most important passages are these:

> The mares that carry me as far as ardor might go
> Were bringing me onward, after having led me and set me down on divinity's many-worded Road,[120] which carries through all the towns (?) the man who knows.
> It was on this road that I was being carried: for on it the much-knowing horses were carrying me,
> * * *
> Towards the light and had pushed back the veils from their heads with their hands.
> That is where the gate of the paths of Night and day is.
> * * *
> The maidens guided the chariot and horses straight along the way.
> And the goddess welcomed me graciously, took my right hand into her own hand,
> And spoke these words addressing me:
> Young man, companion of deathless charioteers, you who
> Have come to our home by the mares that carry you
> I greet you: for it is no evil fate that has sent you to travel
> This road (for indeed it is remote from the paths of men),
> But Right and Justice. It is necessary that you learn everything,
> Both the unshakeable heart of well-convincing (*eupeitheos*) truth.
> And the opinions of mortals, in which there is no true belief (*pistis*).
> But nonetheless you will learn this too: how opinions
> Would have to be acceptable, forever penetrating all things. LM D4 Sextus Empiricus, *Against the Logicians,* tr. LM.[121]

And

> Well then, as for me, I shall say—and as for you, have a care for this discourse when you have heard it—
> What are the only roads of investigation for thought (*noesai*):
> The one, that "is," and that it is not possible that "is not,"
> Is the path of conviction (*peitho*), for it accompanies truth;
> The other, that "is not," and that is necessary that "is not"—
> I show you that it is a path that cannot be inquired into at all.
> For you could not know which is not (for this is impracticable)

---

[120] This language is similar to Plato, *Phaedrus* 246a ff. The motifs are also reminiscent of Homer and Hesiod.

[121] ἵπποι ταί με φέρουσιν, ὅσον τ' ἐπὶ θυμὸς ἱκάνοι,/ πέμπον ἐπεί μ' ἐς ὁδὸν βῆσαν πολύφημον ἄγουσαι/ δαίμονος, ἣ κατὰ πάντ' ἄστη φέρει εἰδότα φῶτα/ τῇ φερόμην, τῇ γάρ με πολύφραστοι φέρον ἵπποι * * * εἰς φάος, ὠσάμεναι κράτων ἄπο χερσὶ καλύπτρας./ ἔνθα πύλαι Νυκός τε καὶ Ἡματός εἰσι κελεύθων * * * ἰθὺς ἔχον κοῦραι κατ' ἀμαξιτὸν ἅρμα καὶ ἵππους./ καί με θεὰ πρόφρων ὑπεδέξατο, χεῖρα δὲ χειρί/ δεξιτερὴν ἕλεν, ὧδε δ' ἔπος φάτο καί με προσηύδα,/ ὦ κοῦρ' ἀθανάτοισι συνάορος ἡνιόχοισιν,/ ἵπποις ταί σε φέρουσιν ἱκάνων ἡμέτερον δῶ,/ χαῖρ', ἐπεὶ οὔτι σε μοῖρα κακὴ προὔπεμπε νέεσθαι/ τήνδ' ὁδόν (ἦ γὰρ ἀπ' ἀνθρώπων ἐκτὸς πάτου ἐστίν),/ ἀλλὰ θέμις τε Δίκη τε. χρεὼ δέ σε πάντα πυθέσθαι/ ἡμὲν Ἀληθείης εὐπειθέος ἀτρεμὲς ἦτορ/ ἠδὲ βροτῶν δόξας, ταῖς οὐκ ἔνι πίστις ἀληθής./ ἀλλ' ἔμπης καὶ ταῦτα μαθήσεαι, ὡς τὰ δοκοῦντα/ χρῆν δοκίμως εἶναι διὰ παντὸς πάντα περῶντα. Sext. Emp. *Adv. Math.* 7.111 (et al).

Nor could you show it. For it is the same, to think (*noein*) and also to be. LM D6 Proclus, *Commentary on Plato's* Timaeus (et al.) + Clement of Alexandria, *Stromata* (et al.).[122]

And

But since all things have been named light and night
  And what belongs to their powers is assigned to these and to those,
  The whole is altogether full of light and of un-gleaming night,
  Both of them equal, since nothing is amidst either of them. LM D13 Simplicius, *Commentary on Aristotle's* Physics.[123]

And

[...] But as for Parmenides, he seems to speak on the basis of more attentive consideration: for thinking that nonbeing is nothing next to being, he believes that necessarily being is one and that nothing else [scil. is] [...] but being obliged to follow the phenomena, and supposing that according to reason the one exists, but according to sensation the multiple does, he posits again that the causes are two and the principles two, the hot and the cold, speaking of them as fire and earth. And among these, he places the hot on the side of being and the other on the side of nonbeing. LM R12 Aristotle, *Metaphysics*.[124]

And

It is necessary that the principle be either one or multiple; and if it is one, either immobile, as Parmenides and Melissus say, or in motion, as the national philosophers say [...]. LM R22 Aristotle, *Physics*.[125]

And

Parmenides, Melissus, and Zeno abolished genesis and destruction because they thought that the whole is immobile. LM R28 Aëtius.[126]

In LM D4, the selection cited from the proem is about a metaphorical journey with divine symbolism about the acquisition of knowledge. The divine symbolism moves

---

[122] εἰ δ' ἄγ' ἐγὼν ἐρέω, κόμισαι δὲ σὺ μῦθον ἀκούσας,/ αἵπερ ὁδοὶ μοῦναι διζήσιός εἰσι νοῆσαι./ ἡ μὲν ὅπως ἔστιν τε καὶ ὡς οὐκ ἔστι μὴ εἶναι,/ πειθοῦς ἐστι κέλευθος (ἀληθείη γὰρ ὀπηδεῖ),/ ἡ δ' ὡς οὐκ ἔστιν τε καὶ ὡς χρεών ἐστι μὴ εἶναι,/ τὴν δή τοι φράζω παναπευθέα ἔμμεν ἀταρπόν,/ οὔτε γὰρ ἂν γνοίης τό γε μὴ ἐὸν (οὐ γὰρ ἀνυστόν)/ οὔτε φράσαις. τὸ γὰρ αὐτὸ νοεῖν ἐστίν τε καὶ εἶναι. Procl. *In Tim.*, 2.105b 13–22; v.3–8a: Simpl. *In Phys.* pp. 116.28–117.1 (et al.); 8b: Clem. Alex. *Strom.* 6.23.3 (et al.).

[123] αὐτὰρ ἐπειδὴ πάντα φάος καὶ νὺξ ὀνόμασται καὶ τὰ κατὰ σφετέρας δυνάμεις ἐπὶ τοῖσί τε καὶ τοῖς, πᾶν πλέον ἐστὶν ὁμοῦ φάεος καὶ νυκτὸς ἀφάντου ἴσων ἀμφοτέρων, ἐπεὶ οὐδετέρῳ μέτα μηδέν.. Simpl. *In Phys.*, p. 180.9–12.

[124] [...] Παρμενίδης δὲ μᾶλλον βλέπων ἔοικέ που λέγειν, παρὰ γὰρ τὸ ὂν τὸ μὴ ὂν οὐθὲν ἀξιῶν εἶναι, ἐξ ἀνάγκης ἓν οἴεται εἶναι τὸ ὂν καὶ ἄλλο οὐθέν [...], ἀναγκαζόμενος δ' ἀκολουθεῖν τοῖς φαινομένοις, καὶ τὸ ἓν μὲν κατὰ τὸν λόγον πλείω δὲ κατὰ τὴν αἴσθησιν ὑπολαμβάνων εἶναι, δύο τὰς αἰτίας καὶ δύο τὰς ἀρχὰς πάλιν τίθησι, θερμὸν καὶ ψυχρόν, οἷον πῦρ καὶ γῆν λέγων, τούτων δὲ τὸ μὲν κατὰ τὸ ὂν τὸ θερμὸν τάττει θάτερον δὲ κατὰ τὸ μὴ ὄν. Arist. *Metaph.* A5 986b27–987a2.

[125] ἀνάγκη δ' ἤτοι μίαν εἶναι τὴν ἀρχὴν ἢ πλείους, καὶ εἰ μίαν, ἤτοι ἀκίνητον, ὥς φησι Παρμενίδης καὶ Μέλισσος, ἢ κινουμένην, ὥσπερ οἱ φυσικοί [...]. Arist. *Phys.* 1.2 184b15–17.

[126] Παρμενίδης Μέλισσος Ζήνων ἀνῄρουν γένεσιν καὶ φθορὰν διὰ τὸ νομίζειν τὸ πᾶν ἀκίνητον.. Aët. 1.24.1 (Ps.-Plut.) [περὶ γενέσεως καὶ φθορᾶς].

us toward *phusis*$_2$[127] as one moves from ignorance (Night) into understanding (Light). The purpose of invoking *phusis*$_2$ symbolism is to put the author's account into a realm that cannot allow for inter-subjective debate based upon empirical principles (common to all parties). Instead, an esoteric situation is created in which the natural philosopher has access to "inside-intuitive information" that the rest of us do not have available to us.[128] This is definitely anti-empiricism. Because the event involves personal revelation via a goddess, it takes on a normative character (via *right* and *justice*). Epistemology that depends upon a single individual's vision of the truth is called *idealism*. This brand of philosophy is epistemologically *anti-real* (respecting an intersubjective understanding of truth). Parmenides' proem states directly, "And the opinions of mortals, in which there is no true belief (*pistis*)." If *belief* is a precondition for *knowing*, and if there can be no true belief by mortals (without supernatural assistance), then knowing is impossible by mortals. This tenet is a fundamental to antirealism.[129]

In LM D6, Parmenides sets out one of his reasons for his antirealistic leanings: the contrast between *being* and *non-being*. Can an "it" either *be* or *not be*? If so, how does the *non-being* ever become? Where does it come from? How can anything become ex nihilo? And if this is impossible, then perhaps nothing *has really become at all*. In such a situation, what we observe to be the case (viz., of there being a multitude of things that are in motion and interaction), is false. Again, we are pushed in the direction of epistemological antirealism.[130]

In LM D13, Parmenides tries to finesse this "being from non-being issue" by using the metaphors of day and night. If day is taken to be what is (*to on*) and night is not-being (*to me on*), and since in our everyday lives we see night pass into day and back into night again, then metaphorically the problem disappears. But does it? *Argument by improper analogy* is a common logical fallacy.[131] Just because day and

---

[127] This position is argued for by Stuart B. Martin, *Parmenides' Vision: A Study of Parmenides Poem* (Lanham, MD and London: University Press of America, 2016).

[128] This is the same strategy that Plato uses in the top level of his "divided line" in which non-discursive *nous* (immediate apprehension) is in the domain of the philosopher, only—Plato, *Republic* 509d-511e; cf. to Parmenides' use of *a priori* reasoning as undermining *phusis*$_1$ see: Herbert Granger, "The Proem of Parmenides Poem" *Ancient Philosophy* 28.1 (2008): 1–20 and José Trindade Santos, "The Role of Thought in the Argument of Parmenides' Poem" in *Parmenides: Venerable and Awesome: Proceedings of the International Symposium* (Las Vegas: Parmenides Publishing, 2011): 251–270.

[129] Often, this skepticism about knowing is undergirded by an assertion that the problem of error, *fallibilism*, is just too great to support any realistic epistemology. For a modern example of such a fallibilistic argument see: Roderick Firth, "Coherence, Certainty, and Epistemic Priority" *Journal of Philosophy* 61.19 (1964): 545–557. For a contrary, realistic position aimed at Firth, see Roderick Chisholm, *The Foundations of Knowing* (Minneapolis: University of Minnesota Press, 1982).

[130] Arguments about privatives (what is not) create logical problems in self-reference (thus leading to epistemological antirealism). For two different takes on this from modern perspectives see: David M. Cornell, "Taking Monism Seriously" *Philosophical Studies* 173.9 (2016): 2397–2415; and Frank A. Lewis, "Parmenides' Modal Fallacy" *Phronesis* 54 (2009): 1–8

[131] For an exposition of this see: Michael Boylan, *The Process of Argument: An Introduction* (New York and London: Routledge): p. 31.

night follow in sequence does not mean that *being* can logically come from *not-being*. There may not be an appropriate analogy here (because it begs the question of *how*).[132]

In LM R12, Aristotle brings the question back to monism. Monism *assumes* that there is something (and only one something—however that is understood). In a sympathetic rendering, Aristotle interprets Parmenides as embracing the physical world, but only parts of it. With respect to the contraries (hot, cold, wet, and dry) which were widely accepted structures of Nature that were also associated with the roots: earth, air, fire, and water,[133] Aristotle offers an interpretation of Parmenides in which this grouping is first narrowed to hot (fire) and cold (earth). From there the hot/fire is associated with *being* (among humans so long as we are warm, we are still alive) and the cold/earth is associated with not-being (among humans when we become cold, we are dead and we are buried in the earth).

Aristotle's reconstruction brings Parmenides' account back into the mainstream conversation of *phusis*—from $phusis_{3/2}$ to $phusis_1$, however; it is dubious whether Parmenides, himself would have accepted this interpretation.

In LM R 22, Aristotle raises a question that Zeno will later address: if all is *one*, then it would seem that motion, also, is impossible. Since we empirically view a reality of many things that are in motion, then in order to reject this we will have to deny the way that most people formulate their beliefs: via empirical experience that suggests that there are many things and that they are in motion. To deny this is to suggest that our epistemological grounds for belief are mistaken and that this flaw leads to epistemological antirealism.

In LM R 28, the concepts of *genesis* and *corruption* are called into question. Aëtius rightly sets out that if there is only one thing that *is*, then it is foundation of all being. It cannot come-to-be or pass away. Such a concept is the basis of most religions—though generally set out as a foundation for this world and sometimes also for a parallel universe of some sort (cf. Plato's realm of the Forms). However, if we fix our attention to the world before us, then we need to think in terms of some persisting substrate (*hupokeimenon*) that would fit this bill.[134]

What is one to make of the monism of Parmenides? One way that some have approached this is through the contemporary philosophical "type-token"

---

[132] Obviously, "begging the question" is another common logical fallacy, *ibid.* And it doesn't get any better in adding new characters to the drama—such as *Persuasion* (as an attendant to Truth—*to on*) or The Other (which represents *me einai*). These added characterizes merely assert the conclusion (a key feature of *begging the question*)—cf. KRS 291, Simplicius, *Phys.* 116, 28: 3–8.

[133] These roots were referred to outside of direct reference to Empedocles. It is also of interest that these contraries were also used in this context by Alcmaeon, according to Aëtius (DK 24 B4) in a medical contest of balancing the quartet with other parallel entities such as "bitter and sweet" as a way to bring about health. This is a predecessor to the humor theory that also rests upon the concept of *balance* as the key to health.

[134] In the ancient and medieval world, this was a position taken by some. In the modern world, the principle of conservation of matter might also fit this requirement.

distinction.[135] Under my reading of this distinction, *types* refer to classes of things. There may be certain sorts of variety within the class, but the class membership affords a sort of unity. For example, within the class of humans some are taller than others and vary in various material ways, but the essential feature (say "rationality" following Aristotle) is the same for all of them. Thus, this most important essential attribute is the same for all and they exhibit *unity* or monism in this sense.

With tokens the emphasis is upon the particularity of individuals. If we were to ascribe token monism to the ontology of Parmenides, then everything would be a part of a singular, particular whole.[136] This is perhaps most understandable from a religious perspective. If everything were part of a Divine presence, then *that* would be primary and all other differences would be accidental and unimportant (bordering upon being illusory). Under this reading all are part of *One* particular. This is the *traditional* interpretation.

While it is difficult to draw this distinction too sharply with the way to understand Parmenides, John E. Sisko and Yale Weiss have set out some insightful distinctions along these lines trying to mark out territory in-between.[137] This would make Parmenides rather more holistic while allowing more for our sensible experience of diversity. (This brings Parmenides a step closer to Heraclitus' *logos* as a unifying factor.)

However, if we accept the traditional interpretation, it would seem that scientific knowledge of nature would be prone to error. Such a barrier would make *knowledge* via sense impressions (a necessary feature for *phusis*$_1$) very problematic—and thus move the author toward epistemological anti-realism, *phusis*$_3$. This is the way I situate Parmenides.

---

[135] The type-token distinction is variously represented. My use of the distinction owes its source to Paul Grice, 'Utterer's Meaning and Intentions' (1969) 78 *Philosophical Review* 147–177 and Willard Van Quine, *Quiddities: An Intermittently Philosophical Dictionary.* (Cambridge, MA: Harvard University Press, 1987).

[136] Reconciling the problems with this traditional approach is the subject of Alex Priou, "Parmenides on Reason and Revelation" *Epoche: A Journal for the History of Philosophy* 22.2 (2018): 177–202. The author moves to Plato's *Parmenides* for a resolution of the particularity problem.

[137] John E. Sisko and Yale Weiss, "A Fourth Alternative in Interpreting Parmenides" *Phronesis* 60 (2015): 40–59. Some readings that might tend toward the reading of Parmenides as "token" are: W.K.C. Gutherie, *A History of Greek Philosophy*, vol. 1 (Cambridge: Cambridge University Press, 1965); Montgomery Furth, "Elements of Eleatic Ontology" *Journal of the History of Philosophy* 6 (1970): 111–132; Alexander Nehamas, "Parmedian Being/ Heraclitan Fire" in V. Caston and D. Graham, eds. *Presocratic Philosophy: Essays in Honor of Alexander Mourelatos* (Farnham, UK: Aldershot, 2002): 45–64; David Graham, *Explaining the Cosmos: The Ionian Tradition of Scientific Philosophy* (Princeton: Princeton University Press, 2006); Stephen Makin, "Parmenides, Zeno, and Melissus" in J. Warren and F. Sheffield, eds. *The Routledge Companion to Ancient Philosophy* (London and New York: Routledge, 2014): 34–48. For an example of reading the monism as a "type" see Patricia Curd, *The Legacy of Parmenides: Eleatic Motion and Later Presocratic Thought.* 2nd ed. (Las Vegas: Parmenides Publishing, 2004): Part II.

Zeno tries a different approach. He is keen on a mathematical worldview which would situate him as *phusis*$_2$ but because his goal, in the use of mathematics, is to undermine the way most people view Nature—and that view is mistaken, thus supporting epistemological antirealism.[138]

Zeno seems to have had his *floruit* in the 460s BCE (based upon Diog. L. ix, 29 = DK 29A 1) or a bit later according to Plato's *Parmenides* (KRS 286). One possible way to reconcile these discrepancies is that they may use other key events to make their case.[139]

The way Zeno fits into this narrative is the way he sets out his antirealism-Parmenides tract via his paradoxes of motion.[140] Here are a few of the key texts on this.

> [*The Grain of Millet*] [. . .] Zeno's argument, which says that any part of a grain of millet makes a sound [. . .]. LM D12a.Aristotle, *Physics* [141]/ [. . .] he [i.e., Aristotle] resolves the argument of Zeno of Elea, which he posed as a question to the sophist Protagoras. For he said, "Tell me, Protagoras, does one grain of millet make a sound when it falls or does the thousandth part of the grain of millet?" When the other answered that it did not, he said, "Does a medimnus[142] of grains of millet make a sound or not when it falls?" When the other answered that it did make a noise, Zeno said, "Well then, is there not a proportion between a medimnus of grains of millet and a single grain the thousandth part of one grain?" And when the other answered that there was one, Zeno said, "Well then, will there not be the same proportions between the sounds with regard to one another? For just as the things are that make a sound, so too are their sounds; and since that is so, if a medimnus of millet makes a sound, a single grain of millet will make a sound too, and so too the thousandth part of that grain." LM D12b. Simplicius, *Commentary on Aristotle's Physics*.[143]

---

[138] For another angle on Zeno's use of mathematics to undermine epistemological realism see: Trish Glazebrook, "Zeno Against Mathematical Physics" *Journal of the History of Ideas* 62.2 (2001): 193–210.

[139] For more on this see: Daniel W. Graham and Justin Barney, "On the Date of Chaerephon's Visit to Delphi" *Phoenix* 70.3/4 (2016): 274–289.

[140] The "traditional" readings of these paradoxes come from: G.E.L. Owen, "Zeno and the Mathematicians" *Proceedings of the Aristotelian Society,* 58 (1957–58): 199–222; J.F. Thomson, "Tasks and Super-Tasks," *Analysis* 15 (1954): 1–13; Gregory Vlastos, "A Note on Zeno's Arrow" *Phronesis* 11 (1966): 3–18; Gregory Vlastos, "Zero's Race Course" *Journal of the History of Philosophy* 4 (1966): 95–108; James Watling, "The Sum of an Infinite Series" *Analysis* 13 (1953): 39ff. For more recent writing on these see: Patrick Reeder, "Zeno's Arrow and the Infinitesimal Calculus" *Synthese* 192.5 (2015): 1315–1335 and Eduardo Noble and Max Fernández de Castro, "The Ancient versus the Modern Continuum" *Formal Sciences and Philosophy: Logic and Mathematics*(2017): 1343–1380; Giuseppe Micheli, "Kant and Zeno of Elea: Historical Precedents of the 'Skeptical Method'" *Revista de Filosofia* 37.3 (2014): 57–64; and Wesley Salmon, ed. *Zeno's Paradoxes* (Indianapolis, IN: Bobbs-Merril, 1970); and Anthony Preus, ed. *Essays in Ancient Greek Philosophy* IV: Before Plato (Albany, N.Y.: SUNY Press, 2001).

[141] ὁ Ζήνωνος λόγος [. . .] ὡς ψοφεῖ τῆς κέγχρου ὁτιοῦν μέρος [. . .]. Arist. *Phys.* 8.5 250a19–22.

[142] A unit of measure around 52 liters.

[143] [. . .] λύει καὶ τὸν Ζήνωνος τοῦ Ἐλεάτου λόγον, ὃν ἤρετο Πρωταγόραν, τὸν σοφιστήν. "εἰπὲ γάρ μοι," ἔφη, "ὦ Πρωταγόρα, ἆρα ὁ εἷς κέγχρος καταπεσὼν ψόφον ποιεῖ ἢ τὸ μυριοστὸν τοῦ κέγχρου"; τοῦ δὲ εἰπόντος μὴ ποιεῖν "ὁ δὲ μέδιμνος," ἔφη, "τῶν κέγχρων καταπεσὼν ποιεῖ ψόφον ἢ οὔ"; τοῦ δὲ ψοφεῖν εἰπόντος τὸν μέδιμνον "τί οὖν," ἔφη ὁ Ζήνων, "οὐκ ἔστι λόγος τοῦ μεδίμνου τῶν κέγχρων πρὸς τὸν ἕνα καὶ τὸ μυριοστὸν τοῦ ἑνός"; τοῦ δὲ φήσαντος εἶναι "τί οὖν," ἔφη ὁ Ζήνων, "οὐ καὶ τῶν ψόφων ἔσονται λόγοι πρὸς ἀλλήλους οἱ αὐτοί; ὡς γὰρ τὰ

And

[*Argument Against the Existence of* Place] Moreover, if it [i.e., place] is one of the things that are, where will it be? For Zeon's *aporia* requires some argumentation. For if every *thing* that exists is in a place, it is clear that there will also be a place of the place, and this will go on to infinity. LM D13 a, Aristotle, *Physics*.[144] / Zeno's argument seemed to abolish the existence of place by asking as follows: "If place is, it will be in something. For everything that is in something; now, in something is also in a place; so the place too will be in a place, and this will go on to infinity. So, place does not exist." LM D13 b, Simplicius, *Commentary on Aristotle's Physics*.[145]

And

[*Called the Dichotomy*] [. . .] the first [scil. Argument] is that there is no motion, because what is displaced must arrive at the half before arriving at the end [. . .] LM D14 Aristotle, *Physics*.[146]

And

[*Called the Achilles*] The second [scil. Argument] is the one called "Achilles." It consists of saying that what is slowest will never be overtaken when it runs by what is fastest. For before that can happen, it is necessary that the pursuer arrive at the place from which the pursued started out, so that it is necessary that the slower one always be somewhat ahead. This argument too is the same as the one by the procedure of dichotomy, but it differs in that the supplementary magnitude is not divided in half [. . .] LM D15 Aristotle, *Physics*.[147]

And

[*Called the Arrow*] If, he says, everything is always at rest when it is in an equal [scil. Space], and what moves is always in the present moment, then the arrow that is displaced is immobile. LM D16 Aristotle, *Physics*.[148]

And

[*Called the Stadium*] The fourth [scil. Argument] is about bodies of the same dimensions that move at an equal speed in a stadium and pass alongside other bodies of the same

---

ψοφοῦντα, καὶ οἱ ψόφοι, τούτου δὲ οὕτως ἔχοντος, εἰ ὁ μέδιμνος τοῦ κέγχρου ψοφεῖ, ψοφήσει καὶ ὁ εἷς κέγχρος καὶ τὸ μυριοστὸν τοῦ κέγχρου." Simpl. *In Phys.*, p. 1108.14–29.

[144] ἔτι δὲ καὶ αὐτὸς εἰ ἔστι τι τῶν ὄντων, ποῦ ἔσται; ἡ γὰρ Ζήνωνος ἀπορία ζητεῖ τινὰ λόγον, εἰ γὰρ πᾶν τὸ ὂν ἐν τόπῳ, δῆλον ὅτι καὶ τοῦ τόπου τόπος ἔσται, καὶ τοῦτο εἰς ἄπειρον. Arist. *Phys.* 4.1 209a 23–26.

[145] ὁ Ζήνωνος λόγος ἀναιρεῖν ἐδόκει τὸ εἶναι τὸν τόπον ἐρωτῶν οὕτως, "εἰ ἔστιν ὁ τόπος, ἔν τινι ἔσται, πᾶν γὰρ ὂν ἔν τινι, τὸ δὲ ἔν τινι καὶ ἐν τόπῳ ἔσται ἄρα καὶ ὁ τόπος ἐν τόπῳ καὶ τοῦτο ἐπ' ἄπειρον, οὐκ ἄρα ἔστιν ὁ τόπος." Simp. *In Phys.*, p. 562.3–6.

[146] [. . .] πρῶτος μὲν ὁ περὶ τοῦ μὴ κινεῖσθαι διὰ τὸ πρότερον εἰς τὸ ἥμισυ δεῖν ἀφικέσθαι τὸ φερόμενον ἢ πρὸς τὸ τέλος [. . .].. Arist. *Phys.* 6.9 239b 11–14.

[147] δεύτερος δ' ὁ καλούμενος Ἀχιλλεύς, ἔστι δ' οὗτος, ὅτι τὸ βραδύτατον οὐδέποτε καταληφθήσεται θέον ὑπὸ τοῦ ταχίστου, ἔμπροσθεν γὰρ ἀναγκαῖον ἐλθεῖν τὸ διῶκον ὅθεν ὥρμησεν τὸ φεῦγον, ὥστε ἀεί τι προέχειν ἀναγκαῖον τὸ βραδύτερον. ἔστιν δὲ καὶ οὗτος ὁ αὐτὸς λόγος τῷ διχοτομεῖν, διαφέρει δ' ἐν τῷ διαιρεῖν μὴ δίχα τὸ προσλαμβανόμενον μέγεθος [. . .].. Arist. *Phys.*6.9 239b 14–20.

[148] εἰ γὰρ αἰεί, φησίν, ἠρεμεῖ πᾶν ὅταν ᾖ κατὰ τὸ ἴσον, ἔστιν δ' αἰεὶ τὸ φερόμενον ἐν τῷ νῦν, ἀκίνητον τὴν φερομένην εἶναι ὀϊστόν. Arist. *Phys.* 6.9 239b 5–7.

dimensions in the opposite direction, the ones starting from the end of the stadium, the others from the middle, in which case, he thinks, one half of a period of time is equal to its double [. . .] LM D18 Aristotle, *Physics*.[149]

A few brief comments on these passages are in order. They raise considerable issues in the philosophy of mathematics and the philosophy of science. First is the Millet Seed. This thought experiment depends upon the notion of *material reductionism*. This has been an historically engaging concept. It suggests that at various levels of material structure—say 1–50 (where 50 is the most organized), that all of the properties that are present in the structure at 50 can be explained by adding up the properties at 1–49 and summing them up together). There is much that is attractive to this mental construct—even to this present day. But there are problems, too, because organizational structures at different layers of the system—say at level 20, 30, and 40—can create emergent properties that only exist in this *particular* organizational scheme at these levels.[150]

To modern readers it is not much of a stretch to accept that one Millet Seed (or a fraction of a Millet Seed) falling to the ground might not make an audible sound to human ears, whereas when 52 liters of Millet Seeds fall to the ground, there will a resounding *thump*. This is because if the standard is "what a group of bystanders can hear" (even if they all have excellent hearing) does not convey a property to all Millet Seeds in any configuration.[151]

However, Zeno does not accept *emergence*. And many to most of his contemporaries probably wouldn't have either. To them material properties are always additive. The given whole is *always* the sum of its parts. To be otherwise creates a logical anomaly, which suggests doubt in reliability of empirical data. Since much in the material world (especially in biomedicine) exhibits emergence at different levels of organization, Zeno's tack here is one step in the direction of epistemological anti-realism, *phusis*$_3$.

In the argument against the existence of place, we are presented with a self-referential context: if there were *place*, then it would have to exist in a place, and that in a place, ad infinitum. Self-referential contexts make one claim in a first order

---

[149] τέταρτος δ' ὁ περὶ τῶν ἐν τῷ σταδίῳ κινουμένων ἐξ ἐναντίας ἴσων ὄγκων παρ' ἴσους, τῶν μὲν ἀπὸ τέλους τοῦ σταδίου τῶν δ' ἀπὸ μέσου, ἴσῳ τάχει, ἐν ᾧ συμβαίνειν οἴεται ἴσον εἶναι χρόνον τῷ διπλασίῳ τὸν ἥμισυν [. . .]. Arist. *Phys.* 6.9 239b 33–240a1.

[150] For some of the background on the dynamics of reductionism and emergent properties in organized systems see: Richard Boyd, Philip Gasper, and J.D. Trout, eds. *The Philosophy of Science* (Cambridge, MA.: M.I.T University Press, 1993): section III. Also see: Boylan (1983): 232–233. See also on emergence: Alexandru Manafu, "How Much Philosophy is in Chemistry?" *Journal for General Philosophy of Science* 45.1 (2014): 33–44; and Peter Corning, "The Re-Emergence of Emergence, and the Causal Role of Synergy in Emergent Evolution" *Synthese* 185.2 (2012): 295–317.

[151] The same would hold true with feathers or other objects that do not appear to create sounds. Fifty-two liters of feathers will also make a sound that could be discerned by most.

logical declaration and then apply the rules of the original first order sentence to that sentence itself. An historically important instance of a self-referential context in the history of philosophy is the *Liars Paradox* (one popular form goes like this "Epimenides, a Cretan, says, 'All Cretans are liars.'" If there are only two truth states: L (liar) and T (truth teller), and if we depict the sentence attributed to Epimenides as α, and if we attribute the speech act of Epimenides as β, then if α is true, β is false and vice versa. This creates a paradox when viewed on the level of first order logic). However, when one views the interaction of the speech act, β according to the rules of α as separate (second order logic), then according to some (such as this author) the paradox takes on new meanings, but is no longer a "flat-out contradiction."[152] This would mean that the force of Zeno's contention that contemplation about whether the *idea of place, as such* is a flat-out contradiction probably doesn't hold the test of time—though it was certainly powerful enough in its historical epoch and moves Zeno another step towards doubting "place" (epistemological antirealism) that would undermine an essential element for *phusis*$_1$ advocates (epistemological realists).

Third are both the Dichotomy and the Achilles which utilize the same notion of measurement for combination in infinite series in both a positive and negative fashion. In the Dichotomy motion is impossible (negative version) because if some body, A, wants to move from her starting point of X to Z, she must first pass-through midpoint Y. But to get to Y she must first pass through the midpoint of the line segment XY. But before that is possible, another midpoint must be traversed, ad infinitum. The result is that motion is impossible (epistemological antirealism).

In the Achilles complementary mathematical series combination is addressed (positive version). For Achilles to overtake the tortoise (who is at unit 5 on his head start), he must move from the starting point of 0 to a point half-way to the tortoise (2.5). But once there, he must get to half-way toward 5 (3.75) and once there to get to 5 he has to get to half again (4.375) and then half again (4.6875), ad infinitum. As many "half-again" instances as you want to put forth, there will always be an interval between Achilles and the tortoise. This means that the speedy Achilles cannot overtake the tortoise and once again, epistemological antirealism is upheld, *phusis*$_3$.

---

[152] Self-referential paradoxes are very crucial to understanding completeness and consistency within axiomatic systems, such as deductive logic and mathematics. For a history of the Liars Paradox (cited above) see Roy A. Sorenson, *A Brief History of Paradox* (Oxford: Oxford University Press, 2003). For the place of self-referential contexts within mathematics and logic, compared, see the useful gloss on Gödel's Theorem: Ernest Nagel and James R. Newman, *Gödel's Proof* (New York: New York University Press, 1958 [1931]). Obviously, since Gödel's Theorem was one of the most influential works on the structure of logic and mathematics, Zeno's use here is very prescient. Two other accounts that situate self-referential contexts as post-Gödel are Richard Heck, "Self-Reference and the Language of Arithmetic" *Philosophia Mathematica* 15.1 (2017): 1–29 and Bas C. van Fraassen, "Perception, Implication and Self-Reference" *Journal of Philosophy* 65.5 (1968): 136–152.

In the modern realm, once again, the issue that Zeno raises gives grist to philosophical analysis on the idea of combinational method. There are two possible sources of the absurdity that Zeno appeals to in these two paradoxes: (a) the impossibility of adding an infinite series; and (b) the impossibility of there being an infinite series or set, in the first place. Due to the influential articles by Vlastos in 1966, *op. cit.* much of the criticism since has examined the problem from the (a) point of view.[153] From the standpoint of both Newtonian and Leibnitzian calculus, a modern could integrate the series and solve the problem. From the standpoint of ancient mathematics Mueller and Heath thought that the method of exhaustion might have offered a solution to the (a) interpretation.[154] Be that as it may, for most contemporaries, these two paradoxes make a strong statement for the antirealist position.

The (b) interpretation would not have been tenable in ancient mathematics and not until the advent of transfinite numbers and Zermelo-Fraenkel set theory in the last century was there any way to respond to this tack. Thus, the (b) interpretation is the most challenging and leads most forcefully to epistemological antirealism, *phusis*$_3$.

In the fourth paradox, the Arrow, there is another attack on the notion of there being a realistic place. What is at stake in this instance is whether space is continuous or whether it exhibits contiguity. The former case is much like naïve empirical experience of space. Things flow from A to B to C without any hindrance. But what if this concept of space were false? If space were not continuous but contiguous? In this case motion would be like movie film: separated by frames. For an object to move, it must go through frames 1, 2, 3, 4, . . . This concept of space would create the conditions "jumping" from one frame to another and that the *ability* to carry out this jump would require additional explanation.[155] As with the other paradoxes,

---

[153] Another interesting way of thinking about the (a) position has been put forward by Charles McCarty, "Paradox and Potential Infinity" *Journal of Philosophical Logic* 42 (2013): 195–219. McCarty argues that even under the (a) interpretation there is no actual infinite set that results, (b) interpretation. Instead, the issue is a multi-level combination problem that can invoke 1st and 2nd order logic. Using Church's Theorem (A. Church "A Note on the *Entscheidungsproblem*" *Journal of Symbolic Logic* (1936):40–41; see also A. Church, *An Introduction to Mathematical Logic* (Princeton, N.J.: Princeton University Press, 1986 [1944]): chapts. 17–18) McCarty argues that these two combinational paradoxes (a) interpretations, point to deep problems of consistency and completeness in arithmetic (the *Entscheidungsproblem*), and that they call for the creation of 2nd order symbolic logic in order to offer a solution. For another look at a not un-related approach to the combination problem of Zeno see: James Levine, "Russell and the Transfinite" *Hermathena: Philosophy and Mathematics* 190 (2011): 53–112. For a more historically centered examination see: Eduardo Noble, "The Ancient versus The Modern Continuum" *Revista Portuguesa de filosofia* T. 73, fac. ¾ (2017): 1343–1380.

[154] Mueller (1981): 234ff.; T.L. Heath "Greek Geometry with Special Reference to Infinitesimals," *The Mathematical Gazette* 11 (1922–1923):252–253

[155] Another take on discontinuity of space is discussed by Constantin Antonopoulos, "Einstein's 'true' Discontinuity with an Application to Zeno" *Theoria* 23.3 (2008): 339–349.

there are some modern paradigms that look at sub-atomic movement in just this way.[156]

In the ancient world, the only way to respond to the Arrow is to take Aristotle's approach and separate the special and temporal elements and assert that both are *continuous* (Aristotle, *Physics* VI 239b 5ff). The arrow never has a "place of its own"—for then it would stop in time and rest in its place. This is what the paradox suggests. Since the arrow (under naïve empiricism) does not rest at any *moment* in its flight, Zeno's analysis is false. This argument in its historical context is weaker than it is in its modern consideration: a weak move for *phusis$_3$*.

The final argument, the Stadium, considers moving bodies (groups of people in α and in β moving in opposite directions inside an artificial space: an enclosed stadium).[157] To make this work one must contrast two competing perceptions of velocity: imagine three observers: A is an observer with a keen sense of velocity and is sitting in the stadium's stands. B is an observer within group α and C is an observer within group β. Observers B and C before the bodies passed each other, would measure their speed against a background—say the stadium. Their assessment would be X miles per hour (assuming that they are moving at the same pace). The assessment of A would also put α and β at X miles per hour.

However, things change when the columns of individuals pass each other. For A (the observer in the stands) there would be no difference in the assessment of velocity, since his/her vantage point has not changed. But for B and C (observers within groups α and β) if they take their assessment against a background in which α and β are in the foreground, then they will observe a velocity of 2X. This is because they are taking into account the velocity of the other column. (This is much of the sensation one has when riding a train and passing another train going in the opposition—there is an appearance of sudden acceleration.)

For contemporaries of Zeno, this reaction would seem to suggest that empirical assessment of velocity under certain, unchanging circumstances is suddenly different. The effect of this would be to lose confidence in naïve empiricism: the foundation of epistemological antirealism, *phusis$_3$*.

All five paradoxes: the Millet Seed, the Self-Referential argument on place, the Dichotomy/Achilles, the Arrow, and the Stadium all use mathematical analysis to generate problems with naïve empiricism (a foundation of epistemological realism). This background will pave the way for Sextus Empiricus and other subsequent *phusis$_3$*, skeptical physicians.

---

[156] For a quick introduction to this see: A.C. Phillips, *Introduction to Quantum Mechanics* (West Sussex, UK: Wiley, 2003): chapts. 3 & 5; and A. Zee, *Quantum Field Theory In a Nutshell* (Princeton, N.J.: Princeton University Press, 2003): part 4.

[157] For a listing of some traditional interpretations of The Stadium see: Kevin Davey, "Aristotle, Zeno, and The Stadium Paradox" *History of Philosophy Quarterly* 24.2 (2007): 127–146.

Summary of All Pre-Socratics

This chapter has attempted to set out a background for the emerging schools of Ancient Medicine. These speculations about Nature reflect a trifurcation from advocating a material basis, to a non-material (theistic or non-material, realistic mathematics), to skeptical antirealistic views of nature. The theories are variously "complete" according to their generating principles. Most are coherent and simple in their exposition, and because of this can be said to be elegant. When Nature is treated as a powerful entity that should be regarded as such, then the maker/artifact dynamic is present. This background can help us better understand the Ancient Greek Medical tradition from the Hippocratic Writers to Galen. That's next!

Chapter 2
The Hippocratic Contribution

Abstract The Hippocratic Writers present a big step forward in the practice of medicine. They present a rationally-based account of material causation of disease and the theoretical framework for confronting patients who are presenting symptoms: diagnosis, prognosis, and treatment. The professional ethics represented in the Oath go a long way towards making medicine a profession.

Keywords Techne · Nature · Diagnosis · Prognosis · Treatment · Psuche

Hippocrates of Cos was said to have lived sometime between 450 and 350 BCE. He was a physician, and the writings of the *Corpus Hippocraticum* provide a wealth of information on biomedical methodology and offer one of the first reflective codes of professional ethics. Though Plato (a contemporary) makes reference to Hippocrates (*Phaedrus* 270a and elsewhere), it is generally believed that most of the writings in the *Corpus Hippocraticum* are actually the work of a number of different writers. By convention of time, place, and general approach a common name of 'Hippocrates' was assigned to the lot (without distinguishing those of the historical Hippocrates).[1]

---

[1] For a discussion of Hippocrates' life and how this is set out with scant textual information see: W.D. Smith, *The Hippocratic Tradition* (Ithaca, N.Y.: Cornell University Press, 1979); R.R.R. Smith, "Late Roman Philosopher Portraits from Aphrodisias" *Journal of Roman Studies* 80 (1990): 127–155. Eric D. Nelson, "Coan Promotions and the Authorship of the Presbeutikos," in P.J. van der Eijk, ed., *Hippocrates in Context: Papers Read at the 11th International Hippocrates Colloquium, University of Newcastle upon Tyne 27-31 August 2002* (Leiden and Boston: Brill, 2005): 209–236; W. Speyer, *Die literarische Fälschung im heidnischen und christlichen Altertum: ein Versuch ihrer Deutung* (Munich: Beck, 1971); M. Lefkowitz, *The Lives of the Greek Poets* (London: Duckworth, 1981); Philip van der Eijk, "On 'Hippocratic' and 'Non-Hippocratic' Medical Writing" in Lesley Dean-Jones and Ralph M. Rosen, eds. *Ancient Concepts of the Hippocratic* (Leiden: Brill, 2016): 17–47; and Jacques Jouanna, *Greek Medicine from Hippocrates to Galen* (Leiden and Boston: Brill, 2012).

© The Author(s), under exclusive license to Springer Nature Switzerland AG 2025
M. Boylan, *A Philosophical History of Western Medicine*,
https://doi.org/10.1007/978-3-031-97806-7_2

On the biomedical methodology side, these writings provide the most detailed biomedical observations to date in the Western world. They also offer causal speculations that can be knitted together to form a theoretical framework for diagnosis, prognosis, and treatment. On the ethical side, their code of professional ethics is so well structured that it continues to stand as a model for other professions. This chapter will examine all four of these contributions by Hippocrates (either the man or the group).

Introduction

Let us begin with a meta-observation about the way patients are viewed by the Hippocratic writers. Going back to the Plato reference from the *Phaedrus* we see the following[2]:

> [269e 4] **Socrates**: All the great arts require endless talk and [270a] ethereal speculation about nature: This seems to be what gives them their lofty point of view and universal applicability. That's just what Pericles mastered—besides having natural ability. He came across Anaxagoras, who was just that sort of man, got his full dose of ethereal speculation, understood the nature of mind and mindlessness—just the subject on which Anaxagoras had the most to say. From this, I think he drew for the art of rhetoric what was useful to it.
> **Phaedrus**: What do you mean by that?
> [270b] **Socrates**: Well, isn't the method of medicine in a way the same as the method of rhetoric?
> **Phaedrus:** How so?
> **Socrates**: In both cases we need to determine the nature of something—of the body in medicine, of the soul in rhetoric. Otherwise, all we'll have will be an empirical and artless practice. We won't be able to supply, on the basis of an art, a body with the medicines and diet that will make it healthy and strong, or a soul with reasons and customary rules for conduct that will impart it to convictions and virtues we want.
> **Phaedrus**: That is most likely, Socrates.
> [270c] **Socrates:** Do you think, then, that it is possible to reach a serious understanding of the nature of the soul without understanding the nature of the world as a whole?
> **Phaedrus:** Well, if we're to listen to Hippocrates, Asclepius' descendant, we won't even understand the body if we don't follow that method (tr. Alexander Nehamas and Paul Woodruff).[3]

---

[2] For a discussion of Plato and the Hippocratics see: Paul Demont, "Remarques sur le Fableau de la médecine et d' Hippocrate chez Platon" Jones and Rosen, 61–82.

[3] ΣΩ: Πᾶσαι ὅσαι μεγάλαι τῶν τεχνῶν προσδέονται ἀδολεσχίας καὶ μετεωρολογίας φύσεως πέρι, τὸ γὰρ ὑψηλόνουν τοῦτο καὶ πάντῃ τελεσιουργὸν ἔοικεν ἐντεῦθέν ποθεν εἰσιέναι. ὃ καὶ Περικλῆς πρὸς τῷ εὐφυὴς εἶναι ἐκτήσατο, προσπεσὼν γὰρ οἶμαι τοιούτῳ ὄντι Ἀναξαγόρᾳ, μετεωρολογίας ἐμπλησθεὶς καὶ ἐπὶ φύσιν νοῦ τε καὶ διανοίας ἀφικόμενος, ὧν δὴ πέρι τὸν πολὺν λόγον ἐποιεῖτο Ἀυαξαγόρας, ἐντεῦθεν εἵλκυσεν ἐπὶ τὴν τῶν λόγων τέχνην τὸ πρόσφορον αὐτῇ.
ΦΑΙ: Πῶς τοῦτο λέγεις;
ΣΩ: Ὁ αὐτός που τρόπος τέχνης ἰατρικῆς ὅσπερ καὶ ῥητορικῆς.
ΦΑΙ: Πῶς δή;
ΣΩ: Ἐν ἀμφοτέραις δεῖ δειλέσθαι φύσιν, σώματος μὲν ἐν τῇ ἑτέρᾳ, ψυχῆς δὲ ἐν τῇ ἑτέρᾳ, εἰ μέλλεις, μὴ τριβῇ μόνον καὶ ἐμπειρίᾳ ἀλλὰ τέχνῃ, τῷ μὲν φαρμακα καὶ τροφὴν προσφέρων ὑγίειαν καὶ ῥώμην ἐμποιήσειν, τῇ δὲ λόγους τε καὶ ἐπιτηδεύσεις νομίμους πειθὼ ἣν ἂν βούλῃ καὶ ἀρετὴν παραδώσειν.

To unpack this quotation, we begin with Socrates' initial speech in which he sets out that the great arts (*technai*) require an understanding of nature (*phusis*$_1$) in order to attain proficiency. The two arts that he compares here are rhetoric and medicine. The example given for rhetoric is Pericles, the great Athenian statesman (acclaimed in superlatives by Thucydides), and for medicine Hippocrates. In the case of Pericles, the vision of nature that made the difference came from Anaxagoras and his doctrines of *nous* and *anoias* (translated above as "mindlessness"). For our purposes here we can view the adaptation of *nous* as a general principle of "intellect within *phusis*$_{1\text{ or }2}$" (metaphysical realism in today's philosophical vocabulary), as opposed to *anoias* which would be *phusis*$_3$ (metaphysical anti-realism in today's philosophical vocabulary).[4] The point of Socrates first speech, above, is to claim that *because* Pericles was able to latch onto something *true* about the human condition, he could use this knowledge of what is true about humankind (*nous*) to be persuasive to the citizens of Athens in order to make his case about various public policies he supported.

In an analogous way, the field of medicine (270b) also grounds its *techne* in large, general principles that can be known—are, in modern jargon, metaphysically real so that knowledge of them is also real (epistemological realism). What is knowable is the nature of the *body* and its ME (material explanation) along with the *soul* (*psuche*—a stand-in for the TE (teleological explanation), it's proper purposes according to its essences). The particulars here are the ME elements—such as a "pain in my head" or a "sour feeling in my stomach"—which cannot be properly understood in this understanding of the *method* without an understanding of what the holistic depiction would be (the TE components).

Under my reading of this passage, the physician (like the rhetorician) must ground a particular clinical interaction with a patient within a general understanding of the operation of the whole body so that variations and their resulting effects might be known within this context.[5] This puts the responsibility of learning as much as possible (of the holistic operation of the entire body) upon the aspiring physician. The underlying assumption is that the parts of the body *do not act entirely independently of each other* but, instead, are a synchronized (symmetrical) whole.[6]

ΦΑΙ: Τὸ γοῦν εἰκός, ὦ Σώκρατες, οὕτως.
ΣΩ: ψυχῆς οὖν φύσιν ἀξίως λόγου κατανοῆσαι οἴει δυνατὸν εἶναι ἄνευ τῆς τοῦ ὅλου φύσεως;
ΦΑΙ: Εἰ μὲν Ἱπποκράτει γε τῷ τῶν Ἀσκληπιαδῶν δεῖ τι πιθέσθαι, οὐδὲ περὶ σώματος ἄνευ τῆς μεθόδου ταύτης.
Plato, *Phaedrus* 269e 4–270c 5.

[4] For an introduction to these terms see: William P. Alston, "What Metaphysical Realism is Not" in W. Alston. *Realism and Antirealism* (Ithaca, N.Y.: Cornell University Press, 2002): 97–116.

[5] Historically, this has been a controversial passage. See: R. Joly, "Le question Hippocratic et le témoinage de Phèdre" *Revue des Études grecques*, 74 (1961):194–223 for historical summaries of discussions; W.D. Smith, *The Hippocratic Tradition* (Ithaca, N.Y.: Cornell University Press, 1979):31–39; G.E.R. Lloyd, *Methods and Problems in Greek Science* (Cambridge: Cambridge University Press, 1991):194–223; and J.H. Solbakk, *Forms and Functions of Medical Knowledge in Plato*, Ph.D. Thesis. University of Oslo. 1993: 170–203.

[6] Again, we see the influence of *symmetry* as a principle of truth in science. It will be the contention of this book, that this is a powerful explanatory principle—not just in medicine, but also in the underlying science, going forward to the present day.

This is an important methodological stance for it puts front-and-center a requirement for the physician to be knowledgeable of the entire human body (if that is understood as the *whole*)—extending to an understanding of basic science (if that is understood as the *whole*). It will be my reading of the passage that both, separately, are to be understood as the *whole*. Such an approach will tend to increase *completeness* and *elegance* (two of our primary categories for theory assessment).

If this is indeed the position of Hippocrates and the Hippocratic writers, in general, then we have an important component of what the methodology of theory should look like. If we were to examine this tenet in light of developments in contemporary medicine, then we would observe the opposite approach, in the main. In contemporary medicine, internists and emergency department physicians are taught a rather general approach, and they are often the first to confront a patient. However, unless the physical complaint is rather general, they are often put into the position of referring a patient to an expert, who is often very specialized and can miss important symptoms that present *outside* of their specialty. This variation away from the Hippocratic principle asserted in this passage can have negative effects upon patients.[7]

Even in the ancient world, there were also potential dissidents who set out that one might be able to take care of joints and dislocations without the extensive holistic training that this passage seems to ascribe to Hippocrates.[8]

Obviously, the holistic approach moves us in the direction of the basic science that the Hippocratic writers assumed in order to move forward to *diagnosis, prognosis,* and *treatment.*

Basic Science Applied to Medicine

Psuche *and the Role of External Air and Other External Factors*

As we have seen earlier with Aristotle, the role of *vitalized air, psuche,* came to be seen as an essential part of biomedical explanation.[9] What is at stake is how much *air* becomes the principal component in regulating and determining the health of

---

[7] My daughter, who is a neurosurgeon, was extensively trained both generally and specialty. This is rather unusual in these modern times. The overspecialization of most surgeons means that they may miss symptoms or improperly read test data outside of their narrow scope. My daughter has related to me many "horror stories" of various medical sub-specialty personnel who have missed critical symptoms outside of their specialty because they never received a more general level of training (in addition to their narrow ability, e.g., to clip an aneurism).

[8] One possible instance of this reported by Galen in his commentary on *Joints* (closely related to *Fractures*) cites such a position by Ctesias of Cnidus, see: W.D. Smith (1979):731; Lloyd (1991): 205.

[9] There is some controversy on this point when one examines the Anonymus Londinensis papyrus, cf. Vivian Nutton, *Ancient Medicine*, 2nd ed. (London and New York: Routledge, 2013): 58–61, cf.

humans. What we have to distinguish is the difference between "air" that is viewed as an *external* source and the "processed" internal air that is turned into *pneuma*.

Certainly, the effect of external input is part of the thesis of *Airs, Waters, and Places*. In a general sense, the idea is that our health can be affected by the quality of air that we take in.

> Whoever wishes to pursue properly the science of medicine must proceed thus. First, he ought to consider what effects of each season of the year can produce; for the seasons are not at all alike, but differ widely both in themselves and at their changes. The next point is the hot winds and the cold, especially those that are universal, but also those that are particular to each particular region. He must also consider the properties of the waters; for as these differ in taste and in weight, so the property of each is far different from that of any other. Therefore, on arrival at a town with which he is unfamiliar, a physician should examine its position with respect to the winds and to the risings of the sun. (*Airs, Waters, and Places,* ll. 1–15, tr. W.H.S. Jones).[10]

In this passage we begin with the exhortation that to inquire the "right way" (*orthos*) into medicine one must consider a variety of external factors (the temperature, the properties of the waters, but most especially *qualities of the air*—including aspects of the winds along with the added input of sunrise and sunset). In the lines that follows, the quality of the waters (including aspects of the soil) creates marshy, brackish conditions that *foul the air*. These exterior factors can create illness (according to the author). If we are to proceed as *dogmatists*,[11] we must know the full-cause (*aitia*) of the disease in order to properly treat it. In this text, causation is a primary concern.

For those physicians who were itinerant, the author suggest that they acquaint themselves with the exterior conditions of air, water, and soil (in each season) so that they might familiarize themselves with these possible exterior conditions that might have cause the disease. For example, unduly hot winds might cause excess phlegm that begins in the brain and runs down into the digestive organs (III. 10ff.). The after-effects are: childhood asthma, maternal spontaneous abortion (miscarriage), childhood epilepsy, male dysentery, diarrhea, ague, fever, eczema, and

---

also Susan Price, "The Peripatetic Hippocrates and other Monists in Anonymus Londiniensis" in Jones and Rosen: 99–116.

[10] Ἰητρικὴν ὅστις βούλεται ὀρθῶς ζητεῖν, τάδε χρὴ ποιεῖν, πρῶτον μὲν ἐνθυμεῖσθαι τὰς ὥρας τοῦ ἔτεος, ὅ τι δύναται ἀπεργάζεσθαι ἑκάστη, οὐ γὰρ ἐοίκασιν ἀλλήλοισιν οὐδέν, ἀλλὰ πολὺ διαφέρουσιν αὐταί τε ἐφ᾽ ἑωυτέων καὶ ἐν τῇσι μεταβολῇσιν, ἔπειτα δὲ τὰ πνεύματα τὰ θερμά τε καὶ τὰ ψυχρά, μάλιστα μὲν τὰ κοινὰ πᾶσιν ἀνθρώποισιν, ἔπειτα δὲ καὶ τὰ ἐν ἑκάστη χώρη ἐπιχώρια ἐόντα. δεῖ δὲ καὶ τῶν ὑδάτων ἐνθυμεῖσθαι τὰς δυνάμιας, ὥσπερ γὰρ ἐν τῷ στόματι διαφέρουσι καὶ ἐν τῷ σταθμῷ οὕτω καὶ ἡ δύναμις διαφέρει πολὺ ἑκάστου. ὥστε ἐς πόλιν ἐπειδὰν ἀφίκηταί τις, ἧς ἄπειρός ἐστι, διαφροντίσαι χρὴ τὴν θέσιν αὐτῆς, ὅκως κεῖται καὶ πρὸς τὰ πνεύματα καὶ πρὸς τὰς ἀνατολὰς τοῦ ἡλίου. *Airs, Waters, and Places,* I. 1–15.

[11] For a discussion of the so-called Dogmatist School of Medicine see: Boylan (2015): 77–98; R.J. Hankinson, "Causes and Empiricism: A Problem in the Interpretation of Later Greek Medical Method" *Phronesis* 32.2 (1987): 329–348; Colin Webster, "Heuristic Medicine: The Methodists and Metalepsis" *Isis* 106.3 (2015): 657–668; and Caroline Petit, "What does Pseudo-Galen tell us that Galen does not?" *Bulletin of the Institute for Classical Studies,* Supplement 114: *Philosophical Themes in Galen* (2014): 269–290.

hemorrhoids (III. 15–30). Cold winds can also create bad effects making people bilious and apt to suffer from pleurisy (IV. 1–20). This is most apparent when the temperature of the winds/air is contrary to the expected at that season: it's unnatural ($phusis_1$).

A similar litany is presented with waters (VII, ff.) and places (where you live— XII ff.).

Together, this analysis blames holistically (from the air, the water, and the geographical location). This etiology is repeated in *Breaths*.

> Now of these obscure matters one is the cause of diseases, what the beginning and source is whence come affections of the body. For knowledge of the cause of a disease will enable one to administer to the body what things are advantageous. Indeed, this sort of medicine is quite natural. (*Breaths*, tr. W.H.S. Jones).[12]

This passage reinforces the method that inquiry into the causes of a disease is necessary to know, in order to be able to understand (diagnose) the *type* of disease which, in turn, allows the physician to administer the appropriate *treatment*. This suggests that the physician must have a grasp of basic science ($phusis_1$) in order to properly diagnose and treat a patient who is sick. Now, again, (as in *Airs, Waters, and Places*) the quality of the inhaled air is the key input (no hint at "skin breathing" here).

The wind (or *pneuma*, III, 3) is said to be one of the three essential elements necessary for men or animals to live (along with food and drink). When the wind or air (*aer*) enters the body, it becomes breath (*phusa*). This treated entity is powerful and necessary for life. Because of this power, it is also a cause of sickness (IV, 1–3). The most common diseases (*koinotatou nosematos*) are fevers: (a) pestilence (that can be epidemic) and (b) sporadic let down by those who lead unhealthy life-styles (VI, 1–5). Both are caused by air (*aer*).

> So, whenever the air has been infected with such *pollutions* [*miasmasin*] as are hostile to the human race, then men fall sick. (*Breaths*, tr. W.H.S. Jones).[13]

Here, is the introduction of a key concept of "polluted air" as a cause of disease: *miasma*. This cause of disease would continue on until the latter part of the nineteenth century C.E. and the advent of *germ theory* (see Chap. 7).

When the etiology of disease is *outside* the patient and it is not measurable quantitatively, then the author must set out some qualitative term of pollution, *miasma*, that affects humans (and sometimes other animals, as well). The problem with this sort of move is that the physician makes reference to a cause that is not in any way measurable as an explanation for disease. This limits the physician's ability to link the cause to create a full *diagnosis*. Unless one knows the cause of the infection, the diagnosis is merely "symptom oriented." When the patient presents with shivering

---

[12] ἐν δὲ δή τι τῶν τοιούτων ἐστὶ τόδε, τί ποτε τὸ αἴτιόν ἐστι τῶν νούσων, καὶ τίς ἀρχὴ καὶ πηγὴ γίνεται τῶν ἐν τῷ σώματι παθῶν; εἰ γάρ τις εἰδείη τὴν αἰτίην τοῦ νοσήματος. οἷος τ' ἂν εἴη τὰ συμφέροντα προσφερέιν τῷ σώματι, αὕτη γὰρ ἡ ἰητρικὴ μάλιστα κατὰ φύσιν ἐστίν. *Breaths*, I, 21–28.

[13] ὅταν μὲν οὖν ὁ ἀὴρ τοιούτοισι χρωσθῇ μιάσμασιν, ἃ τῇ ἀνθρωπείῃ φύσει πολέμιά ἐστιν, ἄνθρωποι τότε νοσέουσιν. *Breaths* VI, 19–21.

just before fevers, and other symptoms also present—such as tremors, the physician is left to basically discuss the process of respiration (*Breaths* VIII). But this is not enough, because it does not give detailed, specific, causal reasons for *treatment* nor does it all for a reasonable *prognosis* for the patient's condition going forward.

A third text along these lines is *The Sacred Disease*. There are several structural similarities between *The Sacred Disease* and *Airs, Waters, and Places*. They both emphasize the roles of the contraries: hot/cold & wet/dry—particularly caused by environmental factors. As was mentioned in the Introduction, this is because the author of *The Sacred Disease* is keen on maintaining a *phusis*$_1$ stance (being very against physicians who rely upon *phusis*$_2$). An example of this comes from chapter 16.

> At the changes of the winds for these reasons do I hold that patients are attacked, most often when the south wind blows, then the north wind, and then the others. In fact, the north and south are stronger than any other winds, and the most opposite, not only in direction but in power. For the north wind contracts the air and separates from it what is turbid and damp, making it clear and transparent. It acts in the same way upon everything as well that rises from the sea or waters, generally. For it separates the moist and the dull from everything, including men themselves, for which reason it is the most healthy of the winds. But of the action of the south wind is the opposite. At first, it begins to melt and diffuse the condensed air, inasmuch as it does not blow strong immediately, but is calm at first, because it cannot at once master the air, that before was thick and condensed, but requires time to dissolve it. (tr. W.H.S. Jones)[14]

In this passage, the writer (like the writer of *Airs, Waters, and Places*)[15] suggests that exterior, environmental factors can be the cause of human disease. Generally, this is a confluence of *normal* factors like the wind: its direction and its composition (dry or damp). But this can also include extraordinary events as well such as earthquakes.[16]

---

[14] Ἐν δὲ τῇσι μεταβολῇσι τῶν πνευμάτων διὰ τάδε φημὶ ἐπιλήπτους γίνεσθαι, καὶ μάλιϲτα τοῖϲι νοτίοιϲιν, ἔπειτα τοῖϲι βορείοιϲιν, ἔπειτα τοῖϲι λοιποῖϲι πνεύμαϲι, ταῦτα γὰρ τῶν λοιπῶν πνευμάτων ἰϲχυρότατα τά ἐϲτι καὶ ἀλλήλοιϲ ἐναντιώτατα κατὰ τὴν ϲτάϲιν καὶ κατὰ τὴν δύναμιν. ὁ μὲν γὰρ βορέης ϲυνίϲτηϲι τὸν ἠέρα καὶ τὸ θολερόν τε καὶ τὸ νοτῶδες ἐκκρίνει καὶ λαμπρόν τε καὶ διαφανέα ποιεῖ, κατὰ δὲ τὸν αὐτὸν τρόπον καὶ τἄλλα πάντα ἐκ τῆς θαλάϲϲης ἀρξάμενα καὶ τῶν ἄλλων ὑδάτων, ἐκκρίνει γὰρ ἐξ ἁπάντων τὴν νοτίδα καὶ τὸ δνοφερόν, καὶ ἐξ αὐτῶν τῶν ἀνθρώπων, διὸ καὶ ὑγιηρότατός ἐϲτι τῶν ἀνέμων. ὁ δὲ νότος τἀναντία τούτῳ ἐργάζεται, πρῶτον μὲν ἄρχεται τὸν ἠέρα ϲυνεϲτηκότα κατατήκειν καὶ διαχεῖν, καθότι καὶ οὐκ εὐθὺς πνεῖ μέγας, ἀλλὰ γαληνίζει πρῶτον, ὅτι οὐ δύναται ἐπικρατῆϲαι τοῦ ἠέρος αὐτίκα, τοῦ πρόϲθεν πυκνοῦ τε ἐόντος καὶ ϲυνεϲτηκότος, ἀλλὰ τῷ χρόνῳ διαλύει. *The Sacred Disease*, XVI, ll. 1–19.

[15] If they are indeed separate or the work of a teacher and his student. For a discussion of the relationship between these two texts see: Philip J. van der Eijk, "'Airs, Waters, Places' and 'On the Sacred Disease': Two Different Religiosities?" *Hermes* 119.Bd., H.2 (1991): 168–176. The two works were often connected in the ancient world. For an example of this, see: Ido Israelowich "The Use and Abuse of Hippocratic Medicine in the *Apology of Lucius Apuleius*" *Classical Quarterly* 67 (2016): 635–644.

[16] For a description of the earthquake parameter see: John Z. Wee, "Earthquake and Epilepsy: The Body Geologic in the Hippocratic Treatise *On the Sacred Disease*" in John Z. Wee, ed. *The Comparable Body: Analogy and Metaphor in Ancient Mesopotamian, Egyptian, and Greco-Roman Medicine* (Leiden: Brill, 2017): 142–169.

In chapter 17, this *phusis₁* approach is continued.

> Men ought to know that from the brain, and from the brain only, arise our pleasures, joys, laughter and jests, as well as our sorrows, pains, griefs and tears. Through it, in particular, we think, see, hear, and distinguish the ugly from the beautiful, the bad from the good, the pleasant from the unpleasant, in some cases using custom as a test, in others perceiving them from their utility. (tr. W.H.S. Jones)[17]

The physical brain that the Hippocratic writer asserts in XX as the seat of the characteristics set out in XVII as the cause of the various intellectual faculties, which are most important to human life, come from a physical organ and not just a *capacity* lingering about the region of the heart (as Aristotle asserts). This clearly sets this Hippocratic writer in the *phusis₁* camp.

The Roles of Interior Factors Causing Disease

Though, exterior causes—such as the seasons and wind are primary causes[18] for disease in the Hippocratic Writers, another possible cause for disease might be *interior causes*. We see this indirectly via failure to follow normal regimen (primarily dealing with diet and exercise which work together to produce health, *Regimen* I. 2, ll. 22–58). But what does this failure do materially (*phusis₁*)?

One possibility is the interaction with the contraries (hot, cold, wet, and dry, *Regimen* I. 3–4). Another theoretical possibility would be the creation of residues (*peritoma*)[19] that we see in Anonymus Londinensis 4, 25–31. Anonymus Londinensis' terminology is Aristotelian and does not occur in the official Hippocratic Corpus, as such. From the Aristotelian perspective it would mean that in the process of digestion (which exists to create nourishment, *trophe*), Because of problems of excess or deficiency in digestion or the needs of the body as the nourishment is distributed, excess material or not-properly-concocted (*pepsis*) material can be created: residues.

However useful this might be for completeness of interior causation, in the accepted Hippocratic Corpus, this process is not set out, as such. However, it may be, there are indirect ties to those who do not follow a healthy regimen. This may throw off the *balance* of the body, which is depicted via the four humors: blood, phlegm, black bile, and yellow bile. The writer of *The Nature of Man* declares,

---

[17] Εἰδέναι δὲ χρὴ τοὺς ἀνθρώπους, ὅτι ἐξ οὐδενὸς ἡμῖν αἱ ἡδοναὶ γίνονται καὶ εὐφροσύναι καὶ γέλωτες καὶ παιδιαὶ ἢ ἐντεῦθεν, καὶ λῦπαι καὶ ἀνίαι καὶ δυσφροσύναι καὶ κλαυθμοί. καὶ τούτῳ φρονέομεν μάλιστα καὶ βλέπομεν καὶ ἀκούομεν καὶ διαγινώσκομεν τά τε αἰσχρὰ καὶ καλὰ καὶ κακὰ καὶ ἀγαθὰ καὶ ἡδέα καὶ ἀηδέα, τὰ μὲν νομῳ διακρίνοντες, τὰ δὲ τῷ συμφέροντι αἰθανόμενοι. *The Sacred Disease*, XVII, ll. 1–9.

[18] Cf. *The Nature of Man*, 9.5; VI, 56 L.

[19] For a discussion of *peritoma* see Michael Boylan, "The Digestive and 'Circulatory' Systems in Aristotle's Biology" *Journal of the History of Biology* 15.1 (1982): 89–118.

> The body of man has in itself blood, phlegm, yellow bile and black bile; these make up the nature of his body, and through these he feels pain or enjoys health. (*The Nature of Man*, tr. W.H.S. Jones).[20]

So, concerning etiology, whatever throws off the balance (symmetry)[21]—whether it be improper digestion or other variations in the regimen—these can be seen as some sort of *interior* focus upon the cause of disease. But what if the cause of the imbalance are exterior factors in the environment (as seen above)? When one considers that the contraries (hot, cold, wet, and dry) are the constituents of the body, then it would seem that whatever alters these would affect the body's balance in a holistic way.[22] For example, draught conditions (exterior factor) might affect the "wet" in the body and alter the creation of and sustaining of bodily fluids. The particular fluid in question might relate back to Aristotle's account of digestion (and the creation of *trophe* and *peritoma*). If the heart is affected, then there will be an alteration in the blood (since the heart was generally thought to be the creator of blood). The liver and spleen were the generators of black and yellow bile. And phlegm has various origins depending on how one assigns the brain the entire endeavor (a cooling agent à la Aristotle or thinking agent taking formed, vital *pneuma* from the lungs and actualized in the brain). Phlegm was generally an internal by-product of exterior cold and resulted in fevers, fatigue, and decreased capacity to breathe (*Aphorisms* IV–V).

Though the role of *peritoma* was vital to Aristotle (based upon existent texts), its role in Hippocratic medicine is not explicit. Rather, the effects of an individual's diet and exercise and their affect upon the balance of the humors is the focus upon interior causation for disease.

The Practice of Medicine

The Fundamentals: Diagnosis, Prognosis, Treatment

The most important practical texts were those that would assist the physician to perform the fundamental roles of medicine: diagnosis, prognosis, and treatment. Now the way that one went about this had something to do with one's epistemology.

---

[20] Τὸ δὲ σῶμα τοῦ ἀνθρώπου ἔχει ἐν ἑωυτῷ αἷμα καὶ φλέγμα καὶ χολὴν ξανθὴν καὶ μέλαιναν, καὶ ταῦτ' ἐστὶν αὐτῷ ἡ φύσις τοῦ σώματος, καὶ διὰ ταῦτα ἀλγεῖ καὶ ὑγιαίνει. *The Nature of Man*, IV. 1–4

[21] One discussion that emphasizes this is: Brooke Holmes, "Proto-Sympathy in the *Hippocratic Corpus*" in Jacques Jouanna and Michel Zink, *Hippocratismes: Médecine, Religion, Société* (Paris: Belles-Lettres, 2014): 123–138.

[22] This can include a version of the humor theory—see: Jacques Jouanna, "The Legacy of the Hippocratic Treatise *The Nature of Man*: the theory of the four humours" from *Greek Medicine from Hippocrates to Galen* (Leiden: Brill, 2012): 355-360, cf. R.J. Hankinson, "Galen on Hippocratic Physics" in Jones and Rosen: 421–444.

If one conceived of the human body as basically interchangeable with other human bodies and if there were general rules governing the nature (*phusis*$_1$) that governed the whole enterprise, then would be drawn to what would come to be later called the Dogmatist school of medicine. However, in the emerging climate the fourth and fifth centuries B.C.E., these three categories were not always discretely set out. There are several issues for this. First, *diagnosis* seems to require that the physician view the realm of diseases as embodied in categories that are already "set-out." Also, treatment options were limited generally to diet and exercise. This often created a situation in which all three categories were lumped together into prognosis. When this combination occurred, the physician was likely to react by turning to the temporal factors of: the past, the present, and the possible future. Obviously, patients would be most concerned with the *future*: will they die or will they live (and under what constraints)?

This temporal element often presented itself in the category of prognosis (encompassing both diagnosis and treatment). Such a temporal scheme was useful to gain the trust of the patient, who would be more likely to believe a medical practitioner who could tell him what his symptoms were at that moment and what they had been days before. For the patient, this would make them think that this physician really knew what he was talking about.

Be that as it may, this presentation will, nonetheless, try to break the three categories apart as it will yield some insight into how medicine is practiced by this group of physicians called the Hippocratic Writers, and how they might be connected to the practice of medicine as it has been laid down in the development to the present era.

Diagnosis and Prognosis

The first of the two groups of writings on diagnosis emphasize certain groups of signs that center the understanding of the symptoms being presented. This semeiotic approach can readily be presented to others via case logs of treated patients and what happened to them and *how*. One such handbook is *Epidemics III*. We start with the patient's name or something about him. This particularizes the account to a single individual who presented with these symptoms to the physician writing the case log. Many of these reports concern patients with fevers and respiratory distress. The writer of the book of cases makes daily reports on symptom signs—such as the color of the skin, the dryness of the mouth, the passage of urine and feces and their composition (hard, soft, dark in color or lighter in color, (sometimes the smell).

Day-by-day these physical descriptions are given until the patient reaches a *crisis*. This is the critical time in which the patient will begin a journey to either recovery or to death. 60% of the patients listed in *Epidemics I* and *III* end up dying. It can be inferred (though rarely overtly stated) that the physician's treatment was largely dietary since there is copious attention to the state of urine and feces. These presumably relate to the process of digestion, so that regulation of digestion can modify these symptoms. However, it seems that in both of these texts the physician is largely impotent in controlling the path of the disease. Instead, the physician looks after the comfort of the patient—doing what he can to modify digestive symptoms.

This sort of approach appeals to either a *phusis$_2$* or a *phusis$_3$* approach. In the former case, one sets out that the material approach is beyond human understanding so that one makes choices "in the dark" (since no one knows the mind of God's will). Therefore, what has worked in the past might be a useful guide to the present. In the latter case, one is skeptical of anyone knowing how the human body works in a compressive way—leading to epistemological skepticism. One can work around the edges, but is actually impotent for the essential work. In the second case, the physician can do a little bit, but the process of recovery is *really out of his hands*. It's rather a palliative approach to medicine.

What is clear in these two texts is that *case-based* approaches can be of some value to another practicing physician because they present *signs* that when they occurred in another individual, led or recovery or to death. In the physician's best personal interest, being pessimistic was rather good for his personal safety (along with a fast horse). Thus, the casebook, in the case of a former patient who recovered, might lead to a sort of ceteris paribus notion of hope, while another case that led to death, would be just fine for the physician because just-in-case the patient recovered, it might enhance the physician's reputation.

One issue that is important to raise in this case-based approach is how one viewed *particularity*. Edelstein thought that many of the Hippocratic writers practiced, just as I have depicted the writers of *Epidemics I* and *Epidemics III*, viz., that the individual depictions were simply that: here's what happened to Mr. X or Ms. X at this time and in these times of the year. There may be *some* indications of *type* (understood very restrictedly), but there was *not* an intent to set out a *classification of diseases* wherein what happened to Mr. x or Ms. X was merely an instantiation (contra to Littré).[23]

The second group of writings on diagnosis emphasizes a certain *systematic* understanding on what humans are like: adult men, adult women, and children. In this case, *diagnosis* consists, in part, in assessing the signs presented before him.[24] Let's examine an example from *Ancient Medicine*. A sick man is different from a healthy man and so requires a different regimen (*Ancient Medicine* III). Learning what symptoms constitute possible sickness is learned from experience as one learns an art (*techne*) rather than a deduction from particular balances of the hot, cold, wet, or dry (*Ancient Medicine* I. 1–18). In the art of medicine practitioners learn what symptoms tend to lead to a disease and what sort of diet and exercise might make them better (*Ancient Medicine* IV). Thus, the physician is one who has discovered the diet and mode of life that is appropriate for the sick (*Ancient Medicine* VII). Both the quality of the food preparation and the amount of food taken in are crucial (*Ancient Medicine* VIII–IX).

---

[23] Edelstein (1967): 67–69.
[24] Cf. *On the physician* (I, 24, 21, ff.): the physician who knows the signs (σημεῖα) will be able to handle the treatment.

The amount of intuition necessary is compared to a pilot of a ship, whose job to get to port is very hard even during calm weather. But when a tempest arises, that job becomes next to impossible. So, it is with the physician. When the patient is only showing mild symptoms, then recovery is often possible when one possesses the art at a rudimentary level. But when a crisis occurs, then the job gets much harder—and in some cases, impossible.

It is intuition and experience that constitute attaining skill in the art of medicine—not the *scientific* determination that the symptoms are caused by an overabundance of the hot, cold, wet, or dry (*Ancient Medicine* XIII).[25] Such a position puts the writer clearly in the camp of those who espouse *phusis*$_3$. There is a skepticism about knowing the underlying material cause of the disease, but there is faith in the idea of matching certain symptoms (diagnosis) with certain possible outcomes (prognosis) that *may* be alleviated by altering diet and activity (treatment via regimen).

Just as the writer of *Ancient Medicine* is skeptical about using the qualities *hot, cold, wet,* and *dry* as tools for diagnosis, prognosis, and treatment, so also is the writer of *On the Nature of Man* skeptical about giving too much weight to a humor theory that prioritizes one humor as more important than another.

> Now about these men I have said enough, and I will turn to physicians. Some of them say that a man is blood, others that he is bile, a few that he is phlegm. Physicians, like the metaphysicians, all add the same appendix. For they say that a man is a unity, giving it the name that severally they wish to give it; this changes its form and its power, being constrained by the hot and the cold, and becomes sweet, bitter, white, black and so on. But in my opinion, these views also are incorrect. Most physicians then maintain views like these, if not identical with them; but I hold that if man were a unity, he would never feel pain, as there would be nothing from which a unity could suffer pain. And even if he were to suffer, the cure too would have to be one. But, as a matter of fact, cures are many. (*The Nature of Man*, tr. W.H.S. Jones)[26]

This is an interesting passage. At the beginning, the author refers to the Pre-Socratic tendency to set out a single principle (*arche*)—such as water, air, the unlimited, etc. From there the author could create a norm (*phusis*) and describe material events—such as the medical status of individuals as going in accord with the norm (*kata phusin*) or going against the norm (*para phusin*). Such "single-principle" accounts

---

[25] Ralph E. Rosen emphasizes this "back-and-forth" with the non-philosophical-universalist approach in "Towards a Hippocratic Anthropology: On Ancient Medicine and the Origin of Humors" in Lesley Dean-Jones and Ralph M. Rosen, *Ancient Conceptions of the Hippocratic* (Leiden: Brill, 2016): 242–247.

[26] εἶναι τὸν ἄνθρωπον, ἔνιοι δέ τινες φλέγμα, ἐπίλογον δὲ ποιέονται καὶ οὗτοι πάντες τὸν αὐτόν, ἐν γὰρ εἶναί φασιν, ὅ τι ἕκαστος αὐτῶν βούλεται ὀνομάσας, καὶ τοῦτο μεταλλάσσειν τὴν ἰδέην καὶ τὴν δύναμιν, ἀναγκαζόμενον ὑπό τε τοῦ θερμοῦ καὶ τῶ ψυχροῦ, καὶ γίνεσθαι γλυκὺ καὶ πικρὸν καὶ λευκὸν καὶ μέλαν καὶ παντοῖον. ἐμοὶ δὲ οὐδὲ ταῦτα δοκεῖ ὧδε ἔχειν. οἱ οὖν πλεῖστοι τοιαῦτά τινα καὶ ἐγγύτατα τούτων ἀποφαίνονται. ἐγὼ δέ φημι, εἰ ἓν ἦν ὤνθρωπος, οὐδέποτ' ἂν ἤλεγεεν, οὐδὲ γὰρ ἂν ἦν ὑφ' ὅτου ἀλγήσειεν ἓν ἐών, εἰ δ' οὖν καὶ ἀλγήσειεν, ἀνάγκη καὶ τὸ ἰώμενον ἓν εἶναι, νῦν δὲ πολλά. *Nature of Man*, II, 1–17.

are easier to deal with because the assumption is that the events to be explain (human health), the explanandum, is ontologically simple. However, it does not give much guidance in how to fix problems that might arise, viz., *disease*.

One possible alternative (also rejected by the author) is positing a static unity of many principles—many of which can be paired with an opposite (to promote symmetry and elegance), such as hot and cold, sweet and bitter, white and black (possibly referring to the colors of bile).[27] Rather, this author has something else in mind. He accepts a principle of pluralism that recognizes all of these principles as existing in the human body and are critical to health *if they are balanced* (*On the Nature of Man*, III. 8–18; IV. 1–20; V. 1–20). Again, we are drawn to *balance* and *symmetry* of a plurality of distinct factors—each of which has an important role to play in health. When one or more become unbalanced by excess or defect (IV. 8–10), then problems occur because of the imbalance/asymmetry. Health is thus calculated against a standard based upon what is "normal" for humans (their *phusis*$_1$).[28] This strategy of designating *health* as fulfilling some sort of "normal" standard has carried on even to our present time, but it is controversial—since health might be best construed as creating or maintaining some particular state that meets an individual's rational life plan.[29]

As I have already stated, the more general term within the triad of *diagnosis, prognosis, and treatment* is *prognosis*. Let us begin with the treatise with that very name.[30]

> I hold that it is an excellent thing for a physician to practice forecasting. For if he discovers and declare unaided by the side of his patients the present, the past, and the future, and fill in the gaps in the account given by the sick, he will be the more believed to understand the cases, so that men will confidently entrust themselves to him for treatment. Furthermore, he will carry out the treatment best if he knows beforehand from the present symptoms what will take places later. Now to restore every patient to health is impossible. (*Prognostic*, translated by W.H.S. Jones)[31]

The author of *Prognostic* sets out the practical side of being able to *forecast*. Now, it is important *how* the physician is forecasting. If it is on the basis of a theory of

---

[27] Often this contrast is with yellow and black bile. It is possible that the author has another contrast in mind here.

[28] Cf. The discussion by Aristotle at the end of Chap. 3.

[29] I discuss some of the most important aspects of this controversy in our current era in Michael Boylan, "Health as Self-Fulfillment" in Michael Boylan, ed. *Medical Ethics*, 2nd ed. (Oxford: Wiley-Blackwell, 2014):44–57—see the Appendix for a version of this.

[30] For a well-known initial starting point for this exploration, see: Ludwig Edelstein, *Ancient Medicine* (Baltimore, MD: Johns Hopkins University Press, 1967): 65–85.

[31] Τὸν ἰητρὸν δοκεῖ μοι ἄριστον εἶναι πρόνοιαν ἐπιτηδεύειν, προγινώσκων γὰρ καὶ προλέγων παρὰ τοῖσι νοσέουσι τά τε παρεόντα καὶ τὰ προγεγονότα καὶ τὰ μέλλοντα ἔσεσθαι, ὁκόσα τε παραλείπουσιν οἱ ἀσθενέοντες ἐκδιηγεύμενος πιστεύοιτο ἂν μᾶλλον γινώσκειν τὰ τῶν νοσεύντων πρήγματα, ὥστε τολμᾶν ἐπιτρέπειν τοὺς ἀνθρώπους σφᾶς αὐτοὺς τῷ ἰητρῷ. τὴν δὲ θεραπείην ἄριστα ἂν ποιέοιτο προειδὼς τὰ ἐσόμενα ἐκ τῶν παρεόντων παθημάτων. ὑγιέας μὲν γὰρ ποιεῖν ἅπαντας τοὺς νοσέοντας ἀδύνατον. *Prognostic* I, 1–11.

disease and a theory of the human body, then it will be a *phusis*₁ affair. If it based upon some divination or connection to god, then it will be a *phusis*₂ affair. Finally, if it based upon an intuition or gut feeling based only upon individual patients the physician has seen in the past with the caveat that all patients are different and we really *can't say for sure*, then it will be a *phusis*₃ affair.

Given, that it is a *confidence-builder* for the patient to hear their physician be able to tell them their symptoms before the consultation (the past), and to enumerate on symptoms of the present (that the patient hasn't yet revealed), it would be useful for the successful physician to obtain these perspectives. These two revelations (past and present) will give the patient confidence about the physician's prediction of the future. Whereas, *diagnosis* per se is only about the present, *prognosis* (in this context contains all three time-frames: past, present, and future).

One way to do this is to refer to case books like the *Epidemics*.[32] If the patient before you *presents* these symptoms (signs), then it is possible/probable that they might continue in the path of the case patient to crisis, and from there, to the outcome of *recovery* or *death*.[33]

A second approach would be to have both theories of how the human body operates and how this day-to-day process could be interrupted. Such an approach would create exterior criteria that applied to *all people* in *all places* on earth. It would move medical theory to an objective, universal standard and away from the radical *particularity* that is evidenced in the *Epidemics* volumes. However, in the Hippocratic writings, it is not clear that this was ever a clear goal.

The method advocated is first to examine the face of the patient: eyes, skin temples, ears, and nose. Is the skin too hot or cold? Make a note. What about the color of the skin and other facial features? Is the upper part of the abdomen (*hypochondrium*) distended or in pain? By answering these questions the physician attends to theoretical completeness and coherence.

Next, what about the excretion system? Are the bowels moving loosely (diarrhea)? Is there a problem with sleep—like sleeping with your mouth open (cf. *Prognostic* III. 12) or seeking to sit up during the worst part of the disease (III. 22)?

In each case, when the answer to any of these questions is "yes," then the physician can turn to his handbook and try to find one or more cases in which the same symptoms (signs) were presented. These could be relayed to the patient—probably emphasizing the most pessimistic picture (for the safety and reputation of the physician). Observing these symptoms and remembering how they played out in other particular individuals will give the physician an advantage in relaying the past and forecasting the fate of the sick individual (*Prognostic* XXV).

---

[32] For an argument that the audience for the *Epidemics* was other physicians, see: Chiara Thumiger, "The Professional Audience of the Hippocratic *Epidemics*: Patient Cases in Hippocratic Scientific Communication" in Petros Bouras-Valhanatos and Sophia Xenophontos, eds. *Greek Medical Literature and its Readers* (New York: Routledge, 2018): 48–64.

[33] Of course, there is the situation of chronic symptoms, but so long as these are manageable, they will be considered to be an *imperfect* recovery.

Now it is important to note that the diseases examined in the *Epidemics* and in *Prognostic* are generally acute diseases with potential lethal outcomes. When we move to "management" (*diaites*) of these diseases, we enter the realm of treatment. Two key texts here are *Regimen in Acute Diseases* and in *Diseases I*.

Treatment

In *Regimen in Acute Diseases* the individual-particularity perspective is brought up.

> The authors of the work entitled *Cnidian Sentences*[34] have correctly described the experiences of patients in individual diseases and the issues of some of them. So much even a layman could correctly describe by carefully inquiring from each patient the nature of his experiences. But much of what the physician should know besides, without the patient's telling him, they have omitted; this knowledge varies in varying circumstances, and in some cases is important for the interpretation of symptoms (tr. W.H.S. Jones)[35]

The reference to *particularity* is a key point—especially when we bring in *treatment*. For treatment decisions by the physicians is never completely a clear fact. Depending upon how much one wants to emphasize *individuality* v. *general nomic classification*[36] it will drive the way the process is characterized. This is a universal issue in medicine up to the current era.[37]

Figure 2.1 is an important worldview background for physicians in all historical epochs (including the contemporary).[38] If we think of each patient's presentation of

---

[34] For a discussion of the general Cnidian methodology toward *particularity* see Boylan (1983): chapt. 1.

[35] Οἱ συγγράψαντες τὰς Κνιδίας καλεομένας γνώμας ὁποῖα μὲν πάσχουσιν οἱ κάμνοντες ἐν ἑκάστοισι τῶν νοσημάτων ὀρθῶς ἔγραψαν καὶ ὁποίως ἔνια ἀπέβαινεν, καὶ ἄχρι μὲν τούτων, καὶ ὁ μὴ ἰητρὸς δύναιτ᾽ ἂν ὀρθῶς συγγράψαι, εἰ εὖ παρὰ τῶν καμνόντων ἑκάστου πύθοιτο, ὁποῖα πάσχουσιν, ὁπόσα δὲ προσκαταμαθεῖν δεῖ τὸν ἰητρὸν μὴ λέγοντος τοῦ κάμνοντος, τούτων πολλὰ παρεῖται, ἀλλ᾽ ἐν ἄλλοισιν καὶ ἐπίκαιρα ἔνια ἐόντα ἐς τέκμαρσιν. *Regimen in Acute Diseases*, part I.

[36] The "nomic" model is akin to the modern Logical Empiricists who create an extensive, axiomatic system—much on the design of Euclid's geometry—to offer *explanation* for scientific phenomena. See: Boylan (2015): 134, n.86.

[37] In contemporary philosophy of medicine, *treatment* is driven by "standard of care" for particular diagnosis and prognosis. This leans toward the nomic, but as my neurosurgeon daughter tells me, a physician's historical memory bank of individual cases can alter diagnosis and prognosis, thus changing the options for treatment.

[38] For example, in the contemporary realm, my daughter, who is a neurosurgeon, has related countless stories on how a particular patient's individual condition is different than the general "textbook" description of a spine or brain pathology which requires her, on the spur of the moment in the operation, to change her initial plan (which was backed-up by all the modern imaging available) and make modifications to get to her goal (which she knows through her experience with countless earlier surgical procedures). In this way, in Fig. 2.1, the balanced approach, Y, is the best for those medical practitioners who have enough experience to move away from the general "textbook" when the textbook approach is clearly wrong in this situation because of individual variance (which is predicted by evolutionary theory, which is at the core of all contemporary medicine).

```
    X                       Y                        Z
(extreme particularity)  (balanced)    (extreme general nomic classification)
```

Fig. 2.1 The stance of epistemic background in diagnosis and treatment

illness (as was alleged of the Cnidian "school"),[39] then each patient's condition is treated as a *causa sui*, unique in its particularity. In the extreme, such a stance would philosophically require that *no cases* should ever be set-out in case books like the *Epidemics*. This is because under "X" mode, the underlying *given* is that there is *utter particularity*—meaning that all *grouping* into (*genos* or *eidos*) will be, at best, *accidental* and, at worst, *misleading*. This is the position of those accepting the *phusis₃* position. Epistemological skepticism about the possibility of *natural kinds* has taken reign. Under these circumstances, a case book would only be useful to assist in a "trial and error" process. In order for the "case book" approach to be effective would be for medical practitioner to be *at least* between Fig. 2.1's "X" and "Y"—preferably closer to "Y."

What makes *Regimen in Acute Diseases* different from the *Epidemics* texts, is that the author is keen to discuss the effects of offering various foods and drinks to patients, who exhibit various symptoms. For example, the use of barley gruel can be calming to the bowels—so long as the patient keeps eating—even at a lower amount (IX). No "Z-Level" account is given, but because a general principle of human reaction to treatment is being put forth, it is minimally-necessary "Y-Level."

Fever-abatement, a common symptom that can have serious effects, can respond to this therapy along with others set out (such as pale or sweet wine, LI-II)—in case this is ineffective.

What we have in this text, is primarily the use of food and drink to combat particular symptoms without a comprehensive account of a nomic account ("Z") in favor of a "Y-Based" approach that assumes that all or most humans will respond to these dietary remedies as opposed to the "X-Based" particularists who cannot admit to any firm, universal rules about humans because they eschew comprehensive, nomic account.

The next text to examine along these lines is *Diseases* I. In this text we definitely move from the "Y-based" epistemological account to the "Z-based" disposition. This mid-point of Y-prime is driven by a desire for *epistemological certainty*.

> Anyone who wishes to ask correctly about healing, and, on being asked, to reply and rebut correctly, must consider the following: first, whence all diseases in men arise. Then, which diseases, when they occur, are necessarily long or short, mortal or not mortal, or permanently disabling to some part of the body or not, and which other diseases, when they occur, are uncertain as to whether their outcome will be bad or good. From which disease there are changes into which others. What physicians treating patients achieve by luck. What good or bad patient suffer in diseases. What is said or done on conjecture by the physician to the

---

[39] For further discussion of this see: Jacques Jouanna, "Egyptian Medicine and Greek Medicine" from Jacques Jouanna, ed. *Greek Medicine from Hippocrates to Galen: Selected Papers* (Leiden: Brill, 2012): 3–20.

patient, or by the patient to the physician. What is said or done with precision in medicine, which things are correct it in, and which are not correct. What starting point of medicine, or end, or middle, or any other feature of this kind has been demonstrated; what truly does or does not exist in medicine. (tr. Paul Potter)[40]

The author of this text puts forth a series of queries that suggests that: (1) There is a common ground among humans being significantly similar, and that diseases are also such that they can be identified, as such. (2) That the physician can become educated in these *facts* so that he might be able to understand both patient and ailment such that *treatment* might be achieved by something other than "luck." "Conjecture" can be replaced by "precision" (understood via completeness and coherence). This is the goal, *phusis*$_1$. The goal is thus reaching Z in the Fig. 2.1. This is an important move within this milieu; it is a decisive move toward *coherence* (one of the criteria set out for theory evaluation).

Under these aspirational standards, the physician should be able to recognize various *sorts* of disease (diagnosis) and understand how they might progress or cross-over into another strain and how long the disease might last on its own (prognosis). Finally, how these might assist in the patient helping the patient to recover (treatment)? If they are clearly stated and not too complicated (*simplicity*—a property of the elegant), then they might be operationally very useful.

Along these lines, is *Regimen in Acute Diseases*. In ancient times this text was sometimes called *Against the Cnidian Sentences* (a work that favored *phusis*$_3$ particularity, cf. the first line of the work). The "acute" refers to fevers (which to this day are considered to be important markers of serious illness). The *symptom* of "fever" is common to many serious diseases. Since this indicates an over-heated body, the logical way to treat this is by making the body cooler. This can include baths and light food (as a beginning treatment) going onto general gastro-intestinal purges and enemas (as a next line). Dietary treatment includes honey and water, and honey and wine. Barley cereal, *ptisan*, and barley water, *chiulos*, were also recommended.

Barley was the grain of choice because of its "smoothness," its "moderate softness," that stimulates *easy evacuation* (IX). The physician must oversee the creation of *ptisan* so that only the best ingredients are used (quality control for ingested ingredients in treatment, (XVI-XVII)). The idea behind this is that if the digestive process from input to output is not performing properly, then there is something

---

[40] Ὃς ἂν περὶ ἰήσιος ἐθέλῃ ἐρωτᾶν τε ὀρθῶς καὶ ἐρωτώμενος ἀποκρίνεσθαι καὶ ἀντιλέγειν ὀρθῶς, ἐνθυμεῖσθαι χρὴ τάδε. πρῶτον μέν, ἀφ' ὧν αἱ νοῦσοι γίγνονται τοῖσιν ἀνθρώποισι πᾶσαι, ἔπειτα δέ, ὅσα ἀνάγκας ἔχει τῶν νοσημάτων ὥστε ὅταν γένηται εἶναι ἢ μακρὰ βραχέα ἢ θανάσιμα ἢ μὴ θανάσιμα ἢ ἔμπηρόν τι τοῦ σώματος γενέσθαι ἢ μὴ ἔμπηρον, καὶ ὅσα, ἐπὴν γένηται, ἐνδοιαστά, εἰ κακὰ ἀπ' αὐτῶν ἀποβαίνει ἢ ἀγαθά, καὶ ἀφ' ὁποίων νοσημάτων ἐς ὁποῖα μεταπίπτει. καὶ ὅσα ἐπιτυχίῃ ποιέουσιν οἱ ἰητροὶ θεραπεύοντες τοὺς ἀσθενέοντας καὶ ὅσα ἀγαθὰ ἢ κακὰ οἱ νοσέοντες ἐν τῇσι νούσοισι πάσχουσι, καὶ ὅσα εἰκασίῃ ἢ λέγεται ἢ ποιεῖται ὑπὸ τοῦ ἰητροῦ πρὸς τὸν νοσέοντα, ἢ ὑπὸ τοῦ νοσέοντος πρὸς τὸν ἰητρόν, καὶ ὅσα ἀκριβῶς ποιέεται ἐν τῇ τέχνῃ καὶ λέγεται, καὶ ἅ τε ὀρθὰ ἐν αὐτῇ καὶ μὴ ὀρθά, καὶ ὅ τι αὐτῆς ἢ ἀρχὴ ἢ τελευτὴ ἢ μέσον ἢ ἄλλο τι ἀποδεδειγμένον τῶν τοιούτων, ὅ τι καὶ ὀρθῶς ἐστιν ἐν αὐτῇ εἶναι ἢ μὴ εἶναι. *Diseases I*, Littré, 140–141.

out-of-balance which has to be corrected by purgative methods and by corrective diet. This occurs against a backdrop of a *normal* human state that, when pushed out of the normal state, demonstrates explicit signs of fever that call for "fever reducing" strategies.

The author of *Regimen in Acute Diseases,* makes great strides in making $phusis_1$ connections between diagnosis and treatment. This is certainly a big step forward in the way we view the practice of medicine from a modern standpoint.

The treatment suggestions should be seen in the context of what is considered to be a healthy lifestyle for day-to-day living, cf. *Regimen in Health*, where diet is important according to the season of the year (air temperature being a key).[41] When the season is hot and dry the body tends toward the burning and parched. These affects can be balanced by more fluids (which leads to cooling and hydration) and less food (which leads to internal heat), bk. I. Walking in the winter should be brisk (to encourage heat) and slow in the summer (to preserve coolness), bk. III.

There is even advice in weight control, Bk. IV for a healthy disposition. Bowels may be artificially emptied with drugs during the winter when constipation is more common, Bk. V. Advice is given for young children and for athletes. The back drop on this is the humor theory supplemented by the contraries.

Basically, the strategy is balance—which also governs treatment options in *Regimen in Acute Diseases.* Primarily driven by the contraries (hot, cold, wet, and dry) within the context of season and location, and by the nascent developing humor theories, there is a prescription for health through diet and exercise (much as there is today—though against a different background of causes). What makes the diagnosis, prognosis, treatment triad different from medicine as practiced today is that the dynamics of the healthy person and the sick person are basically the same except to an excess—which in the extreme can lead to *crisis* which can end up in death.

In medicine today, we view the sick patient as fundamentally different (due to internal causes of germ-theory and other systemic breakdowns and operational imbalances of various sorts (more on this later—see Chaps. 8 and 9).

The Ethical Dimension of Practice: The Oath

The last dimension of Hippocratic practice that I wish to bring forward is The Oath. The dating of this text seems to vary from the sixth century BCE to the first century CE.[42] There are two recognized parts of the Oath: (1) The duties a pupil owes to his teacher and the teacher's family, and (2) Various precepts about what is acceptable

---

[41] One aspect of ancient health was the propagation of dreams. For a discussion of this see: Maithe Hulskamp, "On Regimen and the Question of Medical Dreams in the Hippocratic Corpus" in Jones and Rosen: 258–270.

[42] Two contemporary arguments on this are: Jan N. Bremmer, "How old is the Ideal of Holiness (of Mind) in the Epidaurian Temple Inscription and the Hippocratic Oath?" *Zietschrift für Papyrologie und Epigraphik*, Bd. 141 (2002): 106–108. Christopher A. Faraone, "Curses, Crime Detection and

and unacceptable in medical practice. At least, by the time of Galen, the Hippocratic Oath was referred to as representing positive depiction of the *art* (techne) of medicine.[43]

> By Apollo (the physician), by Asclepius (god of healing), by Hygeia (god of health), by Panacea (god of remedy), and all the gods and goddesses, together as witnesses, I hereby swear that I will carry out, inasmuch as I am able and true to my considered judgment, this oath and the ensuing duties:
>
> 1. To hold my teacher in this art on a par with my parents. To make my teacher a partner in my livelihood. To look after my teacher and financially share with her/him when s/he is in need. To consider him/her as a brother/sister along with his/her family. To teach his/her family the art of medicine, if they want to learn it, without tuition or any other conditions of service. To impart all the lessons necessary to practice medicine to my own sons and daughters, the sons and daughters of my teacher and to my own students, who have taken this oath-but to no one else.
> 2. I will help the sick according to my skill and judgment, but never with an intent to do harm or injury to another.
> 3. I will never administer poison to anyone-even when asked to do so. Nor will I ever suggest a way that others (even the patient) could do so. Similarly, I will never induce an abortion. Instead, I will keep holy my life and art.
> 4. I will not engage in surgery—not even upon suffers from stone, but will withdraw in favor of others who do this work.
> 5. Whoever I visit, rich or poor, I will concern myself with the well-being of the sick. I will commit no intentional misdeeds, nor any other harmful action such as engaging in sexual relations with my patients (regardless of their status).
> 6. Whatever I hear or see in the course of my professional duties (or even outside the course of treatment) regarding my patients is strictly confidential and I will not allow it to be spread about. But instead, will hold these as holy secrets.
>
> Now if I carry out this oath and not break its injunctions, may I enjoy a good life and may my reputation be pure and honored for all generations. But if I fail and break this oath, then may the opposite befall me. (my tr.)[44]

Conflict Resolution at the Festival of Demeter Thesmophoros" *Journal of Hellenic Studies* 131 (2011): 25–44.

[43] Heinrich von Staden, "Division, Dissection, and Specialization: Galen's 'On the Parts of the Medical Techne'" *Bulletin of the Institute of Classical Studies*, Supplement No. 77 *The Unknown Galen* (2002): 19–45. Heinrich von Staden. "In a Pure and Holy Way: Personal and Professional Conduct in the Hippocratic Oath?" *Journal of the History of Medicine and Allied Sciences* 51.4 (1996): 404–437.

[44] Ὄμνυμι Ἀπόλλωνα ἰητρὸν καὶ Ἀσκληπιὸν καὶ Ὑγείαν καὶ Πανάκειαν καὶ θεοὺς πάντας τε καὶ πάσας, ἵστορας ποιεύμενος, ἐπιτελέα ποιήσειν κατὰ δύναμιν καὶ κρίσιν ἐμὴν ὅρκον τόνδε καὶ συγγραφὴν τήνδε, ἡγήσεσθαι μὲν τὸν διδάξαντά με τὴν τέχνην ταύτην ἴσα μενέτησιν ἐμοῖς, καὶ

Within this oath are both a moral code for the profession of medicine and the outlines of a system of "accreditation" for new physicians via a family apprenticeship.[45] These two functions went a long way to establishing medicine as a profession that ordinary people could trust (see below). However, there is another sensibility operating here that is emphasized by Edelstein.[46]

Instead of seeing the oath, primarily, as a paradigm for a code of *professional ethics* (as I mention below), he sees it as a more practical document citing success in families that pass on the family business successfully from one generation to the next. Under this interpretation, the document is more about keeping the family business solvent, which has never been an easy task.[47] When others, outside the family, joined, they became a part of a guild that was certainly ruled by the oldest patriarch.

There is also a tradition that sees the act of "giving an oath" as having a distinctive religious connotation. Under this account, for example, the oath that Asclepiades swore at Delphi undeniably evokes this famous *Oath* preserved in the Hippocratic Corpus.[48]

Further, the uttering of an *oath* or a *curse* with feeling evokes a religious dimension in much of ancient Greece.[49]

βίου κοινώσεσθαι, καὶ χρεῶν χρηίζοντι μετάδοσιν ποιήσεσθαι, καὶ γένος τὸ ἐξ αὐτοῦ ἀδελφοῖς ἴσον ἐπικρινεῖν ἄρρεσι, καὶ διδάξειν τὴν τέχνην ταύτην, ἢν χρηίζωσι μανθάνειν, ἄνευ μισθοῦ καὶ συγγραφῆς, παραγγελίης τε καὶ ἀκροήσιος καὶ τῆς λοίπης ἁπάσης μαθήσιος μετάδοσιν ποιήσεσθαι υἱοῖς τε ἐμοῖς καὶ τοῖς τοῦ ἐμὲ διδάξαντος, καὶμαθητῆσι συγγεγραμμένοις τε καὶ ὡρκισμένοις νόμῳ ἰητρικῷ ἄλλῳ δὲ οὐδενί. διαιτήμασί τε χρήσομαι ἐπ' ὠφελείῃ καμνόντων κατὰ δύναμιν καὶ κρίσιν ἐμήν, ἐπὶ δηλήσει δὲ καὶ ἀδικίῃ εἴρξειν. οὐ δώσω δὲ οὐδὲ φάρμακον οὐδενὶ αἰτηθεὶς θανάσιμον, οὐδὲ ὑφηγήσομαι συμβουλίην τοιήδε, ὁμοίως δὲ οὐδὲ γυναικὶ πεσσὸν φθόριον δώσω. ἁγνῶς δὲ καὶ ὁσίως διατηρήσω βίον τὸν ἐμὸν καὶ τέχνην τὴν ἐμήν. οὐ τεμέω δὲ οὐδὲ μὴν λιθιῶντας, ἐκχωρήσω δὲ ἐργάτῃσιν ἀνδράσι πρήξιος τῆσδε. ἐς οἰκίας δὲ ὁκόσας ἂν ἐσίω, ἐσελεύσομαι ἐπ' ὠφελείῃ καμνόντων, ἐκτὸς ἐὼν πάσης ἀδικίης ἑκουσίης καὶ φθορίης, τῆς τε ἄλλης καὶ ἀφροδισίων ἔργων ἐπί τε γυναικείων σωμάτων καὶ ἀνδρῴων, ἐλευθέρων τε καὶ δούλων. ἃ δ' ἂν ἐν θεραπείῃ ἢ ἴδω ἢ ἀκούσω, ἢ καὶ ἄνευ θεραπείης κατὰ βίον ἀνθρώπων, ἃ μὴ χρή ποτε ἐκλαλεῖσθαι ἔξω, σιγήσομαι, ἄρρητα ἡγεύμενος εἶναι τὰ τοιαῦτα. ὅρκον μὲν οὖν μοι τόνδε ἐπιτελέα ποιέοντι, καὶ μὴ συγχέοντι, εἴη ἐπαύρασθαι καὶ βίου καὶ τέχνης δοξαζομένῳ παρὰ πᾶσιν ἀνθρώποις ἐς τὸν αἰεὶ χρόνον, παραβαίνοντι δὲ καὶ ἐπιορκέοντι, τἀναντία τούτων. The Oath.

[45] The connection to a practicing physician to obtain *knowledge to obtain the techne* connects to modern practice in Western Medicine of "residency" where a medical student who has gone through medical school and passed some key tests, is assigned to specialists who will teach them how to practice a particular sub-speciality of medicine. This pattern is after the ancient Hippocratic-endorsed procedure, as per *The Oath*.

[46] Ludwig Edelstein (1967): 9–63. Vivian Nutton, *Ancient Medicine*, 2nd ed. (London: Routledge, 2013): 69; takes a similar view on the primacy of "practicality" for establishing a going business.

[47] For a discussion of the hypothesis that family businesses fail in the third generation see this link to the Harvard Business Review (accessed December 17, 2021) https://hbr.org/2021/07/do-most-family-businesses-really-fail-by-the-third-generation.

[48] See Jacques Jouanna, "Hippocrates and the Sacred" in Jacques Jouanna, ed. *Greek Medicine from Hippocrates to Galen: Selected Papers* (Leiden: Brill, 2012): ch. 6, 77–118.

[49] Christopher A. Faraone, "Curses, Crime Detection and Conflict Resolution at the Festival of Demeter Thesmophoros," *Journal of Hellenic Studies* 131 (2022): 25–44.

Edelstein focusses upon a population that accepts *infanticide* yet sets out a plank on abortion. These, and other, discrepancies between the life on the ground in ancient Greek-speaking culture and what is put forth seems to be a contradiction. However, if we look at the work of Arthur William Hope Adkins, there are ways to explain this by looking carefully at the normative words and how they were understood via the literature (especially the fictive literature of the drama) of the times.[50] What might appear to be *normative* terms in our "post-Kantian" times, might be more *practical* and *self-serving* as an echo to the *Homeric Ethic*, which praised as *arete* and *agathos* the powerful chieftain who could control his clan. This orientation, for example, changes the contemporary reading of Aristotle's *Nicomachean Ethics* and *Politics* (originally conceived as a single work)[51] from a treatise which modern commentators, like Alasdair MacIntyre, use to construct these works into what is currently called "virtue ethics."[52]

Edelstein points to several tenets of the code that seem to be anomalies. For example, the giving of poison.[53] Since suicide was well-accepted socially, and if a patient was in great pain and one of the handbooks (like the *Epidemics*) suggest that a bad ending was at hand, then wouldn't it be compassionate to give the poison so that a patient might end the inevitable, with less pain?

Concerning abortion, since infanticide was socially acceptable, wouldn't abortion be also? Plato and Aristotle seemed to suggest this.[54]

Therefore, from the point of view of contemporary Greeks the Oath might have been seen to be more of a brief model of how physicians should create a public persona in order to establish a successful family business. Under this reading, the Oath was as much about public *branding* as it was about upholding core ethical principles.

However, that does not take away from the clear influence the Oath had (especially in the Western World) among medical school graduates after World War II.[55] This also roughly corresponded to the rise of medical ethics as a discipline. From

---

[50] For a good depiction of the moral landscape via Adkins' hypothesis see: Michael Boylan, *The Philosophy of A.W.H. Adkins: "Virtue" and "Goodness" in Ancient Greece* (New Castle upon Tyne: Cambridge Scholars Press, 2022).

[51] Boylan (2022): ch. 9.

[52] For an introduction to this viewpoint see: Alasdair MacIntyre, *After Virtue*, 3rd ed. (Notre Dame, IN: Notre Dame University Press, 2007), cf. Michael Boylan, *Basic Ethics*, 3rd ed. (New York and London: Routledge, 2021).

[53] Some would make a link here between the prohibition concerning abortion and the prohibition on giving a poison. For a discussion of some links to more contemporary discussions see: Heike Baranzke, "'Sanctity-of-Life' —A Bioethical Principle for a Right to Life?" *Ethical Theory and Moral Practice* 15.3 (2012): 295–308 and Gerald Dworkin, "Patients and Prisoners: The Ethics of Lethal Injection" *Analysis* 62.2 (2002): 181–189.

[54] Certainly, abortion was not then performed as a surgical procedure, Edelstein (1967): 18–20.

[55] See: https://historynewsnetwork.org/blog/5278 (accessed December 19, 2021).

that modern point of view, let me make a few comments on how the structure of the Hippocratic Oath has influenced modern practical ethics.[56]

From the modern point of view, let me begin by saying that I think that among professional codes, the Hippocratic Oath is a good one. It balances between very specific prohibitions such as not administering poison or not having sexual relations with one's patients, to more general principles such as "I will concern myself with the well-being of the sick." and "do no harm." These general principles are very useful because they govern a larger domain than simply prohibiting a particular action. These principles are not set out without context. Instead, they are put into the context medicine's mission.

Beginning in #1 (from my translation, above) the tone is set that medicine is an art that is "given by the gods." It is an esoteric art that is to be reserved for those who are willing to commit to the provisions of the code. Thus, it is not open to everyone. This fulfills the condition of specialized knowledge mentioned earlier. It is for the sake of doing good to others and always avoiding harm. This fulfills the condition of providing a service for others.

Next, the code ties itself to the larger moral tradition, "I will commit no intentional misdeeds." Whereas "harm" has a direct link to manner in which medicine is practiced, "misdeeds" links the physician to the larger moral tradition. There is no possible hiding in the shared community perspective alone.

These three factors are the basis of any good professional code (Fig. 2.2).

Where codes of professional ethics fail, is in overemphasizing one of these elements too highly or in ignoring an element entirely. If codes of ethics exist in order to remedy the "inward perspective" problem described above, then they must create links to more general "shared worldviews." This would put them in the realm of common morality.

This is the most important point from my perspective. So often the "practice" of the profession defines its excellence in an introspective way such that the achievement of these functional requirements is all that matters—divorced from any other visions, viz., moral visions.

In the modern arena, many professional codes have evolved from a legal perspective. The practitioners of the profession do not want to go to jail or to be sued. Thus, they create certain codes that will make this possible situation less probable. These sorts of codes are defensive in nature and stand at the opposite end of the spectrum from the Hippocratic Oath. Their mission is not to set internal standards and link to common morality, rather they seek to "shave" as close as possible to maximizing an egoistic bottom line at the expense of the pillars of professionalism: one's specialized education and one's mission to serve others.

Any code that takes as its basis merely a negative approach designed to protect the practitioner from going to jail or being sued is fundamentally inadequate. This is not where we should set our sights. Rather, we should dream about what the

---

[56] For example, in lectures and in the first edition of my book, *Basic Ethics* (Upper Saddle River, N.J.: Prentice Hall, 2000) I set it as a paradigm for the use of "codes of ethics" which were being evaluated among the professions.

A Good Professional Code Should Contain

1. A specific listing of common abuses.

2. A few general guidelines that tie behavior to the mission of the profession.

3. A link to general theories of morality.[1]

Fig. 2.2 Elements in a normative professional code. [a]As mentioned earlier with reference to A.W.H. Adkins, this is a tortured question concerning the ancient Greeks and how they are interpreted today. However, for our purposes here: modern reconstruction of an ancient text in a modern context, we may adapt this understanding to a post-Kantian understanding of ethics: theory and applied

profession may be—in the best of all possible worlds. This properly sets the mission that should drive all codes of ethics.

The Hippocratic Oath in its application today, does just that and is a driving influence in countless generations of physicians in the Western World—especially since 1945 (the end of the Second World War).[57]

---

[57] My friend, the late Edmund Pellegrino, used to have first year medical students recite the Oath as they began and when they ended their medical studies at Yale and later at Georgetown.

Chapter 3
Plato's *Timaeus*

Abstract Plato offers an account of disease that fits into his "likely story" natural philosophy and his moral theory that no one consciously does evil (meaning evil actions are a disease). In many respects, Plato's account is consistent with natural philosophers who came before and after him. His realistic account asserts that exercise and diet are keys to health—putting him in alignment with most Hippocratic writers.

Keywords Plato · Timaeus · Ion · Divided line · Nature · Forms · A likely story · Stoicheia · The four elements · Liver · Inflammation · Madness

Plato is interesting to a history of medicine because of his acquaintance with contemporary medical writers and his critical interest in the philosophy behind this space in natural philosophy. This chapter will focus upon Plato's most concentrated exposition on natural philosophy, *Timaeus*.

In a general, there are two distinct approaches to *Timaeus*: (a) as solely a metaphorical exposition that is about creation and origins; and (b) a set of speeches that vary in intent from *a likely story about creation* to *realistic accounts of natural philosophy*.[1] This author thinks that both approaches are correct within certain con-

---

[1] The *Timaeus* is variously read as a *literal* account—primarily of creation and basic cosmology or as a *figurative, literary exposition* to be interpreted much as one might interpret a narrative story. Proponents of the first group include: Gregory Vlastos, *Platonic Studies* (Princeton, N.J.: Princeton University Press, 1981); W.K.C. Gutherie, *A History of Greek Philosophy: The Later Plato and the Academy*, vol. 5 (Cambridge: Cambridge University Press, 1978); Jean Pépin, *Mythe et Allégorie: les Origines Grecques et les Contestations Judéo Chrétiennes* (Paris: Études Augustiniennes, 1976); and Richard Mohr, *The Platonic Cosmology* (Leiden: Brill, 1985). The second group includes: Svetla Slaveva-Griffin, "'A Feast of Speeches:' Form and Content in Plato's *Timaeus*" *Hermes* 133 (2005): 312–327; Matthias Baltes, "Die Weltentstehung des Platonischen Timaios nachden antiken Interpen" 2 vol. *Philosophia Antiqua* 30 & 35 (Leiden: Brill, 1976–1978); David

© The Author(s), under exclusive license to Springer Nature
Switzerland AG 2025
M. Boylan, *A Philosophical History of Western Medicine*,
https://doi.org/10.1007/978-3-031-97806-7_3

texts.² I will try to make clear my perspective as we examine several key arguments in the text. In this presentation, I will highlight both Plato's general approach to natural philosophy and a few insights into how the human body operates with a nod or two to directions for health.

I agree with G.E.L. Owen that the *Timaeus* is among Plato's last dialogues (in a sequence of three following *Republic* and preceding *Critias*) and therefore represents his most mature thoughts on the issues raised.³

In keeping with the methodology of the book, I will first examine some general principles that Plato expresses in the general philosophy of science and then focus on selected, specific points relating to biology and medicine.

Plato's General Philosophy of Science

The way I parse the dialogue, there are 13 principal arguments.⁴ In this section I will bring up arguments from most of the dialogue, as I believe this is the primary focus for Plato here.

Runia, *Philo of Alexandra and the Timaeus of Plato* (Leiden: Brill, 1986); J.M. Dillon, "The Riddles of the Timaeus: Is Plato Sowing Clues?" in Mark Joyal, ed. *Studies in Plato and the Platonic Tradition: Essays Presented to John Whittaker,* (London: Routledge, 2017): 25–42; Ralf Becker, "Zahlen vom Mythos zum Logos und Zurück" *Allgemeine Zeitschrift für Philosophie* heft 44.1 (2019): 45–60; Andrea Wilson Nightengale, *Genres in Dialogue: Plato and the Construct of Philosophy* (Cambridge: Cambridge University Press, 1995); Kathryn Morgan, "Designer History: Plato's Atlantis Story and Fourth Century Ideology" *Journal of Hellenic Studies* 118 (1998): 101–118, and Luc Brisson, *Plato the Myth Maker* tr. B. Naddaf (Chicago: University of Chicago Press, 1982).

²I set out my rules for being on one side or the other in my book, *Fictive Narrative Philosophy: How Fiction can Act as Philosophy* (New York and London: Routledge, 2019).

³G.E.L. Owen, "The Place of the Timaeus in Plato's Dialogues" *The Classical Quarterly* 3.5 (1953: 79–95).

⁴The conclusions of these 13 are: Argument One (27d–29d): Humans cannot provide the best accounts and must settle for likely stories grounded in belief.
  Argument Two (30a–30b): The world (universe) possesses soul and intelligence.
  Argument Three (30c–31b): There is only one world and one perfect type for animal creation.
  Argument Four (31b–32c): The primary qualities owe their combination potential to the existence of a 3rd entity, thus exist the four elements: earth, fire, air, and water.
  Argument Five (33b–36c): [Geometry and arithmetical proportions are useful to explain creation].
  Argument Six (37d–38b): Time is an outcome of creation (necessary due to the possibilities open) and came to be just as creation occurred.
  Argument Seven (41b): Three sorts of humans have been created.
  Argument Eight: (46d–48b): *Nous* and *ananke* struggle together in creation to set out purposive good whenever possible.
  Argument Nine (48e–52b): Being, space + generation (*on te kai xoran kai genesin einai*) are [regulative] principles that exist in heaven.
  Argument Ten (57e–58a): There is motion in nature.
  Argument Eleven (58a–58b): There is no void.

Plato's General Philosophy of Science

In Argument #1 Plato argues:

1. There are two sorts of thing that *are*: A. the eternal thing that has no becoming (*ti to on aei, genesin de ouk exon*, 27d 6), B. that which is always becoming but never gets there (*ti to gignomenon men aei, on de oudepote*, 28a1)—A[ssertion] (27d–28a)
2. That which is apprehended by reason (*noesis* and *logon*) is always in the same state—A (28a)
3. That which is apprehended by *doxa* and *aesthesis* always *becomes* but never *gets there*—A (28a)
4. [There is a link between ontology and the mode of knowing]—A
5. [Eternal things link to reason and logic while transitory things (that come-to-be and pass away) are linked to sensation and belief]—1–4
6. All created things are created by a cause—A (28a)
7. The *demiourgos* [a paradigm for understanding creation] looks to eternal patterns that are noble/beautiful/symmetrical (*kalon*)—A (28b)
8. Created patterns are not wholly noble/beautiful/symmetrical—A (28b)
9. The universe is as noble/beautiful/symmetrical as possible—A (29a)
10. We know that the universe was created by a *demiourgos* based upon eternal patterns—4–9 (29a)
11. Words are akin to matter—A (29b)
12. Books or arguments can imitate the eternal or they can imitate other books—A (29b)
13. *Being* is to truth as *becoming* is to belief—A (29c)
14. [The irrefutable and invincible are better than the refutable and vulnerable]—F[act]
15. [The best account is one that flows from eternal truth]—11–14
16. Humans only have limited access to the eternal forms—F (29e)

---

17. Humans cannot provide the best accounts and must settle for likely stories grounded in belief—15–16 (29d)

In premise #1 of this argument Plato follows pre-Socratic writers such as Heraclitus who also set out the realm of becoming and the realm of logos. In Plato's Divided Line in the *Republic* this is the division which separates both the *ways of knowing* and *the things that are*.[5] The eternal objects exist in another realm that is not visi-

---

Argument Twelve (68d): The experimental verification of color will at best be incomplete.
Argument Thirteen (90a–90c): Humans should seek Truth because it is lasting (immortal) and superior to satisfying transitory longings.

[5] The Divided Line—*Republic* 509d–511e
Level One—**The Intelligible:**

(A) Forms known through dialectical intellection (*noesis*)

(B) Mathematical objects known through hypothetical reflection (*dianoia*)

Level Two—**The Empirical:**

(A) Empirical objects known through grounded belief or trust (*pistis*)

(B) Empirical facsimiles known through image arranging (e.g., advertising, rhetoric, et al., *eikasia*)

The Argument

ble—often called the *realm of the forms*. These are partially known by propositional knowledge (*dianoia* and *episteme*) while the full knowledge only comes through intellectual intuition—direct knowledge (*nous*). For a theory of discovery (*genesis*), this makes a lot of sense—not just in Plato's time but throughout all history of humankind. Those who *discover* have an inspiration that is immediate and is akin to Plato's depiction of the poet in *Ion* 535d 5–533e. This is because what is *being discovered* are eternal, invisible truths (*phusis$_2$*).

Plato contrasts this higher form of knowledge with the information garnered through the senses that is always changing, premises 2–3. This means that, though they may provide practical information for a particular circumstance, they may have rather limited application over the long haul—thus being rather problematic for medical principles (within a medical handbook). Such limited application would, in this context, align Plato with *phusis$_3$*—belief can never be anything more. And, in principle (because of the Divided Line), empirical explorations are bound within this category.

However, in premise 6, Plato sets out a version of the *principle of sufficient reason*.[6] This is a very important principle in the general philosophy of science because it gets rid of accounts that claim that event x "just happened." *Every* natural event occurs because of some logically or empirically prior events/characteristics. Thus, the natural philosopher (including medical practitioners) cannot just throw their hands up in the air when asked to answer the question, "why." They can say that they don't know, but they cannot say that that the resolution is, in principle, inscrutable. There is a cause, whether or not the practitioner can fathom it or not.

Premises 7–10 talk about the existence of a creator who acts according to patterns that are generally *to kalon*, but may not always be so. I take this use of *to kalon* to refer to what is noble/beautiful/symmetrical/elegant. As mentioned earlier, I take this component to be important in the history of science and a fortiori of medicine. However, it does reflect *phusis$_2$*. Thus, the creator (however this is construed—an

---

1. The sources of knowing are empirical or mental—A
2. Empirical sources are easy to come by—A
3. [That which is easy to come by is trivial]—A
4. Empirical sources are trivial—1–3
5. [Trivial and non-trivial are logical contradictories]—F
6. Mental sources are non-trivial—4, 5
7. Direct propositional knowledge is relatively easy to come by—A
8. Direct propositional knowledge is relatively trivial—7, 3

---

9. The highest form of knowledge comes from pure (dialectical and non-propositional) contemplation—4, 6, 8

---

[6]The argument that all events in Nature have a cause, the principle of sufficient reason, is often ascribed to Gottfried Wilhelm Freiheer von Leibniz, *Discourse on Metaphysics,* ed. Albert R. Chandler, tr. George Montgomery (LaSalle., IL: Open Court, 1924 [1902]).

active intelligence or a purely material (almost mechanical) entity.)[7] Depending upon how one views Plato here, the generating force can move on a continuum between *phusis*$_{1-2}$ (the key terms here are *kata phusin archen* (29b3) and *peri te eikonos kai peri tou paradeigmatos* (29b 4–5)). These suggest a matching against a standard and a copy of that standard. The standard may be physical or superphysical, which would constitute the poles.

In either case, the analogy to words suggests the *stoicheion* position of elements creating larger wholes[8] This is an important point that links Plato to the pre-Socratics. The power of the creator clearly puts this entity into the "maker/artifact dynamic."

In premise 13 *being* is associated with truth while *becoming* is associated with belief. This assertion also goes back to the recognition that *what* one observes *here* and *now* exists in a foundational way that is different from speculations about causation that are inherently more complicated—especially in medical matters.[9] The fact that in medicine, especially, it is *becoming* that this dynamic is most important (prognosis). Belief is certainly subject to counterfactual conjectures. Thus, Plato's pronouncement here is more in tune with *phusis*$_3$. For this reason, we have only "educated guesses" that will guide our judgment in these matters based on how we, in our limited capacity (as humans), have access to the absolute truth in the forms. That means that what we are left with is only a likely story (*ton eikota muthon*, 29d 2).[10] This clearly situates Plato, with respect to understanding the process of discovery into the *phusis*$_3$ group.

In Arguments #3 and #5, Plato argues in what might be termed the *second* and *fourth stage* of the creation story, respectively, that what comes-to-be on earth has some relation to the forms. But because parts cannot be accurate copies of *what is*, there is nothing on earth that will fit the *type* for that sort of thing (30c–31a). In the case of animals, this means that there is variety in the individual, whole organisms. Since man is an animal, it would mean that individual persons have some various differences from the "perfect" homo sapiens. This poses a potential problem for medicine in diagnosis, prognosis, and treatment. Again, *phusis*$_3$ is affirmed.

---

[7] I explore the dimensions of material & mechanical (such as it might exist in this time) in my article: Michael Boylan, "Mechanism and Teleology in Aristotle's Biology" *Apeiron 15*.2 (1981): 96–102 and Boylan (1983): ch. 3.

[8] This position is most apparent in Plato's Theaetetus. For a discussion of this see: T. A. Druart, "Le 'Stoicheion' dans le 'Théétèete' de Platon" *Revue Philosophique de Louvain* 91 (1968): 420–436.

[9] The sense of epistemological *foundationalism* I am using follows my late friend, Roderick Chisholm, *Theory of Knowledge* (Englewood Cliffs, N.J.: Prentice Hall, 1989). Though Plato does not mention medical knowledge in this context, I am making an application by subsumption from the enunciated proposition.

[10] I term this sort of discourse "abductive logic" see Michael Boylan, *Fictive Narrative Philosophy: How Fiction can Act as Philosophy* (New York and London: Routledge, 2019): ch.5, cf. Michael Boylan, *The Process of Argument: An Introduction* (New York and London: Routledge, 2020): ch. 6.

In a move to find a connection between the imperfect parts with their combination into wholes, Plato turns to geometry.[11] This is perhaps the key methodological turn for Plato's philosophy of science. As set out by Ian Mueller, the geometry of the time was largely based upon construction and this abstract, nascent axiomatic approach would permit a more rigorous presentation than the Pre-Socratics provided for connecting parts to constructed wholes.[12]

Plato had a similar interest in constructing wholes from parts in the Theaetetus, 182c–183b 5. In that instance the *stoicheia* as element has a logical and linguistic sense, but here in the *Timaeus* the implication is material.[13]

This move toward geometry, as a way to explain Natural phenomena, begins in argument #5 where it attaches to astronomical events—such as the orbits of the planets.[14] The planets are spheres and their orbits are some deviations from a perfect circle.[15]

---

[11] This move to geometry as an overarching principle in Plato's general philosophy of science has had some broad discussion. For an introduction to some of this see: S. Glenn, "Proportion and Mathematics in Plato's Timaeus" *Hermathena* 190 *Philosophy and Mathematics* (Summer 2011): 11–27; John Cleaery, "'Proclus' Philosophy of Mathematics" in G. Bechte and D.J. O'Meara, eds. *La Philosophie des mathematiques de l'Antiquité tardive* (Fribourt: Editions Universitaires Fribourg Suisse): 85-101); Ronald F. Kotrc, "The Dodecahedron in Plato's 'Timaeus'" *Rheinisches Museum für Philogie, neue folge* 124, Bd. H ¾ (1981): 212–222; Ernesto Paparazzo, "It's a World Made of Triangles: Plato's 'Timaeus' 53b–55c" *Archiv für Geschichte der Philosophie* 97.2 (2015): 135–159; _____. "A Note on the Construction of the Equilateral Triangle with Scalene Elementary Triangles in Plato's 'Timaeus 54 A&B'" *The Classical Quarterly* 54-n.s. 65.2 (2015): 552-558; and Jan Opsomer, "Plutarch on the Geometry of the Elements" from Michiel Meeusen and Luc van der Stock, eds. *Aspects of Plutarch's Philosophy of Nature* (Leuven: Leuven University Press, 2015): 29–56; D.R. Lloyd, "Triangular Relationships and Most Beautiful Bodies: 'On the Significance of ἄπειρα at Timaeus 57d5 and the Number of Plato's Elementary Triangles'" *Mnemosyne* 62 (2009): 11–29—which supports Francis Cornford, *Plato's Cosmology* (London: Routledge, 1935) and Charles Muggler, *La physique de Platon* (Paris: C. Klincksie, 1960).

[12] Ian Mueller, *Philosophy of Mathematics and Deductive Structure in Euclid's* Elements (Cambridge, MA: MIT Press, 1981):286–306.

[13] For a more complete discussion on this connection see: Mary Louise Gill, "Matter and Flux in Plato's 'Timaeus'" *Phronesis* 32.1 (1987): 34–53.

[14] For a discussion of some of these dynamics see: Andrew Gregory, "Eudoxus, Callippus and the Astronomy of the 'Timaeus'" *Bulletin of the Institute of Classical Studies,* Supplement no. 78: *Ancient Approaches to Plato's "Timaeus"* (2003): 5–28.

[15] IV. Fourth Stage= **Argument Five**

1. The most perfect geometric figure is the sphere—A (33b)
2. The various realms of the universe (stars, planets, and earth itself) must be spheres arranged by concentric orbits—1 (36b–c (cf. I, 1))
3. [The order of creation follows the order of excellence]—A
4. [The soul is more excellent than the body]—A
5. The soul was created before the body—3,4 (34c)
6. Arithmetical proportions relate parts to a whole—A (35b)
7. The universe is to be viewed primarily through the whole—A (II, 7)
8. The parts of the universe should be seen as arithmetical properties—6, 7 (35b–36c)

Between arguments Plato tries to integrate parts of his material theory with an assumption that everything in nature is integrated within a symmetrical whole. He says: (39e–40a) that the *four types of beings* relate to the *four elements*: the gods = fire; birds = air; fish = water; land creatures = earth. This is a way that Plato goes about things—taking one problem at a time as a means of addressing well-defined issues within a separate, established context. The four roots of Empedocles are a given and there are four ontological entities he wishes to put forward—so he links them. This sort of eristic mindset is served by his use of *division* as his mode of classification, cf. *Statesman* 266e. This sort particularism is reminiscent of the Hippocratic, *Epidemics III*[16]: various insights are made on data points without integrating them within the larger explanatory system (see next section).

We can therefore characterize Plato's general philosophy of science based upon the *Timaeus* as: Searching for the necessity of mathematics—especially geometry—which it provides. Another move towards certainty, is his use of combining various small, distinct, eristic points together so as to appear to have constructed a whole account. (It should be noted that the entire *crucible/receptacle* episode in creation pitting pure symmetry of design versus what is possible (the countering force of *Ananke*) makes the account less a priori assertive.[17]

Plato's Non-systematic Insights into Biomedicine

In this section we will examine several cases in which Plato makes various unconnected speculations on biomedicine that are not connected by any systematic apparatus, but are rather a function of Plato's eristic method.[18] This would impact the *completeness* criterion.

---

9. [Geometry and arithmetical proportions are useful to explain creation]—2, 5, 6–8

   note: the proportions of 36c are spelled out in 36d. Six planets with seven unequal circles with 2 and 3 and 3. I surmise the following ordering: A-group = Sun, Mercury, Venus; B-group = Moon, Saturn, Mars, Jupiter; 37a → the soul is invisible; heaven is visible. The world soul stretches everywhere and partakes (*meta-exein*) in reason and harmony.

[16] For an example of this see my analysis of case 7, Littré VII in Boylan (2015).

[17] Argument Eight: (46d–48b): *Nous* and *ananke* struggle together in creation to set out purposive good whenever possible.

[18] The characterization of Plato as an eristic philosopher (as opposed to being a systematic philosopher like Aristotle) lies in his interest in using the *elenchus*—that seeks to refute a proposition put before him by another philosopher and stopping there. For some discussion on this method broadly understood see: Gail Fine, "Aristotle and the 'Aporema' of the 'Meno'" *Bulletin of the Institute of Classical Studies,* supplement no. 107, *Aristotle and the Stoics 'Reading' Plato* (2010): 45–71; Raphael Woolf, "Consistency and *Akrasia* in Plato's *Protagoras*" *Phronesis* 47.3 (2002): 224–252; Alexander Nehamas, "Eristic, Antilogic, Dialectic: Plato's Demarcation of Philosophy from Sophistry" *History of Philosophy Quarterly* 7.1(1990): 3–16.

In *Timaeus* 64a 2 Plato suggests that he will give a general description of the human body and how it works. In a little over a page Plato sets out the following principles. First, he sets out the common parts of the whole body and the major affections, viz., pleasure and pain. These come about from the following internal causes: being by nature (*phusis*$_1$) *easily moved* (*eukinetou*) or by nature (*phusis*$_1$) *difficult to move* (*duskinetou*). The former is according to nature (*kata phusin*), 64b 3; while the latter is the opposite. Thus, when the common parts work together, things move smoothly. There is harmony. When there is intransience, then there is cacophony. If *pleasure*, as an affection of the body, can be linked to *health* and *pain* to *disease*, then Plato is setting forth a theory of medicine that expresses balance and harmony of the respective body parts to be constitutive of health and the opposite to disease.

This move toward harmony and symmetry as being conducive to health and disharmony and asymmetry to disease is the essence of the humor theory. At roughly the same time period the Hippocratic writers were asserting something similar without the general connection to the cosmos as well. We have no definite dating of texts nor quotations that would allow any firm assertion of influence one way or the other.

What we can say is that Plato's writings on the general philosophy of science use geometry and formal symmetry as critical macro principles. Thus, it is not surprising that the same holds true with looking at living organisms such as humans. This also lends elegance to the theory due to the formal symmetry.

At *Timaeus* 65b 4 Plato turns his attention to particular parts and functions. These range from *taste* (65c), to *smell* (66d), to hearing (67a 7), to sight (67c 4)—which takes up some space and is one of Plato's more engaging discussions over time—especially as it explores color recognition theory, its causes and verification.[19] At 68e 6 Plato sets out that there are two sorts of causes: the necessary (*to anagkaion*) and the divine (*to theion*). I interpret these to be similar to *phusis*$_1$ and *phusis*$_2$.

At 69c Plato seems to assign the head (brain?) as the receptacle of the superior (eternal/divine) *psuche*. Between 69c and 70a 1 Plato talks about two sorts of souls[20]: one that is akin to the immortal (and thus akin to reason and the top-two levels of the Divided Line?) and one which is mortal which is inferior to the first. The "neck" is placed between, to separate these (69e). Plato believes that they operate upon different principles and must be kept apart.

---

[19] The key point here is whether sight and color recognition is an active or passive process on the part of the agent. For some general comments on this question, in general, see: Todd Gansom, "Are Colors Representational?" *Philosophical Studies* 116.1 (2013): 1–20; Pendaran Roberts, "An Ecumenical Response to Color Contrast Cases" *Synthese* 194.5 (2017): 1725–1742; and Nicholas Gaskill, "Epilogue after the Color Sense" from *Chromographia: American Literature and the Modernization of Color* (Minneapolis: University of Minnesota Press, 2018): 239–250.

[20] For modern readers *psuche* need not refer to some abstract entity that is the essence of someone apart from fulfilling some actual function. In the *Phaedo*, Plato does treat *psuche* in this manner in his argument for the immortality of the soul. But in this context within the *Timaeus*, I would contend that something approximating some biological function is what is intended.

To put this into context, there was a debate in this time period between whether the rational capability was located in the brain or in the region around the heart.[21] The Hippocratic writers were generally for the brain and Aristotle (later) was for the region of the heart. It is possible that Plato wants to split the difference, as he creates two divisions: (a) between the head and lower body and (b) between lower gut and upper gut (69e–70a). This could allow Plato to answer the question both ways. However, he does not elaborate enough on these distinctions and how they relate to function.

At 70b Plato moves to some more vital organs: the heart, the stomach, and the liver. It has been this author's opinion that in the history of medicine these three organs and the systems in which they are said to operate are essential to understanding the origins of ancient medicine.[22]

Concerning the heart, Plato sets it out as the source of blood and blood vessels[23] We can assume that since the blood "races through the limbs" (70b) that it does so for a somatic purpose. Plato sets out that blood conveys the whole power of feeling in the body—much as we might attributed to the nervous system today. Since the blood goes out from the heart, and is continually being made by the heart, we may assume that it gets *used up* in the body in carrying out these neurological functions (though the mechanism for this is not described).

Plato does recognize that since he accepts the elements of earth, air, fire, and water, in general (69b–c), that it would make sense to associate "work done" with heat. (Afterall, when humans do work, they get *hot*.) Since *too much* heat would violate the symmetry/balance principle just enunciated, there needs to be a balancing principle: enter breath. The lung is a collection device for cooler air, it can alleviate the heat (70c–d).

In the middle-to-lower gut Plato first assigns to the stomach-region as a manager of food (70e). "Managing food" is important, though Plato gives us no details except to say that it is done with making "as little noise and disturbance as possible" (71a).[24] Plato goes on to create a hierarchy in the body that is not dissimilar to the hierarchy

---

[21] One argument against linking the mind to an organ such as the brain is set out in Amber D. Carpenter, "Embodied Intelligent (?) Souls: Plants in Plato's *Timaeus*" *Phronesis* 55.4 (2010): 281–303.

[22] Andrés Pelavski, "Physiology in Plato's *Timaeus*: Irrigation, Digestion, and Respiration" *The Cambridge Classical Journal* 60 (2014): 61–74; Michael Boylan, "The Digestive and 'Circulatory' Systems in Aristotle's Biology" *Journal of the History of Biology* 15.1 (1982): 89–118; _____. "The Hippocratic and Galenic Challenges to Aristotle's Conception Theory" *Journal of the History of Biology* 15.1 (1984): 83–112; _____. "Galen on the Blood, Pulse, and Arteries" *Journal of the History of Biology* 40.2 (2007): 207–230; _____. *The Origins of Ancient Greek Science: Blood—A Philosophical Study* (London and New York: Routledge, 2015).

[23] *Phlebes*, should not be viewed as "veins" in these early contexts, but rather as "blood vessel." There is no distinction between veins and arteries yet.

[24] There are also hints in this region of exploring conception theory: James Wilberding, "Plato's Embryology" *Early Science and Medicine* 20.2 (2015):150–188; and Sara Brill, "Animality and Sexual Difference in the Timaeus" from *Plato's Animals: Gadflies, Horses, Swans, and other Philosophical Beasts* (Bloomington, IN: Indiana University Press): 161–176.

of the three sorts of citizens in the Republic, guardians, auxiliaries, and the *hoi polloi*. In each case it is *imitation* of a higher class that marks the realm of the lower class(es). Here, also, the gut receives the likeness of the upper trunk (and head?)—wherever the mind is located.[25]

The liver is set out as the principal organ of the lower gut and even given the power of prophecy (though the efficacy of this prophecy is dubious), (72b). What we notice in these passages in the dialog is a lack of any systematic approach that connects various organs using common function (as we will see in Aristotle). This is in keeping with Plato's general eristic argumentative approach that often looks at the truth of particular propositions in isolation—often using a reductio ad absurdum strategy to defeat that single claim only.[26]

For our purposes, the most important part of this discourse is the origin of disease (a principal concern in the history of medicine). At 81e 6 Plato says that there are four classes (*genon*) out of which the body is composed: earth, fire, water, and air. An excess of any of these is *para phusin* (contrary to Nature). Also, if their natural position in the body is changed, this can also be a cause for disease.

Also, when the composition of these is amiss (82c), this is also a cause of disease. For example, blood should create flesh and sinews. But when flesh is created out of residues and becomes discomposed, its waste goes through the blood vessels and blocks the blood's normal actions so that the blood does not get "used up" but accumulates creating a surplus of blood which then becomes polluted by bile, serum, and phlegm (82e).[27] This stops the blood from fulfilling its proper purpose of giving nourishment to the body (83a). The flesh does not renew itself and becomes old and grows black and bitter. This bitterness is due to acidity.

Another class of diseases comes about from the outer causes of the wind,[28] and the inner causes of phlegm and/or bile (84c). Normally, this would be purged by the air brought in by the lungs, but when there is some obstruction in this process or when too much air gets in, then the result is an extreme which, by definition, is against Nature (*phusis*$_1$) 84c–d.

---

[25] Plato does mention in 73c that the "brain" might be connected to bone marrow in some fashion. However, the reference is not specific enough to draw any conclusions about any connection to the "mind." For more on this see: John E. Sisko, "Cognitive Circuitry in Pseudo-Hippocratic 'Peri Diaites' and Plato's 'Timaeus'" *Hermathena* 180 (2006): 5–17.

[26] The reductio ad absurdum style affects the way an argument is presented and its scope. For a bit of background see: Catarina Dutilh Novaes, "Reductio ad absurdum from a Dialogical Perspective" *Philosophical Studies* 173.10 (2016): 2605–2628; Brendan O'Sullivan, "The 'Euthyphro' Argument 9d–11 b" *Southern Journal of Philosophy* 44.4 (2006): 657–675; and Elliot Welch, "Self-Predication in Plato's 'Euthyphro'?" *Apeiron* 41.4 (2008): 193–210.

[27] These are, of course, three of the four Hippocratic humors (though sometimes phlegm and serum are combined and bile is distinguished by "black" and "yellow"). Blood, a key Hippocratic humor is not mentioned here—but was above at 82c as a kind of huper-agent that might create and interact with flesh and sinews.

[28] Cf. the Hippocratic work, *Airs, Waters, and Places*, which shares some concerns about exterior effects and in particular, *places*.

At 85b 6 Plato brings up the issue of *inflammation* (*phlegmainein*)[29] as a cause of disease. Unlike the writer of *On the Nature of Man*, this inflammation who used the word in connection with phlegm, Plato attributes this to bile giving off fibers that congeal.

At 86b ff. Plato sets out that diseases of the mind (madness and stupidity) come about via acid and briny phlegm and other bitter and bilious humors wander about the body and find no exit or escape. Again, this is a reference to imbalance of humors that can cause ill temper, rash action, and melancholy (87a). Because this account is rather vague and does not attach itself to exact triggers and systemic effects, it is useless for diagnosis and prognosis (essential for medicine). One is inclined to think that this really comes in the context of Plato's ethical position that no person commits evil voluntarily (86e). Thus, Plato here offers a "disease account" of evil human action.[30] Since Plato claims that no one does evil voluntarily, all evil action must be the result of ignorance—brought on by illogical reasoning or by disease.[31] This, of course, brings "disease" to another level.

Plato's prescription for a prophylactic against disease is exercise (89a ff.). This is shared by most of the Hippocratic writers who also suggest a regulated diet.

In general, we can walk away from the *Timaeus* with some general ideas that relate to science and medicine and a few particulars that will be advanced by various other writers in physicians within the ancient Greek world as we progress eventually to Galen.[32]

---

[29] The word, φλεγμαίνειν, has a connotation of inflammation due to heat caused by phlegm, cf. the Hippocratic Work, *The Nature of Man*, 4, cf. Plato, *Republic* 564b.

[30] This position is similar to one modern psycho-analytic position set out by Karl Menniger, *The Crime of Punishment* (New York: Viking, 1968).

[31] Colin David Pears, "Congruency and Evil in Plato's 'Timaeus'" *Review of Metaphysics* 69.1 (2015): 93–113, cf. C.M. Chilcott, "The Platonic Theory of Evil" *Classical Quarterly* 17.1 (1923): 27–31.

[32] Galen viewed the *Timaeus* as a source of medical doctrine and used it in the following works: *The Faculties of the soul* (following the mixtures within the body); *On the Powers and Mixtures of Simple Drugs*, and *On the Use of the Parts*. For a discussion of this see: Aileen R. Das, "Re-evaluating the Authenticity of the Fragments of Galen's 'On The Medical Statements in Plato's Timaeus' (Scorialensis grac. ф-III 1. 11ff. 123r–126v)" *Zeitschrift für Papyrolgie und Epigrapik*, Bd. 192 (2014): 93–103.

Chapter 4
Aristotle

Abstract This chapter outlines some of the general strategies of Aristotle's philosophy of science found in the *Organon*—particularly *Prior and Posterior Analytics* and squares them with his biological writings stemming from *The Parts of Animals*.

Keywords Aristotle · Nature · *Nicomachean ethics* · *Eudemian ethics* · *Organon* · *Prior analytics* · *Posterior analytics* · *Syllogisms* · *Divided line* · 4 scientific questions · Induction · Material explanation · Teleological explanation · *The parts of animals* · Empedocles · Dissection · *Ancient medicine* · Pneuma theory · Blood temperature · Galen · Digestion · Conception

Aristotle's father was the court physician to the King of Macedon, Amyntas II. In my short story "Aristotle the Outsider,"[1] Aristotle is always an outsider: the *other*. *First*, he doesn't go into medicine (as was the path at the time to keep things close—either among family or to surrogates who became "like family"), but goes more generally into biology. In a real sense (since a third of his extant works are in biology), he is the founder of this discipline in the West. The big differences are two: (1) As a biologist and not a physician, he broader and more theoretical than most physicians, and (2) He covers the anatomy and physiology of more organisms than just humans. Both of these points are important as we explore his role in general science and biological science—especially as it might be relevant to the practice of medicine. *Second*, Aristotle goes to live in Athens (which had been controlled by the hated Macedonians including Aristotle's pupil, Alexander). As an outsider, Aristotle couldn't own land nor become a full citizen. These are important points to help us

---

[1] Michael Boylan, "Aristotle the Outsider" in Michael Boylan and Charles Johnson, eds. *Philosophy: An Innovative Introduction—Fictive Narrative, Primary Texts, and Responsive Writing* (Boulder, CO: Westview, 2010).61–72.

understand Aristotle. His perspective as an outsider rather freed him at being comfortable at being at the outside of conventional attitudes. It helped him formulate his unique positions on natural philosophy.

General Philosophy of Science Trying to set out what Aristotle's general philosophy of science is constitutes a controversial exercise. One approach (perhaps one that Aristotle intended)[2] was that Aristotle intended to create a logical structure that would guide all investigations into nature. By doing this Aristotle may have intended coherence and symmetry (and by extension, elegance). This sort of structure would be teleological—such was Aristotle's disposition, cf. the opening lines of the *Nicomachean Ethics* in which he links all *making* to a method that is governed by an orientation that is end-oriented.[3]

Aristotle's logical structure for a general philosophy of science is found in the *Organon*. The logic part of the method is set out in the *Prior Analytics* and its application to Nature is found in the *Posterior Analytics*.

In the *Prior Analytics* Aristotle structures formal, deductive logic. Each argument will have two premises and a conclusion and three terms: the subject and predicate of the conclusion and the middle term (which exists only in the premises in order to connect the subject and predicate of the conclusion—also called *the extremes*, *APr.* 1.5). According to whether the middle term acts as the subject in one premise and predicate in the other, and according to the quality of the premises (universal affirmative, universal negative, particular affirmative, or particular negative),[4] the syllogism will have a "figure." Some figures are universally valid (meaning the structure of argument's premises will always bring about the argument's conclusion). An example of this is the AAA—first figure (Fig. 4.1).[5]

Of course, there is more to logic than mere formal validity. The premises, themselves, must be *true*. If the premises are true and the argument is valid, then the conclusion possess deductive certainty. It's not just *probable* or the *best* among the available alternatives.[6]

So, what is the *purpose* of creating a formal logical presentation? One purpose might be the ability to defend a conclusion of such a procedure as being *apodeictic* (certain). There are certainly epistemological and metaphysical advantages for

---

[2] I look at this approach in my book, Michael Boylan, *Method and Practice in Aristotle's Biology* (Lanham, MD and Oxford: UPA/Rowman and Littlefield, 1983): 18–29.

[3] *NE* 1094a1–2: Πᾶσα τέχνη καὶ πᾶσα μέθοδος, ὁμοίως δὲ πρᾶξίς τε καὶ προαίρεσις, ἀγαθοῦ τινὸς ἐφίεσθαι δοκεῖ, διὸ καλῶς ἀπεφήναντο τἀγαθὸν οὗ πάντ' ἐφίεται.

[4] "All S is P" is universal affirmative; "No S is P" is universal negative; "Some S is P" is particular affirmative; "Some S is not P" is particular negative—generally called "A," "E," "I," and "O" propositions, respectively.

[5] The "M" refers to the middle term; its function is to connect the "S" to the "P." The "S" refers to the subject term of the conclusion and the "P" is the predicate term of the conclusion. For longer arguments, sorites, the conclusion of argument #1 becomes the major premise (first premise) of the second argument. The result rather resembles pearls on string.

[6] For clarification of these other sorts of logical goals (induction and abduction) see: Michael Boylan, *The Process of Argument* (London: Routledge, 2020): chapts. 5 & 6.

All M is P

All S is M

All S is P

Fig. 4.1 The universally valid AAA first figure logical syllogism (The second figure is: P-M/ S-M, therefore; S-P/ The third figure is: M-P/M-S, therefore: S-P/ The fourth figure is: P-M/M-S, therefore: S-P. [Note that Aristotle did not recognize the fourth figure.] Each proposition also possesses a quality as per above. There are 64 different moods to be enumerated, though Aristotle does not go through all this detail. For a classic work on Aristotle's logic as set out in the *Prior Analytics* see: H.W.P. Joseph, *An Introduction to Logic* (Oxford: Oxford University Press, 1906).)

being able to claim that the conclusion of an argument is *certain*. However, when Aristotle moves over to the *Posterior Analytics,* he immediately contextualizes this. "All teaching and learning that involves the use of reason proceeds from pre-existent knowledge. This is evident if we consider all the different branches of learning, because both the mathematical sciences and every other art are acquired in this way" (tr. Trendennick).[7]

Of course, this sentence makes use (in a different context) of Plato's contention that all knowledge is pre-existent so that "education" is not teaching, but stimulation for recollection.[8] This is connected to Plato's theory of reincarnation and how we forget what we're exposed to (regarding forms) between lives.[9] Aristotle does not accept Plato's theory of forms, his theory of reincarnation, nor his theory of recollection as how we learn. Thus, his reference here is clearly *ironic*.

However, what it does show in the *APo.* Passage is that though we have a syllogism that because of its structure is *valid* (meaning that the structure of the premises as stated lead to a necessary conclusion (as in an AAA-First Figure Syllogism)) it does *not* mean that we are finished. The truth of the premises is still in question. This is what is necessary to complete the project (to make the argument *true* and *complete*—and thereby *sound*).[10] What we should take away from this, is that *formal method* is highlighted by Aristotle in the *Prior Analytics* in such a way that the presentation of what it *means to study nature* (become a natural philosopher) takes on a structure beyond previous practitioners. Interacting with the explanandum to

---

[7]*APo.* 71a 1–4: Πᾶσα διδασκαλία καὶ πᾶσα μάθησις διανοητικὴ ἐκ προϋπαρχούσης γίγνεται γνώσεως. φανερὸν δὲ τοῦτο θεωροῦσιν ἐπὶ πασῶν, αἵ τε γὰρ μαθηματικαὶ τῶν ἐπιστημῶν διὰ τούτου τοῦ τρόπου παραγίγνονται καὶ τῶν ἄλλων ἑκάστη τεχνῶν.

[8]*Meno:* (81 c–d): 1. All "learning" is *anamnesis*—A/ 2. *Anamnesis* is a process within us—F/ 3. Teaching is learning from without—F ∴/There is no learning via teaching—1–3 (82)

[9]For an account of this process recollection see: Dorothea Frede, "Not in the Book: How does Recollection Work?" Proceedings of the Second Symposium Platonicum Pragense. (Praha: Oikúmené. 2001).

[10]It should be mentioned here that though "soundness" is a developed concept of Aristotelian logic, Aristotle only addresses this indirectly in the *Prior Analytics*, cf. Joseph, *loc. cit.*

create an explanans now requires rigid rules of procedure as opposed to inspired acts of intuition (*nous*)—as was the case with Plato (cf. *The Divided Line*'s highest level).

With these formal, structural, tools of logic in hand, Aristotle seems to suggest in *Posterior Analytics* that we still need some additional logical categories to ascertain what is going to be necessary in order to fulfill the essential entities for analysis (such as the difference between the fact (*hoti*) and the reasoned fact (*dioti*) and existential import (*ei esti*) and the essential nature [definition] of the object under observation (*ti esti*)).[11] These are meant to interact organically so that the reasoned fact connects the fact to its explanation (definition) in the *ti esti*. This organic connection commits Aristotle (in theory) to giving an underlying account that is organic and not ad hoc (see Chap. 6).

The purpose of this structure is to create a logical *conceptual* framework upon which to present scientific accounts. Some have criticized Aristotle here as creating one structure in the *Prior* and *Posterior Analytics* and then not following his own prescription when he is engaged in his biological investigations (his principal area of field research).[12] It has been my approach that instead of contradicting himself, Aristotle is involved in two different projects when he is engaged in his biological investigations and when he is setting out as what the presentation standards are for a finished science. The *Analytics* are concerned with the latter.

Of these crucial four questions that define scientific inquiry for Aristotle (and ensure *completeness*—the fact, the reasoned fact, the mode of existence, and the definition) the first two are a pair and make it possible to set-out the last two. For recognition of *the fact*, we need sense perception of the particular (*APo.* 99b 35) which leads to a developed power of retention from the persistence of sense perceptions (*APo.* 99b 36–100a 1). This power of retention is memory (*APo.* 100a 3–4). The addition of multiple memories in conjunction with continued sense input creates experience, *empeiria* (100a 5).[13]

The process can be reconstructed as follows:

1. Sense perception of the particular (99b 35).
2. A developed power of retention from the persistence of the sense perceptions (99b 36–100a 1).
3. This power of retention is memory (100a 3–4).
4. The addition of multiple memories in conjunction with continued sense input creates experience, *empeiria* (100a 5).
5. From experience (i.e., being able to identify the common element in the memories and to match it against sense input) arises a universal through induction (100a 5–8).

---

[11] *APo.*, 89b 23–25: Τὰ ζητούμενά ἐστιν ἴσα τὸν ἀριθμὸν ὅσαπερ ἐπιστάμεθα. ζητοῦμεν δὲ τέτταρα, τὸ ὅτι, τὸ διότι, εἰ ἔστι, τί ἐστιν.

[12] These are set out in Boylan (1983): ch. 1.

[13] Three studies of note: J.H. Lesher, "'Just as in Battle?' The Simile of the Rout in Aristotle's 'Posterior Analytics' II 19" *Ancient Philosophy* 30.1 (2010): 95–105; David Bronstein, "The Origin and Aim of 'Posterior Analytics' II.19" *Phronesis* 57.1 (2012): 29–62; and Greg Bayer, "Coming to Know Principles in 'Posterior Analytics II 19'" *Apeiron* 30.2 (1997): 109–142.

6. Induction admits the universal into the soul by the power of *nous* (in the case of unqualified knowledge) and by the power of *doxa* (opinion, in the case of general knowledge, *gnosis*)—1–5 (100b 5–100b 9).[14]

Argument 3.1: The Operation of Induction

What is clear in Argument 3.1 is that induction, *epagoge*, operates via *aesthesis* (sensation), *mneme* (memory), *empeiros* (experience), and *nous* (intellectual intuition). When a particular group of common facts are present in the mind, then intellectual intuition arranges these common elements into a universal. It is not clear just how many cases are necessary for an induction. In *APr.* 68b 26–30, it seems as if all the cases must be considered (commonly called "enumerative induction" in the history of science) while in *Topics* 105a 13–16, one or two cases seems to be enough.[15] Perhaps, it would be fair to conclude from these two passages that one must go through some *appropriate* number of cases when practically employing induction since complete enumeration of all the cases is rarely possible. But what number of cases is "appropriate"? This would depend upon the judgment of the practitioner. But is this *good enough*? In the history of science, some practitioners are eager to *prove* their hypothesis and will bend over backwards to make the data fit the explanation (often called "confirmation bias").[16]

The actual apprehension of the universal is an intellectual process which was first occasioned by an act of sensation. It would be incorrect to demand (as some of the empiricists did) that there be no ratiocinative act in an empirical investigation. Such a demand would give immediate recognition of the "fact" (*hoti*, *APo.* 1.31) total sway in comprehending nature. Likewise, knowledge of only the "reasoned fact" (*dioti*) that is tied to no actual phenomena (as the uncritical rationalists would have it) could lead to *incoherence*. These are the two sides of pre-Aristotelian natural philosophers who focus upon *phusis*$_2$ or *phusis*$_3$. Ultimately, they are not robust enough to support the theoretical science that must support medicine.

Aristotle also cites some key distinctions necessary for the logical presentation of nature within the *Posterior Analytics*. These include the distinction between accidental relations (*sumbebekos*) as opposed to actual causal relations that will lead to unqualified knowledge (*apodeixis*). One key hurdle that must be met is that alleged causal connection (*dioti*) is "prior" and "primary."

---

[14] This is not a logical argument, but a "process checklist."

[15] Sometimes, enumerative induction is called "perfect" induction by Aristotle and the use of fewer examples "imperfect" induction. For a discussion of these see Boylan (1983): 20–28.

[16] The modern research on scientific "confirmation bias" is large. Some contemporary representative samples include: Olav B. Vassend, "Confirmation measures and Sensitivity" *Philosophy of Science* 82.5 (2015): 892–904; Kevin T. Kelly, Konstantin Genin, Hanti Lin, "Realism, Rhetoric, and Reliability" *Synthese* 193.4, Special Issue: Causation, Probability, and Truth—The Philosophy of Clark Glymour (2016): 1191–1223; Jan Sprenger, "A Novel Solution to the Problem of Old Evidence" *Philosophy of Science* 82.3 (2015): 383–401; and Boas Miller, "When is Consensus Knowledge Based? Distinguishing Shared Knowledge from mere Agreement" *Synthese* 190.7 (2013): 1293–1316.

By "prior" or "more knowable" in relation to us I mean that which is nearer to our perception, and by "prior" and "more knowable" in the absolute sense I mean that which is further from it. The most universal concepts are furthest from our perception, and particulars are nearest to it.[17] And these are opposite to one another.

To argue from primary premises is to argue from appropriate first principles; for by "primary premises" and "first principles" I mean the same thing.[18] (tr. H. Tredennick and E.S. Forster).

To be prior in an ordinary sense, the source of the cognitive recognition must be the result of a sensory experience. It is something we've *seen, heard, smelled, touched,* or *tasted.* Our senses don't lie to us. They are epistemologically *simple* and the touchstone to truth.[19] However, given that, and the process of induction (II.19) including memory and the power of creating generalizations, there is another (more philosophical) sense of prior that refers to the universals (that exist at various levels, as per the chart of the eight categories of thought, *Categories* IV).

I've always contended that the best way to test *prior* in the absolute sense of universals is to submit the proposition to the logical test of convertibility: "All S is P & All P is S." For example, "All Humans are Rational and All Rational Beings are Humans."

To be *primary,* the proposition at hand must be an *arche.* That is, it must be the fundamental causal factor. This is a stricter criterion than being *prior.* For example, take *risibility* (the ability to laugh). One may properly say "all humans are risible" and "all [animals] that are risible are humas." This logical convertibility shows that the trait in question is prior because of its universal applicability (as shown through logical convertibility).

But this is not enough. We need to know more. We don't laugh (coming from the ability to see the disparity in two propositions in an intentionally discordant fashion) because we are *human,* but rather because we are *rational.* Thus, logical convertibility is not enough. We must get to the fundamental *why.* Only *then* have we a window to *dioti* (the reasoned fact).

These two distinctions of *dioti* underline all four of Aristotle's causes. The material and efficient causes are *discovered* in real time first. Then, upon reflection we become aware of that *out of which materially* (material cause) something comes-to-be through the motions of these material agents (efficient cause). This pairing corresponds to the sense of *prior* which is, at first, known via senses and then moves to a more universal representation (as per above). I have called this duo the *material explanation* or ME.[20] Throughout the biological works, Aristotle uses ME both in

---

[17] Cf. *Met.* 1029b 3 ff.

[18] λέγω δὲ πρὸς ἡμᾶς μὲν πρότερα καὶ γνωριμώτερα τὰ ἐγγύτερον τῆς αἰσθήσεως, ἁπλῶς δὲ πρότερα καὶ γνωριμώτερα τὰ πορρώτερον. ἔςτι δὲ πορρωτάτω μὲν τὰ καθόλου μάλιςτα, ἐγγυτάτω δὲ τὰ καθ' ἕκαςτα, καὶ ἀντίκειται ταῦτ' ἀλλήλοις. Ἐκ πρώτων δ' ἐστὶ τὸ ἐξ ἀρχῶν οἰκείων, ταὐτὸ γὰρ λέγω πρῶτον καὶ ἀρχήν. *APo.* 72a1–7.

[19] This is an important distinction in the history of philosophy. Are the senses always reliable? Aristotle says, "yes." Later philosophers, like Descartes in the *Meditations* I, suggests that the senses may be sources of deception.

[20] Boylan (1983): 125, 128, 146, 157–158, 165–167, 234–237.

his mode of discovery and his presentation of his findings. It is usually associated with necessity.

The second mode combines the causes of formal cause (that sets out the essences of some subject according to what is primary, above) and the final case (that provides the ultimate criteria of why one account might come before another in priority (as per the risibility example above)). I call this pairing TE or teleological explanation.[21] This is usually connected with the overall *reason why* from a system standpoint.

Both ME and TE are used extensively in Aristotle's biological investigations. They spring naturally from the distinctions made in the *Posterior Analytics*. This combination goes a long way to refute those who contend that Aristotle's pronouncements in the *Posterior Analytics* are divorced from his actual method when engaged in biological research.

The last nod to theory takes us to the beginning of *The Parts of Animals*.

Aristotle's Biomedical Applications

This sub-section will examine Aristotle's biomedical applications using his theoretical structure set out in the previous section. At the beginning of *The Parts of Animals*, Aristotle says:

> In all study and investigation, be it humble or noble, there appear to be two types of proficiency. One is scientific knowledge of the subject; the other is an educated person's competence. For the latter has the power to judge what is put forth properly and what is not.[22] (tr. mine)

Here, Aristotle distinguishes between the specialist (who has *episteme*) and the generalist who has a general education (*paideia*) about the subject matter. This is important, for in biomedical investigation, one must have competence with some particular area of medicine (for example), but that has to be grounded within a wide-ranging theory of science. It has been the practice of this book to begin with the wide-ranging theory of science and then examine how this may have helped the practitioner make advances that affect the practice of medicine.

What is significant about this passage is that Aristotle allows for the situation in which a practitioner may have expert knowledge of biomedicine, for example, without being well-versed in other subjects that may relate to general science, like mathematics or logic. This contrasts with Plato, who thought that one had to have the entirety of natural science to have any of it.

---

[21] Boylan (1983): 93, 101–103, 235–236.

[22] Περὶ πᾶσαν θεωρίαν τε καὶ μέθοδον, ὁμοίως ταπεινοτέραν τε καὶ τιμιωτέραν, δυο φαίνονται τρόποι τῆς ἕξεως εἶναι, ὧν τὴν μὲν ἐπιστήμην τοῦ πράματος καλῶς ἔχει προσαγορεύειν, τὴν δ'οἷον παιδείαν τινά. πεπαιδευμένου γάρ ἐστι κατὰ πρόπον τὸ δύνασθαι κρῖναι εὐστόχως τί καλῶς ἢ μὴ καλῶς ἀποδίδωσιν ὁ λέγων. *PA* 639a 1–7.

Aristotle, with his first-hand empirical knowledge of marine zoology—particularly from tide pools—could write on these matters with authority.[23] But, since one cannot have thorough first-hand knowledge of all species, sometimes a general practitioner must critically interview those who have—in Aristotle's case one prominent example was bee-behavior from beekeepers.

Of course, the problem with this is that the beekeepers might have one thing on their mind: producing honey for sale. Whatever relates to this, they observe, and whatever does not relate to this, is merely "noise" to them. They may even make-up details that they might relate to Aristotle upon his queries.

But since one person cannot (contra Plato) know the entire realm of zoology, this approach is all that is open to Aristotle. However, it should be clear, that Aristotle (like Plato) is moving toward universals. He doesn't just want to talk about natures of particular species—such as Man, Ox, or Lion—but rather the common character that they possess between species and genera (P.A. 639a 15–19).

It is these general universals between species and genera that allow Aristotle, and those who follow him later, to apply certain shared structures and their activity (anatomy and physiology in modern parlance). And though Aristotle did not hold Empedocles' ideas on evolution, he does create "quasi-homology" based upon his taxonomic depiction of animals.[24]

This allowed for the practitioner to examine a non-human species and dissect it (since such was not legally allowed for humans) and then to make comparisons-speculations on similar organs in humans—at least according to anatomy. (Dissection requires killing the animal (or using a dead animal) for anatomy and comparative anatomy, but because the body is dead (in either case), physiology was rather speculative.)

A second question is also set out at the beginning of *PA*:

> Should the student of Nature, just as the mathematician does when he is doing astronomy, first consider the datum of the animals and each of their parts and only then go on to state the *reasoned fact* and the causes, or should he follow another procedure? (my tr.)[25]

To unpack this, we have to ask how the mathematicians act while doing astronomy. First, we may rule out astrology (*phusis$_2$*). Instead, the mathematicians seek to *predict* the positions of the planets against their "fixed star" backdrop. So, also should the student of biological Nature. They also require a causal account of the *phainomena* that will approach the certitude of the mathematician examining the super-lunar realm. This certitude can create an attitude toward Nature of causal power that could support the maker/artifact dynamic.

---

[23] For a detailed account of what Aristotle *did* and *did not* know "first-hand" see Boylan (1983): ch. 4.

[24] See Boylan (1983): pp. 42, 62–63; 225–229.

[25] νῦν γὰρ οὐ διώρισται περὶ αὐτοῦ, οὐδέ γε τὸ νῦν ῥηθησόμενον, οἷον πότερον καθάπερ οἱ μαθηματικοὶ τὰ περὶ τὴν ἀστρολογίαν δεικνύουσιν, οὕτω δεῖ καὶ τὸν φυσικὸν τὰ φαινόμενα πρῶτον τὰ περὶ τὰ ζῷα θεωρήσαντα καὶ τὰ μέρη τὰ περὶ ἕκαστον, ἔπειθ' οὕτω λέγειν τὸ διὰ τί καὶ τὰς αἰτίας, ἢ ἄλλως πως. P.A. 639b 8–11.

Many of these answers are very controversial and "have not been decided" (639b 7). Aristotle sees himself as part of an ongoing debate just as the writer(s) of *Ancient Medicine* did, but here, Aristotle spends more time on methodology in order to frame just what the solutions might look like.[26]

Aristotle's innovation to the history of medicine—beyond methodology and comparative zoology—rests in five areas: (1) Pneuma Theory; (2) The balance of blood temperature; (3) An account of the digestive and "circulatory" systems, and (4) Conception theory, (5) And his doctrine of comparative anatomy based upon dissection and analogy. Each of these deserves a tract of its own, and I have contributed to this in the past.[27] Here, I intend to summarize these findings as they paint a picture of Aristotle's contribution to the History of Medicine.

Pneuma Theory First there is *pneuma*. When assessing what the human body in its everyday activity is able to do, one of the most prominent principles is that it is capable of physical activity. We want to lift our left arm and then we are able to lift it, ceteris paribus. There is some connection between our rational resolve and our bodily response. What is the ME of this? There must be a connection between *thought* and *physical activity*. To get at this connection further, it is important to note the role of blood. Through blood vessels (no distinction between veins and arteries at this juncture) we have a mechanism to transfer an active, potent fluid throughout the body.

How much is included in blood? Most bodily functions.[28] We need a transport system for providing nourishment to the body, getting rid of wastes, creating active materials for reproduction, and for giving *pneuma* a system of connection between thought and action. Now, in this period of history there was a debate on whether the brain was the source of cognition or whether cognition occurred somewhere else. The answer to this question is important to physicians because many physiological maladies include issues involving our mental state. The source of the mental state is a vital first step in being able to create a diagnosis that will drive treatment.

Aristotle is on the side of "somewhere else." He puts this region near the heart (the creator of blood). As blood is created two vital elements are added in its composition: *trophe* (nourishment) and *pneuma* (the vital power within the blood that is the result of lung breathing that mixes air with the right amounts of heat and water). Thus, as the blood is created, it becomes the conduit of both nourishment and

---

[26] These two questions arise again in *PA* I.4 & 5.

[27] For *pneuma* see Michael Boylan, *The Origins of Ancient Greek Science: Blood—A Philosophical Study* (New York and London: Routledge, 2015): 34–36, 54, 59, and blood 87, relationship to air 118, nerve-like activity 81, 112, types of *pneuma* 118. For the *balance of blood temperature see:* Boylan (2015) 61–62; Boylan (1983): 187–189. For an account of *the digestive and "circulatory" systems see:* Michael Boylan, "The Digestive and 'Circulatory' Systems in Aristotle's Biology" *Journal of the History of Biology* 15.1 (1982): 89 118. And *Aristotle's conception theory in context see:* Michael Boylan, "The Galenic and Hippocratic Challenges to Aristotle's Conception Theory" *Journal of the History of Biology* 17.1 (1984): 83–112.

[28] This is the central thesis of Boylan (2015).

"animal spirits."[29] Because the heart is so central to this creation, Aristotle assigns the source of cognition to the general region of the heart in the newly *pneuma*-enriched blood.[30] This means that, as such, "thinking" occurs in no actual organ (*phusis$_2$*), but rather in the region of the heart by through the material agency of *pneuma* + blood (*phusis$_1$*). Though this account employs both *phusis$_1$* and *phusis$_2$,* the balance leans clearly in the direction of *phusis$_2$*. Since blood is created via the digestion of food-process, and *pneuma* is produced via lung breathing and transported by the pulmonary vein to the heart, this combination is set forth with an ME account that brings with it necessity.

"Necessity" is an important concept--though Aristotle's account of "*ex ananke*" can be puzzling. According to the context it can refer to different aspects of the *dioti* of nature/Nature.[31] Finding the appropriate role of material necessity (within the ME) is very important for the history of medicine because it can make a significant difference in diagnoses and prognosis. It is also important within any account of material necessity that the teleological element (TE) is not left out. Though it isn't stated in medical terms, this combination will offer a good perspective for the practicing clinician for understanding what *ought to be happening* as opposed to the symptoms presented and the *force* of these particular symptoms.

Aristotle tries to combine these two needs in his account of conditional necessity (*PA* 642a2)

> There are then two causes: *that-for-the-sake-of-which*[32] and the cause from *necessity*[33]—for many things are generated from necessity. Perhaps one might ask which necessity is meant by those who say *from necessity*. For neither of the two modes defined in our philosophical treatises can be meant here.[34] In things generated, however, there is a third kind. It is seen in things which come-to-be; as when nourishment is necessary, it is not according to either

---

[29] Aristotle never uses a phrase like "animal spirits" but later writers up to the nineteenth century take *pneuma* theory and use this translation. They are trying to account for what we moderns would describe as "neurological activity."

[30] Boylan (1983): 170–172, 184–185, 190–191, 195–199, 201–202.

[31] For example see: David Balme, translation of *De Partibus Animalium I* (Oxford: Oxford University Press, 1972); Wolfgang Kullmann, *Wissenschaft und Methode: Interpretationem zur aristotelischen Theorie der Naturwissenschaft* (Berlin: De Gruyter, 1974), Anthony Preus, *Science and Philosophy in Aristotle's Biological Works* (New York: G. Olms, 1975); Richard Sorabji, *Necessity, Cause, and Blame* (Ithaca, New York: Cornell University Press, 1980); Mariska Leunissen, Alan Gotthelf, "'What's Teleology Got to Do With It?' A Reinterpretation of Aristotle's 'Generation of Animals V'" *Phronesis* 55.4 (2010): 325–356; and James Lennox, "Aristotle on the Emergence of Material Complexity: *Meteorology* IV and Aristotle's Biology" *HOPOS* 4.2 (2014): 272–305.

[32] TE (formal and final causes combined).

[33] ME (efficient and material causes combined).

[34] See *Metaphysics 1015a 20ff*—mode one and *Politics* (1282b 19) and *Eudemian Ethics* (1217b 23)—mode two. The meaning listed in mode one is "absolute" necessity and in mode two "coercive" necessity. In the *PA* passage a third mode is listed "conditional necessity."

of the other modes [absolute or coercive necessity], but we mean that without nourishment no animal can live. This is *conditional necessity*. (my tr.)³⁵

In absolute necessity a → b, a is essential in order to obtain b. In coercive necessity, a → b, means that the presence of condition "a" tends to bring about "b." In conditional necessity, a → b means that from the point of view of "b," "a" must occur. Thus, the focus changes from the antecedent (in the former) to the consequent (in the latter).³⁶ This emphasizes teleology so that though conditional necessity combines TE and ME, the TE element is primary.

Thus, in the case of pneuma theory, conditional necessity enters the picture starting from the vantage point of the consequent—our body parts move (by themselves and by voluntary directive). This "b" requires an antecedent "a." The antecedent is pneuma. The other two highlighted biomedical principles act in the same way.

The Balance of Blood Temperature This principle has two parts to it: the idea of *balance* as TE and the ME of achieving that. First, there is the notion of balance. As mentioned in the Introduction, *symmetry* (the concept upon which "balance" finds its origin) is important in Aristotle's biological practice.³⁷ *Balance* and *symmetry* are important concepts to Aristotle's biological writings. He uses extensive reference to bi-lateral symmetry of the body (e.g., if you look at almost any animal's face and create a lateral line, the left and the right are generally identical). This occurs also with the entire body. Embedded within the concept of *symmetry* is the concept of *balance*—the occurs not only with the "non-uniform parts" (like arms, organs, and other bone structures) but also with the "uniform parts" (blood, serum and phlegm).³⁸

The symmetry between the uniform parts will prove to be an influential concept among the Hippocratic writers and then to Galen, who expands upon this concept so that it becomes a key measure of health going forward—even to today.³⁹

---

³⁵ Εἰσὶν ἄρα δὺ αἰτίαι αὗται, τό θ' οὗ ἕνεκα καὶ τὸ ἐξ ἀνάγκης, πολλὰ γὰρ γίνεται, ὅτι ἀνάγκη. ἴσως δ' ἄν τις ἀπορήσειε ποίαν λέγουσιν ἀνάγκην οἱ λέγοντες ἐξ ἀνάγκης, τῶν μὲν γὰρ δύο τρόπων οὐδέτερον οἷον θ' ὑπάρχειν τῶν διωρισμένων ἐν τοῖς κατὰ φιλοσοφίαν. ἔστι δ' ἕν γε τοῖς ἔχουσι γένεσιν ἡ τρίτη, λέγομεν γὰρ τὴν τροφὴν ἀναγκαῖόν τι κατ' οὐδέτερον τούτων τῶν τρόπων, ἀλλ' ὅτι οὐχ οἷόν τ' ἄνευ ταύτης εἶναι. τοῦτο δ' ἐστὶν ὥσπερ ἐξ ὑποθέσεως. *PA* 642a 2–10.

³⁶ In the nineteenth century version of Aristotelian Logic, this change in emphases is captured by the difference in *modus ponens* (a → b /a/ ∴ b) and *modus tollens* (a → b/ ~b ∴ ~a), see Joseph (1906).

³⁷ *PA*. 652b 35, 686a 10, *GA* 719b 10, 723a 30, 727b 10, 729a 15, 739b 1, 742b 25, 743a 25f, 767a 15ff, 772a 15ff, 775b15, 777b25, 779b 25, 780b 20, 786b 5.

³⁸ Boylan (1983): uniform—181–191; non-uniform—191–202.

³⁹ This is also a principle embraced in east Asia—thus creating a common theme among Chinese (and other east Asian thinkers), as well—see: G.E.R. Lloyd, *Ancient Worlds: Modern Reflections—Philosophical Perspectives on Greek and Chinese Science and Culture* (Oxford: Oxford University Press, 2004) and more broadly G.E.R. Lloyd, *Intelligence and Intelligibility: Cross Cultural Studies of Human Cognitive Experience* (Oxford: Oxford University Press, 2020). An important book that explores balance and symmetry in Plato and Aristotle is: Theodore Tracy, *Physiological Theory and the Doctrine of the Mean in Plato and Aristotle* (The Hague: Mouton, 1969).

Now in the case of the temperature of the blood, there is a problem.[40] Blood, according to Aristotle, is formed in the heart, and held within *like water in a jar* (*PA* 650b 10). From there, the blood passes nourishment to the body from very small vessels that are practically (though never completely) continuous with the skin. Using analogy to various sorts of "making" in the human material world, such processes require *heat* to get things done. Thus, the heart creates blood that is very warm—too warm for its principal purposes: to nourish the body and to carry *pneuma*. This imbalance must be countered because of the principle of symmetry. The "cooling element" for the blood is the brain. Thus, the blood goes from the heart to the brain via the carotid artery and there is cooled (as the brain is depicted as a cooling agent, devoid of its own blood) and descends to the body via the vena cava—fully balanced and ready to perform its two functions.

So important is the principle of balance to Aristotle, that he makes the brain, a major body part, the balancing device for blood temperature.[41]

Of course, both the concepts of "balance" and "symmetry" assume some standard by which "the more and the less" can play a part in getting the middle (the right balance), which is the best course. But when one comes upon some condition, that by nature are sometimes good, there is a necessity of having a standard to measure *excess* and *defect*. Aristotle famously set this out in the ending of the *Eudemian Ethics* in his analogy between moral choice and the deliberations a doctor has to make in his practice.

> But since a doctor has a certain standard by referring to which he judges the healthy body and the unhealthy, and in relation to which each thing, up to a certain point, ought to be done and is wholesome, but if less is done, or more, it ceases to be wholesome, so in regard to actions and choices of things good by nature, but not laudable, a virtuous man ought to have a certain standard both of character and of choice and avoidance; and also in regard to large and small amount of property and of good fortune (tr. H. Rackham).[42]

This analogy presents several interesting points. First, just like moral reasoning, there are *facts* that contribute to a standard upon which a decision might be made that has grounds for "correctness." In the case of the doctor, the "correctness" will help the patient to live. In the case of the moral decision-maker, his excellent (*arete*)

---

[40] I set this out in more detail in Michael Boylan, "The Digestive and 'Circulatory' Systems in Aristotle's Biology" *Journal of the History of Biology* 15.1 (1982): 89–118.

[41] A good discussion of balance in Plato and Aristotle can be found in Theodare James Tracy, *Physiological Theory and the Doctrine of the Mean in Plato and Aristotle* (Chicago: Loyola University Press, 1969).

[42] Ἐπεὶ δ' ἔστι τις ὅρος καὶ τῷ ἰατρῷ πρὸς ὃν ἀναφέρων κρίνει τὸ ὑγιαῖνον σῶμα καὶ τὸ μή, καὶ πρὸς ὃν μέχρι ποσοῦ ποιητέον ἕκαστον καὶ ὑγιεινόν, εἰ δὲ ἔλαττον ἢ πλέον οὐκέτι, οὕτω καὶ τῷ σπουδαίῳ περὶ τὰς πράξεις καὶ αἱρέσεις τῶν φύσει μὲν ἀγαθῶν οὐκ ἐπαινετῶν δὲ δεῖ τινὰ εἶναι ὅρον καὶ ἕξεως καὶ τῆς αἱρέσεως καὶ φυγῆς, καὶ περὶ χρημάτων πλῆθος καὶ ὀλιγότητα καὶ τῶν εὐτυχημάτων. Aristotle, Eudemian Ethics VIII.3 1249a 22–1249b 4.

choices (using facts in the world)[43] will help him to become (*agathos*) which will promote his human flourishing (*eudaimonia*).[44]

In each case, we are examining a *phusis*$_1$ dynamic. There is a standard and the actions that follow from deliberate choices by the doctor/moral agent. These choices bring about results. To be competent is to be judged by the outcomes of these choices. What is still needed is how we can come up with such a standard (in the case of medicine)? To answer this query we must turn to the *ergon* or the functional criteria by which we might begin to create such standards.

The Digestive and "Circulatory" Systems In order to create standards for health, the physician needs to know more about how the body works. As far as influencing the practice of medicine, the digestive and "circulatory" systems are perhaps the most influential.[45] Concerning *digestion*: food is taken in the mouth, masticated, swallowed, and then goes to the stomach where it undergoes a transformation/concoction (*pepsis*) and travels via the hepatic portal system to the liver where it undergoes a second *pepsis*. Then it travels via the venous system to the heart where it undergoes a third *pepsis* in which nutrified/pneumatized blood is created that must go up to the brain to become balances and ready to perform its two primary functions for the body (nourishment and "neurological function").[46]

The word "circulatory" is put in quotes because the concept of circulation refers to a *going out* and *a coming back*. Aristotle's system has no *coming back*. Blood leaves the heart to the brain and then goes to the rest of the body via blood vessels (no distinction is made between arteries and veins). That blood carries out a particular function and then is "used-up." So, there is a "one way" circulation only.

This is an important diagnostic and treatment tool because it involves checking temperature and pulse as a way of determining health.[47]

---

[43] This might make Aristotle a moral realist—though this problem is a little more complicated: see Boylan (2020b): ch. 1.

[44] This process is a bit more complicated than it is often set out to be by proponents of virtue ethics. For a more rounded view of these dynamics, see: Michael Boylan, ed. *The Philosophy of A.W.H. Adkins: "Virtue" and "Goodness" in Ancient Greece* (New Castle Upon Tyne: Cambridge Scholars Publishing, 2022).

[45] Boylan (1982).

[46] Much of the material part of the "neurological function" (the animal spirits driven by *pneuma*) are in the fluid part of the blood that Aristotle calls *ichor*. For mor discussion on this see Boylan (2015): 1, 2, 8, 9, 10, 13, 14, 17, 18n11, n12, 33, 56–59, 65, 72n52, 75n98, cf. Boylan (1983): 184ff.

[47] See Michael Boylan, "Galen on the Blood, Pulse, and Arteries" *Journal of the History of Biology* 40.2 (2007): 207–230. Some of the key features of pulse that have been set out by Galen and later developed by Avicenna. These categories of the pulse include: Quality of expansion (length, width, depth); Quality of impact (strong, weak, moderate); Duration of cycle (fast, slow, moderate); Duration of pause (successive, different, moderate); Between beats (full, empty, moderate); Compressibility (hard, soft, moderate); Pulse perspiration (full, empty, moderate); Regularity; Order and disorder (ordered, irregular, irregularly disordered); Rhythm (similar, different, out of rhythm)—for a further depiction of the use for medical diagnosis and prognosis see: Joseph E. Pizzorno and Michael T. Murray, *Textbook on Natural Medicine* (Pittsburgh, PA: Churchill Livingston, 2020).

Because, in this context, digestion has the causal role of affecting the process of "one-way circulation" delivering nourishment and "neurological" function, and because digestion is partially controllable by what we eat, "nutrition" (as an area of controlling health) is opened up by Aristotle's work on these underlying biological systems. The writers who come after Aristotle use this information to create a program of health that relies heavily on these distinctions in order to create practical, trial-and-error manuals for diagnosing and treating disease using *nutrition* as a key component.

Conception Theory The penultimate category that I set out as Aristotle's innovation in medicine is his work on conception theory.[48] In the ancient world, "conception theory" was largely thought to be a "woman's problem." Since the society was very andro-centric, it was assumed that any problems with pregnancy were *of course* a problem with the woman who had accepted perfectly fine seed, *spermatos*, into a closed growing system that many had likened to seeds planted in the field. If all the seeds were fine, then all problems with crops were due to problems with the soil, watering, or weeding the plants (the woman's womb). The mechanics (ME) here follow from the digestive system. The digestive system, as outlined above, is built upon *pepsis* (key physical changes analogous to cooking food).[49] When blood passes through testes in men and the ovaries in women, a *pepsis* occurs that energizes the blood to be capable of vitality (reproductive potential). The entire process can be seen, again, from the analogy to cooking.

In cooking, which is governed by providing the proper amount of heat to what is being prepared, getting the right mixture at the right temperature is the name of the game. If (1) conception theory is like cooking and if (2) (as Aristotle incorrectly conjectured) adult women are colder than men (because of the loss of menstrual blood and the heat it holds, *GA* 727a 15),[50] and if (3) being "hotter" means food is properly prepared and "colder" means the food is inferior, then (4) infants that are "cooked" within the uterus and have the proper amount of heat, end up being "male" while those that are "undercooked" become female. This means that females (in conception theory) are *deformed males*—meaning they were "undercooked."[51] This argument is obviously subject to "confirmation bias" via the social disregard of the role of women. But it also creates a warning sign for those in the history of medicine (including today) about how social prejudices can affect the integrity of medical research.[52]

---

[48] For more details see Boylan (1983).

[49] Boylan (1983): 151, cf. *GA* 750b 25, 752b 15, 30, 753a 15, 775a 15, 786a 15.

[50] Women, on average, have a higher temperature than men (contra Aristotle) though there are a number of critical factors to examine, see: H. Kaciiuba and R. Grucza, "Gender Difference in Thermoregulation" *Curr Opin Clin Nutr Metab Care.* 4 .6 (2001): 533–536.

[51] *GA* 737a 25, 767a a 35, b5, 775a 15; see also Boylan "The Galenic and Hippocratic Challenges to Aristotle's Conception Theory" *Journal of the History of Biology* 17.1 (1984): 83–112.

[52] For an exploration of modern sex bias against women as research subjects see: Edward E. Barlett, "Did Medical Research Routinely Exclude Women? An Examination of the Evidence" *Epidemiology* 12.5 (2001): 584–586. See also: Ann Hanson's work on how social prejudices can

What Aristotle does *right* here is that he tries to create a model on embryo development based on a *phusis*₁ model (cooking as a transformative mode for matter: both in the materials and in the transformative process). This is a better strategy for a medical understanding of embryonic development and the health of mothers than a *phusis*₂ account that might require *prayers* and special amulets created by *holy figures—such as priests.*[53]

Those who advocate *phusis*₃ would resist "theories" with their ME accounts and focus on desired outcomes, TE-leaning. The purpose (desired outcome) would be clear, but the material causes would be held up in the air—in a sort of *skepsis* (keeping lots of options open and fighting closure). Rather, the *phusis*₃ advocate will lean toward a "best practices" attitude so that they cannot be blamed for bad outcomes. One good way to do this is by getting a copy of a medical practitioner's clinical notebooks (more on this in the next chapter on the Hippocratic writers).

Other critical questions in conception theory include the "pre-formation" vs. "epigenetic" theory of fetus development and inheritance. In the first case, the pre-formation notion of fetal development suggests that at the moment of the mixture of male *sperma* seed with female *katamenia* (fortified menses) a very small, fully formed homunculus was created. After that, all that was needed was for the homunculus to grow larger.[54]

Also, connected to the pre-formation strategy is the *pan-genesis* theory that the reproductive fluids (*sperma* [male seed = semen] and *katamenia* [female seed = menses) derive their potency (*dunameis*) from all parts of the body. The evidence for this is that the sexual orgasm is felt generally throughout the body. If this event is experienced generally, then this pleasurable event is connected to another general event—such as the discharge of defining aspects of that person's uniform and non-uniform parts (delivered to the blood stream). These bits of defining characteristics from all the parts go into the gonads (male and female) to represent a potentiality (*dunameis*) of the male/female combination engaged in intercourse. They then combine and, in short order, create the homunculus. This theory was advocated by several among the Hippocratic writers.

Aristotle, on the other hand, depicted the process differently. Instead of the pan-genesis posit of pulling identifying material from all the body parts to reproduce into a homunculus, Aristotle takes a rather different approach in having the male seed (*spermatos*) produced in the testes and the female factor (*katamenia*) produced in the ovaries. From there, the embryo is produced one part at a time (*epigenesis*), beginning with the heart (the first organ to come-to-be designating life and the first

---

affect the integrity of ancient medical research, "Papyrology: Minding Other People's Business" *Transactions of the American Philological Association* 131 (2001): 297–313.

[53] Even with the *phusis*₁ accounts, a woman who wanted a female child might be inclined to go out at night and expose her abdomen to the cold winds so that her uterus might be cooler and thus produce the "undercooked" female baby.

[54] It is my opinion that this debate still exists today among the non-medical community: it is at the basis of much of the worldview behind the anti-abortion popular argument.

organ, when it fails, designating death).[55] Why this approach is important to medicine is that life (and death) do not occur *all at once*. The heart is first (based upon an observation of chicken eggs)[56] and thus is the *arche* (which both means "first" and "most important principle"). Aristotle uses this dual meaning to his advantage to signify both. Thus, care of the heart becomes a fortiori the principal concern of medicine: it is both the principle of *life* and *death*.

Comparative Anatomy Based upon Dissection and Analogy There are some who do not believe that Aristotle was actively engaged in animal dissection and using this to make application to humans—with further speculation as to physiology.[57] I take the opposite position contending that not only did Aristotle engage in animal dissection, but also that he used principles of comparative anatomy to make applications to humans via analogy.

We can start this short examination by making reference to the *standard account* that sets out that dissection of animals was permitted in the ancient Greek World, and some human dissection of corpses (though there were some taboos about cutting up dead bodies). In Alexandria, there was said to be a tradition of dissection on convicts and other miscreants who were still alive (vivisection) without any anesthesia (which only became developed in the nineteenth century—see Part II).[58]

Though it is certainly true that some of the reports about animal behavior and body parts (including both anatomy and physiology) are Aristotle's second-hand discussions with those who dealt with those sorts of animals in their economic jobs,

---

[55] On epigenesis see *GA* 734a 25ff. and my comments on a number of these passages in Boylan (1983)—see especially: 142, 152, 170–172, 184–185, 190–191, 195–199, 201–202. The reader should also look at the forthcoming book by P.N. Singer concerning Galen's writing on health: https://www.cambridge.org/core/books/galen-writings-on-health/4DE48C8D5BCA07F9B13DF76D31F17DCF.

[56] Actually, it is the spinal cord that is first, but this observation would require staining techniques not available to Aristotle.

[57] For example, Daryn Lehoux, "Observation Claims and Epistemic Confidence in Aristotle's Biology" *Isis* 108.2 (2017): 241–258.

[58] For a background on dissection see: Ludwig Edelstein, *Ancient Medicine* (Baltimore: Johns Hopkins University Press): 247–302; F. Kudlien, "Antike Anatomie und menschlicher Leichnam" *Hermes* 96 (1969): 78–94; Peter Marshall Fraser, *Ptolemaic Alexandria,* 3 vol. (Oxford: Clarendon Press, 1972) I, 335–356; James Longrigg, "Superlative Achievement and Comparative Neglect: Alexandrian Medical Science and Modern Historical Research" *History of Science* 19 (1981): 155–200; Heinrich von Staden, *Herophilus: The Art of Medicine in Early Alexandria* (Cambridge: Cambridge University Press, 1989); _____. "The Discovery of the Body: Human Dissection and its Cultural Contexts in Ancient Greece" *Yale Journal of Biology and Medicine* 65 (1992): 223–224; Rebecca Flemming, "Empires of Knowledge: Medicine and Health in the Hellenistic World" in A. Erskine, ed. *A Companion to the Hellenistic World* (Oxford: Blackwell, 2003); Michael Boylan, *The Origins of Ancient Greek Science: Blood—A Philosophical Study* (London: Routledge, 2015) 81, 111, 130 n15.

e.g., bee keepers,[59] still most of his works recorded in the *Parts of Animals* and *The History of Animals* are the result of careful observation, and dissection of animals.[60] Some specifics on this include Aristotle's pioneering work on cuttlefish reproduction (that continued to be cutting edge even into the twentieth century).[61] Note also that Aristotle's depiction of "intermediates" (*epamphoterizein*) suggests anatomical knowledge that surely came from dissection.[62]

Other general concepts in his theory of anatomy include bilateral symmetry (which implies both anatomical studies of various animal species and the insertion of a concept of biological analogy which allows the researcher to make inter-species assumptions.[63] Further, Aristotle suggests that the best way to prepare an animal for dissection is by strangulating them first.[64]

Aristotle discusses procedure for dissection[65] and concrete details that aspire to descriptive diagrams that certainly suggest that he had perfected techniques that yielded this detailed account.[66] His explicit use of analogy between various species' anatomy is also clear both for the uniform (blood and tissue) and the non-uniform parts (bones and organs).[67]

Analogy is a powerful tool in general zoological explanation. It enables the biologist to unite the animal kingdom in a powerful way—but it is only possible with careful observation that would clearly require dissection. Aristotle, himself, refers to a work that he composed called "On dissection,"[68] Clearly, some of his controversial conclusions depend upon not only animal but some access to human dissection (most likely of cadavers).[69]

---

[59] Michael Boylan, *Method and Practice in Aristotle's Biology* (New York and Oxford: Roman and Littlefield (UPA) 1983): 142–155.

[60] Boylan (1983): 158, 179, 195, 235.

[61] Boylan (1983): 156. See also H.A. 524a 4–5; 541b 9–15; cf. H.A. 544a 12, G.A. 720b 33.

[62] Boylan (1983): 179, cf. discussion of the segmented spine P.A. 640a 35.

[63] See P.A. 666b 26–35 on horns, chambers of the heart, and blood vessels.

[64] H.A. 513a 13–27.

[65] H.A. 511b 11–23.

[66] H.A. 510a 30–35; 511a 13, 525a 9, 550a 25, 566a 15.

[67] A. non-uniform parts—H.A. 486b 19; 489a 14, 29; 497b 20, 33; 502b 32; 503b 31; 530b 33/ B. uniform parts— H.A. 486b 19; 487a 5, 9; 489a 22; C. Sameness and Difference—H.A. 486b 19, 488b 31, 491a 18, 497b 11

[68] See P.A. 650a 30, H.A. 495b 10, 497a 32, 509b 22, 511a 13, 514b 10, 525a 9, 530a 31, 565a 12, 566a 15.

[69] See his conclusions on embryos at P.A. 666a5, and blood vessels P.A. 668b 30.

Conclusion

Aristotle was not a physician (though he was the son of a physician), and thus, he does not fit into the history of medicine as a practitioner, but rather as someone who has set out: (1) General principles in methodology of science, in general—particularly on what counts as good reasons for causation, (2) General principles in methodology in biomedicine, in general (TE and ME) that would help both physicians and zoologists, (3) Observations (some based upon animal dissections that, by analogy, are applied to humans); (4) These biological observations (guided by a systemic methodology) are critical to physicians who seek to help patients via *pneuma* theory, the balance of blood temperature, the digestive and "circulatory" systems, conception theory, and anatomy (and speculations on physiology). These core accounts have direct impact upon the practice of medicine and are crucial in any historical account. They create a more *complete* account of biology than had hitherto been brought forward, and it is systematic (and thus largely coherent). Because *teleology* (TE) and the *material account* (ME) are used so frequently in a theoretically consistent manner, this gives Aristotle's account a simplicity that yields *elegance*. Together, these create a *standard of care* that is just as important in medicine as it is in moral decision-making. The notion of a standard of care to guide the practice of medicine among a community of health professionals, has grown as a critical component to the development of medicine as a science that lay people can trust and rely upon—continuing until the present day.

Chapter 5
Galen's Resolution of *Nature* in Medicine

Abstract This chapter examines Galen along with his contemporaries in the Pneumatist, Empiricist, and Methodist schools and argues that Galen presents an account that is more complete and coherent than his contemporaries. In this approach he employs the three senses of nature that have been outline in this book.

Keywords Galen · Empiricists · Methodists · Dogmatists · Soranus · Pneumatists · Thessalus

Galen was a pre-eminent figure in Ancient Medicine.[1] Writing some 400 years after the Hippocratic writers. Galen was born in 129 CE in Pergamum (Bergama, W. Turkey) into a prosperous family. His influence upon medicine was monumental.[2] Because vhe read and reflected upon the thinkers before him, his commentary on them enabled Galen to act via synthetic discourse to contemplate a unified theory of medicine that systematically brought together much of went before, as well as his attempt at creating a theory that made the best of the Pneumatists, Empiricists, Methodists, and Dogmatists (the prominent, competing medical *schools* at the time).

---

[1] One of the very best sources of information on the historical significance of Galen in his context is Vivian Nutton, *Ancient Medicine*, 3rd ed. (London: Routledge, 2024). I have had contact with Dr. Nutton for 40 years and have learned a lot in the exchanges.

[2] One recent work that documents Galen's historical influence is: Petros Bouras-Vallianatos and Barbara Zipser, eds. *Brill's Companion to the Reception of Galen* (Leiden: Brill, 2019).

Introduction

It has been my contention that one of the best ways to understand the period between the Hippocratic writers and Galen is through advances in anatomy and in the examination of the role of blood, along with how these affect conception theory.[3] It's not so much that the basic science has had a monumental change over this time period as it is how that science is applied to process of understanding the patient's condition so that a diagnosis might be rendered that is based upon facts that present themselves during the initial consultation with the physician.

The underlying understanding of human anatomy can help in this $phusis_1$ approach because it expands the context of the explanandum which the physician observes in order to render his *explanans*. His big work on anatomy, *Anatomical Procedures*, was based upon his dissections.[4] This empirical, "fact-based" approach came to be termed the Dogmatist approach in medicine. Under this grouping Galen chooses to identify his methodology with Aristotle and Hippocrates[5] —though concerning several over-arching structural principles, he often identifies with Plato.

Part of the reason that Galen was adept at anatomy was both through his possible study in Egypt[6] and his work with wounded gladiators in 157 CE upon his return to Pergamum. Galen's duties were to keep the gladiators alive.[7] This background experience may have given Galen more anatomical experience than his contemporaries.

In 162 CE Galen left Pergamum for Rome with an established reputation. After some scuffles with the establishment, he was able to obtain the favor of Marcus Aurelius and Lucius Verus in northern Italy as they prepared for a war campaign. Verus died soon thereafter and Galen accompanied Aurelius to Rome. These events helped to promote Galen's position within the empire.

In Galen's principal works on anatomy, *On the Anatomy of Muscles*, and *Anatomical Procedures*,[8] he rejected Aristotle's assertion the that heart was the primary organ of the body and embraced the Platonic notion that the brain, heart, and liver were tripartite principles for the functioning of the human body. The ways

---

[3] See my articles: Michael Boylan, "The Galenic and Hippocratic Challenges to Aristotle's Conception Theory" *Journal of the History of Biology* 17.1 (1984): 83–112; and "Galen's Conception Theory" Journal *of the History of Biology* 19.1 (1986): 44–77; and in my book, *The Origins of Ancient Greek Science: Blood—A Philosophical Study* (London and New York: Routledge, 2015): chapter 5.

[4] For a discussion see: J. Rocca, "Anatomy" in R.J. Hankinson, *The Cambridge Companion to Galen* (Cambridge: Cambridge University Press, 2008): 242–262.

[5] For a discussion of this identification see Inna Kupreeva, "Galen's Theory of Element" *Bulletin of the Institute of Classical Studies,* No. 116 *Philosophical Theories in Galen* (2014): 153–196.

[6] Since Egypt was a center for the use of vivisection and comparative anatomy (with animals) Galen's writings suggest an acquaintance with these techniques—see: Julius Rocca, "The Brain beyond Kühn: Reflections on 'Anatomical Procedures' Book IX" *Bulletin of the Institute of Classical Studies, Supplement* No. 77 *The Unknown Galen* (2002): 87–100.

[7] See John Scarborough, "Galen and the Gladiators" *Episteme* 2 (1971): 98–111.

[8] The latter work contains anatomical discussion on bones, nerves, veins & arteries, and muscles.

these were unified was via blood, which was formed from nutriment (*trophe*) in the liver via a "concoction" that included vitalized air from the lungs and skin to the level of *pneuma* and was carried throughout the body.[9] This vitalized blood acted both for giving the parts the fuel needed for current activity and growth, but also functioned as an agency for the actualization of movement and bodily function (carrying on many of the tasks that modern theories assign to the nervous system (both autonomic and somatic)).

Some students of this era of medical history might be surprised that dissection was more embraced by the so-called Dogmatic practitioners (who make *theory* an important part of their presentation) whereas the Empiricist practitioners were often skeptical of conclusions derived from dissection. The reason for this is that when one views the anatomy via dissection (generally of a corpse (human or otherwise) or the vivisection of an animal (also done on humans in Egypt) there may be significant *phusis*$_3$ reasons for doubting what you see before you as being *understandable* or *useful*.[10]

In *On the Use of the Parts* and *On the Opinions of Hippocrates and Plato*, he sets out positions on physiology. It is here that he provides some more of the details of his teleology under the general umbrella that *Nature does nothing in Vain*.[11] This is both a principle of teleology (and as such admits to a *phusis*$_2$ slant) and a broad explanatory principle: each body part has a function: if I don't see the function, the problem is with me because nothing is uselessly attached. (This supports the attitude of the maker/artifact dynamic.) Aristotle said much the same thing as did Plato at times. It is also an embedded background to affirming that all biological events are caused by biological causes—because there is an underlying teleology to the entire endeavor.

Galen set out four *natural faculties* within the body that accounted for how processes (*phusis*$_1$) occurred: attraction, assimilation, excretion, and growth.[12] This could be further structured by the categories of: (1) Bodily function about to be undergone (*dunamis*); (2) Bodily functions that were ongoing in carrying about their tasks (*energeia*); and (3) Completed bodily functions (*ergon*). Obviously, this structure is very Aristotelian in its composition and execution. Because of this overarching structure that describes the parts of the body and how they operate (with some variation for individuality, of course), this group emphasized in a strong way the *phusis*$_1$ approach. Thus, the label "dogmatists" is appropriate.

---

[9] Boylan (2015): ch. 5.

[10] See my discussion of this problem in Boylan (2015): 83–85.

[11] Again, we confront here how we understand the activity of "N/nature"—See Boylan (1984): 89–109.

[12] Galen, *On the Natural Faculties*, tr. A.J. Brock (Cambridge, MA: Harvard University Press, 1916): 2, 9–24.

Galen and His Contemporaries

This section will examine three groups of thinkers that vied for recognition, against the Dogmatists, as the proper way to do medicine in Galen's time.[13] The first of these are the Pneumatists.[14] They put stock in *pneuma* as the essential ingredient of health. The origins of this sect are unclear. Nutton suggests that their founder may have been Athenaeus of Attaleis (S.W. Turkey), who was a pupil of Posidonius—though probably not the famous Stoic philosopher.[15] The principal tenet of the Pneumatists was that source of health was balancing *pneuma* in the body through the regulation of hot and cold. Illness occurred when this balance went too much to an extreme, and it could be remedied via a treatment toward the opposite extreme.[16] For the most part, illness was more often occasioned by an excess of heat so that cold baths were a preferable life-style in all seasons to keep the tendency toward the *hot* in check. Because the causes of disease are all sensible, this seems to lean toward *phusis*$_1$, but because the underlying explanandum is not empirically open for inspection, viz., that the vital air, *pneuma*, is affected in its ability to act as it should by an imbalance that is not empirically scrutable; therefore, there is a leaning here towards *phusis*$_2$. In the end, it would be best to describe the Pneumatists as a mixture: *phusis*$_{1+2}$.

A second group of physicians are termed as Methodists. In the clinical setting what is the proximate center of attention? There is, of course, a patient who has come to you for help. But how do you create an epistemological framework to deal with your patient? One way is to focus on the disease. Of course, this was also the strategy of the Dogmatists. They focused upon the disease in a manner that emphasized the *causes* of disease with reference to a relatively standard characterization of the human body, *phusis*$_1$.

However, another approach is to streamline the process—bypassing anatomy/physiology and speculation upon the material structure of human beings and how they might get out of sync. In order to avoid the philosophical debates of the Dogmatists (that skeptically-oriented physicians found tedious and unhelpful), and

---

[13] Philip von der Eijk takes a similar position that I will take that Galen tries to merge "top-down" and "bottom-up" approaches in assessing his methodology that tries to use what his predecessors and contemporaries got right and wrong (much as Aristotle did), see: Philip van der Eijk, "Galen on the Nature of Human Beings" *Bulletin of the Institute of Classical Studies, Supplement 114* (2014): 89–134.

[14] See M. Wellmann, Die pneumatische Schule bis auf Archigens (Berlin: Weidmann, 1895), c.f. also Aretaeus, *Acute Diseases* 1, 5; 2, 3 and discussed by Karl Deichgräber, "Aretaeus von Kappadokein als medizinischer Schriftsteller" *Abhandlungen der sächsichen Akademie der Wissenschaften,* 63 (1971): 3 and A. Roselli, "Areteo di Cappadocia lettore di di Ippocrate" in Philip van der Eijk, *Hippocrates in Context: Papers Read at the 11th International Hippocratic Colloquium, University of Newcastle on Tyne 27-31 August 2002* (Leiden: Brill, 2005): 413–432.

[15] Nutton (2024): 207.

[16] This notion that too much hot or cold causes illness continues among the general populations on earth even to the present day.

to avoid the Empiricist complicated clinically-oriented therapeutic metrics, the Methodists offered a simpler, handbook approach based upon the commonalities of diseases.[17]

Disease is presented as constricted, un-constricted, or mixed.[18] Diseases themselves were either chronic or acute (with the former being too constricted and the latter too un-constricted).[19] This led to treatment strategies that accepted the *diatritus*, ongoing 3 day cycles, in order to track the path to recovery.[20] This streamlined approach could be franchised on a large scale rather efficiently (which is why learning the therapeutic approach called "the method" (*methodos*) was a path that could be completed in as little as 6-months!).[21] This was the quickest apprentice model for physicians among the various schools. It was no wonder that such efficiency was greatly appreciated in the Roman Empire, that valued the combination of simplicity and results.

The emphasis upon results over theory is also a characteristic of one form of scientific pragmatism. Pragmatism is outcome oriented. Whether some posit is true or not is dependent upon whether using that posit in a particular context gets you what you want: i.e., it works. So, if we use the constricted/un-constricted/mixed model and get results, then that is truth. This goes against the correspondence theory of truth model of the Dogmatists.[22] The longstanding argument against scientific pragmatism is that something may "work" but not for the reasons given, but for other reasons unknown to the practitioner—this is called a *sumbebekotos*, accidental, inference by Aristotle (*Meta.* I, 4, 8). However, such distinctions were of no practical interest to the Methodists. [23]

---

[17] Most take the school of the method to have been founded by Themison of Laodicea, a pupil of Asclepiades of Bithynia. Then Thessalos put everything together around the time of Nero adding the doctrine of *metasynkrisis*, on modifying the pores of the body toward tightness or looseness and the doctrine of three-day intervals for evaluating disease, *diatritos* (cf. Nutton (2024): 197). Since there are few texts, this book will follow this account. It should be noted, that Edelstein very strenuously denies that Themison be viewed as the founder of the Method—leaving that role to Thessalos (Edelstein, 1967): 174–179. A source for some of the fragments of the Methodists can be found in M. Tescusan, ed. and tr. *The Fragments of the Methodists. Methodism outside Soranus.* Vol. 1. (Leiden: Brill, 2004).

[18] Edelstein (1967): 181 calls these moist and dry and they affect the whole body. It should be stated that the Methodists were not keen on strict definitions. This is because that approach characterized the dogmatists, against whom they sought professional differentiation.

[19] Nutton (2024): 195.

[20] D. Leith (2008) "The *diatritus* and Therapy in Graeco-Roman Medicine" *Classical Quarterly* 58: 581–600.

[21] Pseudo-Galen, Kühn XIV, 684.

[22] For a discussion on how the various theories of truth models work see: Boylan (2007): chapter 4.

[23] Though Soranus said that reading natural philosophy might be a good discipline for the mind in the form of juvenile education, this theoretical part is "useless for our purpose" (Soranus): 4.

Since there are so few existent complete texts by Methodists (which is due to Galen's dominance),[24] we must turn to Soranus to see how this theory plays out in explanations about two cases concerning blood. In the *Gynecology* Soranus addresses the female discharging too much blood due to a difficult labor, leading to miscarriage (*ektrosis*) and various ulcerations. And secondly irregular, heavy menstrual flow leading to fatigue.

In both cases we have a discharge of too much blood. In order to get some background on the discharge of blood in women we must briefly return to *ichor*. As I have written elsewhere, *ichor* was set out as a connection to the blood of the gods. It provided support for *phusis*$_2$ explanations.[25] I also set out some nuance as some common usages of *ichor* as a thin fluid associated with animal meat juice stood side-by-side with other usages even as the tragedians as still refer to it being the blood-of-the-gods.[26] In the writings of Soranus (98–138 CE), the bestial (lower-than-human) meaning is primary and is used to explain disease[27]:

> Re: Menstrual blood v. *ichor* (Soranus I. 4. 19)
> Menstrual blood is depicted as pure while *ichor* is corrupted [like] animal meat. When the menstrual flow comes from the uterus, it is pure blood. When it comes from the vagina, it is tainted and displays characteristics of *ichor* (my tr.).
> Re: Fertility and the proper moisture of the uterus (Soranus I. 9. 34)
> Blood is set out as providing the proper moisture while *ichor* is an ersatz moisture that will not prepare the uterus properly (my tr.).
> Re: Spontaneous abortion (Soranus I. 18. 59)
> "a watery discharge appears. . . an *ichorous* or sanguineous fluid—like the water in which meat has been washed." When this appears in a pregnant woman, a miscarriage is about to occur. Once again *ichor* is a sign of *para phusin* (my tr.).
> Re: Skin exanthemata and itching (Soranus II. 25 [45]. 52 [121]) "When the *rheumatismos* [flow of bodily fluids] and the great discharge of *ichors* have ceased . . . we resort to olive oil . . . and with the white of an egg beaten and combined with a moist wax salve [an astringent]" (my tr.).
> Re: Uterine Prolapse (Soranus IV [XX]. 36 [85] "Euryphon makes the patient hang by her feet from a ladder for a whole day . . . . Euenor inserts ox meat into the vagina not knowing that the *ichors* produced by the putrefaction will, by their pungency, cause ulcerations" (my tr.).

In each of these cases we see *ichor* to be an agent of the un-constricted. The remedy is to balance the un-constricted with various constricting agents such as a vinegar and papyrus potion that is rubbed on the labia and the orifice of the vagina (vinegar

---

[24] There are, of course, the fragments and testimonials collected by Tecusan (2004)—which are substantial. However, I also think that there is also a non-theoretical-skeptical bent to the Methodists that give a reason for why Galen became to dominant. My analysis goes against Edelstein's analysis of the skeptical connection with the Methodists: Edelstein (1967): 186–189.

[25] Boylan (2015): ch. 1–2.

[26] As per Jouanna (2012): 197–209.

[27] This does not totally dismiss the magical *phusis*$_2$ sense. For example, after this very passage Soranus brings up the use of magical amulets saying that though he does not use them, he doesn't forbid them either: "Yet one should not forbid their use; for even if the amulet has no direct effect, still through hope it will possibly make the patient more cheerful" (Soranus III.10.42).

is an astringent and papyrus is part of the delivery mechanism meant to keep the mixture in place).[28]

In severe cases, a vaginal suppository is used (consisting of ashes of a sea sponge soaked in raw pitch).[29] Ashes are astringent and pitch is an effective device to hold the suppository in place.[30] These treatments are meant to counteract the un-constricting agents (such as *ichor*) which are *causing* the miscarriage, ulcerations, or heavy menstrual flow. (We have to be careful about characterizing these agents as *causing*. This is because "causation" is a by-product of the Dogmatists' approach. But the Methodists *are* asserting three *type-relations*: the constricted, the unconstricted, and the mixed. Type-token relations are explanatory so they function causally—whether the Methodists admit it or not. The Methodists want to say that their categories are merely self-evident commonalities that are revealed during the 3-day cycles of disease. But this is incoherent (as Galen and Celsus point out).[31] It ignores the distinction between what is physically evident in the body (Empiricism and Dogmatism) and what might be hidden (Dogmatism). It also brings in a notion of types by the backdoor (Dogmatism). For this reason, it seems useful to categorize the Methodists as in-between the Empiricists and the Dogmatists, i.e., $phusis_{1+2}$.

After the Dogmatists, Pneumatists, and Methodists, the last major medical sect that I will highlight are the Empiricists. What were the Empiricists (*empeirikoi*) about? Well, to begin it has to do with the root term, *empeiria* which means experience, but what does "experience" mean? Does it refer to uncritical interaction with the world or something more robust along the lines of Aristotle's critical

---

[28] The Galenic doctrine of *mixtures* is critical to his biomedical presentation. In the recent book, Peter N. Singer and Philip J. van der Eijk, and Piero Tassinari, *Galen: Works on Human Nature*, vol. 1 "Mixtures (De temperamentis)" (Cambridge: Cambridge University Press, 2019), Singer and van der Eijk set out a pattern: Book One details the different sorts of mixtures of the *hot, cold, wet, and dry*. Book Two is on their presentation in the human body, and Book Three depicts how they operate, in what I call a $phusis_1$ manner within the human body. This depiction of mixture meets my evaluative category of elegance as it makes use of artistic metaphors such as Galen's citation of the "Canon of Polyclitus" which is a metaphor based upon a statue of a well-balanced human form. Certainly the elegance of symmetry—here set out as a well-balanced human form works positively in my assessment of Galen's biomedical science.

[29] Soranus, III, 10.41

[30] Soranus rejects bloodletting (contra Themison) because it does not contract, but further loosens the blood flow and can therefore threaten the life of the woman, Soranus III. 10. 42.

[31] Galen, Walzer and Frede (1985) *On the Sects for Beginners*, ch. 6, and Celsus, *De Medicina*, "Prooemium" 62–73. R. J. Hankinson argues in "Causes and Empiricism" *Phronesis* 32.3 (1987): 329–348, that the "hidden" was confronted by the Empiricists via inference from common observation and by associative signs—e.g., if a woman is lactating, then she has conceived. The empirically verifiable lactation is a sign of pregnancy. One may infer the latter from the former—though, strictly speaking, it is not an instance of causation (à la Hume), cf. David Sedley, "On Signs" in Jonathan Barnes, et al. ed. *Science and Speculation* (Cambridge: Cambridge University Press, 1982): 239–266. It should be noted that Thessalus, a Methodist physician from Tralles (70–95 CE) who was a Roman physician to Nero, who claimed he could teach Methodism to anyone in six months. Because of such claims, he is spoken of by Galen with scorn, Galen, *De Meth. Med.I .1*, vol. x Kuhn. He is not to be confused with an earlier Thessalus (fifth century BCE), who is sometimes put into the Hippocratic Corpus as a contributor on *Epidemics* and *On Nutriment*).

empiricism? This is a very big question.[32] The answer goes back to Celsus quoted above: the evident and the hidden. The Dogmatists were keen on discovering the hidden, underlying mechanisms that had broad explanatory power. But the Empiricists were reticent because they were advocates of a sort of skepticism that is "token-oriented."[33] Their response to the Dogmatist theories was, "Not so fast." This gives rise to an attitude that refuses closure on *phusis*$_1$ or closure on the possibility of discovering the universal nature of the human organism nor of its challenges in injury or disease. This reticence to closure can be termed skepticism. I have claimed elsewhere that this lies on a continuum between critical questioning, *skepticism*$_1$ to radical avoidance to general theory *skepticism*$_3$.[34] Different thinkers in each medical sect might vary where they lie on this continuum, but only the Empiricists claim *skepticism*$_3$, and thus earn a solid positioning in *phusis*$_3$.

I follow those who situate the Empiricists in the Hellenistic Medicine period (largely as a reaction to the Dogmatist anatomists, Herophilus and Erasistratus) in the person of one of Herophilus' students, Philinus of Cos around 260 B.C.[35] The Empiricists did not believe in the hidden causes because of epistemological reservations.[36] Instead, they relied upon *peira* and *empeira* (the former being more connected to trial and error). In this way von Staden likens them to B.F. Skinner's behaviorism which used operant conditioning to create trial and error physically evident results that belied "hidden" psychological motivations of the so-called folk-psychologists (those in the Freudian tradition).[37] Since there were so many possible

---

[32] In modern French, this duality is maintained in "experience" which can refer to uncritical (naïve) single perturbation style of experiment along with the more critical reflections on such, the hypothetical-deductive method.

[33] In this context the "type-token" distinction refers to the relationship between large categories of understanding (universals) and individuals. This is a big distinction in contemporary philosophy, see: Paul Grice, "Utterer's Meaning and Intentions" *Philosophical Review* 78 (1969): 144–177; W.V. Quine, *Quiddities: An Intermittently Philosophical Dictionary* (Cambridge: Harvard University Press); and David Papineau, *Philosophical Devices: Proofs, Probabilities, Possibilities and Sets* (Oxford: Oxford University Press, 2013). In the ancient world, some of the dynamics of this can be found in "the One and the Many" dispute in Plato.

[34] Boylan (2015): 88–93.

[35] In support of this approach see: Karl Deichgräber, *Die griechische Empirikerschule,* 2nd ed. (Berlin: Weidmann: 1965); Michael Frede, "The Ancient Empiricists" in *Essays in Ancient Philosophy* (Oxford: Clarendon Press, 1987): 243–260; and R.J. Hankinson, "The Growth of Medical Empiricism" in D. Bates, ed. *Knowledge and the Scholarly Medical Traditions* (Cambridge: Cambridge University Press, 1995): 41–59, and Nutton (2024). For a contrary opinion see Hankinson (1995): 226. Another contrary view pushes the date much earlier see: Edelstein (1967): 195–203.

[36] Some of these heavy epistemological requirements are set out by M. Vegetti, "L'Épistémologie d'Érasistrate et la Technologie Hellénistique" *Ancient Medicine in its Socio-Cultural Context; Papers read at the Congress held at Leiden University, 13–15 April 1992.* 28 (1995): 461–472.

[37] Von Staden (1992): 235. On criticisms of folk psychology along this and affiliated arguments see: Paul Churchland, "Eliminative Materialism and the Propositional Attitudes" *Journal of Philosophy* 78.2 (1981): 67–90; and Jerry A. Fodor, "The Mind-Body Problem" *Scientific American* 244 (1981): 114–123.

epistemological objections to the hidden via *skepticism*$_3$, the alternative, viz., *the evident* was chosen. This was done for clinical reasons of efficiency. After Herophilus and Erasistratus, this approach became dominant (partially also due to *phusis*$_2$ advocates and those who had religious and moral objections to dissection).[38]

If one were to create a general approach for the Empiricists it might be the so-called *tripod*: (1) keen observation of the patient by the physician, *autopsia*; (2) some lore of what had worked in the past (either in the practitioner's first-hand experience or via second-hand experience—the writings of some other practitioner) *historia*; and (3) a skepticism-*phusis*$_3$ approach directed toward the attitude of *similarity of connection*.[39] These deserve a little more attention. First is the keen observation of the patient (first hand observation). This includes listening to the reports by the patient of his own condition along with the physician's own observation. However, this is also a bit controversial concerning blood. This is because the pulse is the one of the primary "evident" inputs that is available to the physician. However, it is impossible to totally expunge theory from observation (even if the theory is not properly recognized, as such). Herophilus had a rather complicated account of the pulse that distinguishes the beginning of the heart's systole from its quiescence (which varies greatly from new born babies to old men). In counter distinction, the Empiricists do not consider the diastole (because it is not observable with the fingers). Instead, they simplify by measuring only beat and interval.[40] This a conscious choice they make to eschew the more complicated theory of pulse set out by Herophilus (that arose from his visually-based understanding of the heart's activity). But if you reject dissection in either of its forms, then you will likewise reject this more nuanced account of the pulse as a diagnostic tool in favor of a simpler model. This is because those who view the body as an entity that, in principle, can be understood mechanically will be drawn to anatomy and physiology. Those who demur do so because they view this asserted precision as illusory: better to concentrate upon the patient before him and the macro symptoms. Interior diagnostic tools (such as the pulse) lead the physician, by necessity, to theories of the hidden—Dogmatism.

The second leg of the tripod refers to the past. First in line would be the attending physician's own personal first-hand experience and his knowledge of what worked and what didn't work in *similar* circumstances. This evidence-based medical approach[41] was highly reliant upon drug/nutrition treatment (which along with

---

[38] Von Staden (1992): 236.

[39] Deichgräber (1965): 83, 165 describes Glaucias' work *The Tripod* that sets out this doctrine.

[40] Harris (1973): 193.

[41] Of course, evidence-based medicine is still in combat with theory-based medicine. See, for example, the following exchange in the *British Medical Journal*. F. Davidoff, B. Hayne, D. Sackett, R. Smith, "Evidence-Based Medicine" *BMJ* 310 (1995): 1085–1086; and a reply in the letters section—N.J. Pearson, J. Sarangi, R. Fay, "Evidence-Based Medicine" *BMJ* 312 (1996). 380; and finally, David Sackett, William M.C. Rosenbert, J. A. Muir Gray, R. Brian Haynes, W. Scott Richardson "Evidence-Based Medicine: What it is and What it isn't" *BMJ* 312 (1996): 71–72. For a version of this based upon the Empiricists v. Dogmatists see Mohan Matthen, "Empiricism and

lifestyle choices, regimen, constituted the general purview of the physician). Since regimen is a largely a preventative medicine technique, it was not of value in situations in which the patient might be in crisis. The only other technique, manual manipulation (including cutting—sometimes called surgery) is largely constrained for the Empiricists to versions of massage since invasive cutting inclines one to the acceptance of humor theory.[42]

However, when the attending physician is either new at the job or encounters a case that is novel to him, then there is recourse to second-hand trial-and-error accounts. These might be the more clinically-oriented accounts from earlier writers[43] or via oral transmission (much as folk-medicine accounts were later transmitted in medieval Europe generally via the women in the family).[44]

The final leg of the tripod refers to *similarity*. This is a very tortured philosophical concept. It refers to a correspondence theory of truth. In the correspondence theory some instance, x, is true just in case x resembles some model that is assumed to be a proper instance.[45] There are two basic flavors of this theory: (a) the one-many interpretation of Plato's Theory of Forms where when some token x is compared to the Form of X—the more similar it is, the more true it is; and (b) natural typology where when some token x is compared to an accepted natural type of x, then x is more true as it approaches verisimilitude to the accepted type. In the first instance, we must posit a separate realm of nature along the lines of $phusis_2$. This, of course, was not the direction of the Empiricists because it included the inscrutable. In the second instance, one is comparing some instance of a fever (a token) with all the

---

Ontology in Ancient Medicine" *Apeiron* 21 (1988): 99–121. Matthen claims that because Empiricists are not committed to logically-based theories that may belie any particular instance of falsification, the Empiricists were really in a better position to jettison treatments that simply did not work. This is because they did not have a general theory that they have bought into. This is essentially the same dynamic in the modern context.

[42] Of course, reference to humors is not a "stand alone" position. But since cutting that is more invasive will entail the loss of considerable blood, surgeons who do not take this into account will be largely unsuccessful. This is why I suggest that surgeons who attempt more invasive techniques will be inclined to the Dogmatic approach.

[43] These extend before the Hippocratic writers and include the accumulation of written texts / short guides, summaries etc.

[44] Two contemporary examples of folk medicine practiced by women within a family/social setting are: Jewel Babb and Pat Littledog, *Border Healing Woman: The Story of Jewel Babb as told to Pat Littledog*. 2nd ed. (Austin: University of Texas Press, 1999) and Michael Mucz, *Baba's Kitchen Medicines* (Edmonton: University of Alberta Press, 2011). Another interesting case concerns modern malaria treatments that embrace folk herbal medicine in Africa in what is termed "reverse pharmacology" in which folk herbal treatments are used—such as argemone Mexicana tea for Malaria in Mali (that has an 89% recovery rate—compared to 95% with ACT) see: Bertrand Graz, et al. (January, 2010) "Argemone Mexicana Decoction versus Artesunate—Artesunate-Amodiaquine for the Management of Malaria I Mali: Policy and Public Health Implications" *Transactions of the Royal Society of Tropical Medicine and Hygiene.* 104.1: 33–41; and Merlin Wilcox (March, 2012) "Improved Traditional Medicines in Mali" *Journal of Alternative and Complementary Medicine.* 18.3: 212–220.

[45] For a more complete account see Boylan (2008): 78–84.

tokens of fever one has experienced (directly or second-hand). In the judgment of the attending physician, the token is tied by common properties to some other token instance. This act of identification constitutes the similarity leg of the tripod.

This situating of medical practice into a tripod is a nod at symmetry (elegance) which is one of the three categories of evaluating scientific theories set out earlier.[46] In either of the two interpretations set out above either a form (type) or a representative instance of token represents the beautiful because of the fidelity of what is before the scientist and *standard* that the physician is looking for. This is true of Methodism and the Dogmatism represented in the *Epidemics* volumes which calls for such epistemological maneuvering in order to ascertain whether the handbook and the patient before him match well enough to be an instance of identification leading to a proper diagnosis.[47]

However, the act of identification and the epistemological barriers of *skepticism*$_3$ also are the grounds for *discarding* the work of the anatomists, Herophilus and Erasistratus. For starters, a live body is not the same as a dead body. For all one knows these major dissimilarities might lead to major error. Even vivisection creates a very artificial context that might lead to no useful information. Thus, if there are moral and religious objections to dissection and vivisection, and if there is a very real possibility that the end product is *inaccurate*, then these factors incline the Empiricist practitioners to avoid this data. Roughly speaking, this application of the careful scrutiny maxim allies itself to the "do no harm" tenet in medicine (cf. the Hippocratic Oath).[48]

But implicit within the notion of *similarity* is a notion of connection. A token "x" is similar to a token "y" on the basis of some connection, F (where "F" is a property that they both share). Galen, in his work, *An Outline of Skepticism,* suggests that the Empiricists put forth a five-part continuum by which x may be similar to y: (1) when the shared trait always occurs; (2) when the shared trait occurs for the most part; (3) when the shared trait occurs half the time; (4) when the shared trait occurs rarely; (5) when the shared trait never occurs.[49] This mix of qualitative and rough-quantitative criteria provides some substance to what is meant by *similarity* which fits into the "symmetrical" (elegance) criterion.

However, because the Empiricists do not accept type-oriented Dogmatic pronouncements on hidden truths (scientific laws), even in the first category in which

---

[46] The other two were completeness & coherence and the Maker/artifact dynamic that honored Nature and its outputs.

[47] This process of identification carries on to the present. There are often differences of opinion on how to read various tests and scans that are performed on patients. Such differences give rise to contentious discussions in M&M (morbidity and mortality) meetings in the days following.

[48] Of course, the words *do no harm* do not occur in the Oath, but have come to be the common interpretation of "I will help the sick according to my skill and judgment, but never with an intent to do harm or injury to another" ll. 18–19)—see Boylan's translation (2014): 77 and the discussion in Chap. 4.

[49] Galen, "An Outline of Empiricism" in *Three Treatises on the Nature of Science,* tr. R. Walzer and M. Frede (Indianapolis, IN: Hackett, 1985): chapters 2, 6, cf. Hankinson (1995): 227–228.

the connection occurs all-the-time, there is no necessity to this connection.[50] The *sign (sema)* can have power of self-validation (epistemologically).[51] This validation of the sign is either natural (based upon the first form of correspondence theory of truth) or conventional (based upon the second form of correspondence theory of truth).[52] It is clear that the Empiricists intend the latter (since the former would force them to be Dogmatists).

However, Galen (a Dogmatist) argues in *On Medical Experience* that every patient presents in a unique way. Given this, how can the Empiricists claim to find *actual* properties that x and y share? Often two patients will share many properties, but most of these will be accidental and not medically relevant. How can the theory-eschewing Empiricists authentically assert which similarities are relevant and which are not—without their own Dogmatic theory?[53]

Despite this rebuke, Galen did recognize Heraclides of Tarentum (an Empiricist) as a learned practitioner.[54] However, the great reliance upon drugs for treatment in a trial-and-error fashion also inclined many Empiricists toward $phusis_2$ in the use of charms.[55]

The reason that the Empiricists were not so keen to delve more into blood is because blood is part of *the hidden* and so is really out of bounds. Only the pulse (in a simplified account) is given to the clinical physician practitioner. However, there is the treatment of bloodletting that was practiced by some Empiricists.

Finally, there were some practitioners who used empiricist trial and error methodology in order to determine what often worked in certain circumstances, but did not buy into the $phusis_3$ skeptical methodology feeling instead that if a trial and error finding in pharmacology (the use of plants and herbs—sometimes prepared in a particular fashion), for example, worked in certain situations, then that was

---

[50] Cf. David Hume's argument on this exact point (*Enquiry*—1977 [1748]):

1. "Causation" implies non-logical connection between cause and effect—Fact (p. 49)
2. In matters of fact sentiments are formed through direct experience—Fact (p. 49)
3. When two sentiments are often linked together the ideas that are formed from them will also be linked—Assertion (p. 50)
4. The constant conjunction of objects will give rise to an idea of the objects being linked—2, 3 (p. 51)
5. There is no necessary relation between these two objects—Fact (p. 51)
6. The two habitually joined objects are not, strictly speaking, cause and effect—1, 5 (p. 51)
7. [What is not necessary is contingent]—Fact

---

8. What is commonly called cause and effect is really two contingently linked objects and not properly called causation—4, 6, 7 (p. 52)

---

[51] For a discussion of this in a general scientific context see: Elizabeth Asmis, *Epicurus' Scientific Method* (Ithaca, NY: Cornell University Press, 1984).
[52] Galen, Walzer and Frede (1985) *An Outline of Empiricism*: ch. 8.
[53] Galen, Walzer and Frede (1985) *On Medical Experience*: chapts 3–6.
[54] Galen, K: 12, 534, 989; 13, 462.
[55] Nutton (2024): 152.

enough to justify its being tried again with similar clinical circumstances presented themselves. One prominent example of such a practitioner was Pedanius Dioscorides (40–90 CE).[56] Dioscorides' work was readily available to Galen and other medical writers/practitioners of the period.

A Brief Selection of Key Galenic Texts

As in the previous chapters, I wish to put the cap on this era of medical history on Galen by examining what I consider to be a brief, select series of texts that illustrates some of what I have contended in this chapter. Galen was an ardent philosopher who used Plato or Aristotle's philosophical principles as he saw fit to help create a theoretical structure that paid attention holistically to the symmetrical (elegance), analytically to the systemic properties of completeness and coherence, and with a sense of intellectual reverence to Nature as a teleological creative force (maker/artifact distinction) in *On the Use of the Parts*.

Galen, as a Dogmatist, believes health (*hugieia*) is a positive state within the individual and is not merely the absence of disease (*nosos, nosema*). This positive state comes-to-be when the functions (*energeiai*) of the body are in accord with nature (*kata phusin*). This is Galen's functional understanding of health. But there is also another understanding: health exists when the constitution (*kataskeue*) and/or condition (*diathesis*) of the organs of the body function properly (naturally—*kata phusin*).[57] In the former case if a person has a good diet and exercise regimen, then they will, ceteris paribus,[58] be healthy.

Since disease is a negative condition it comes upon the individual in a non-voluntary fashion, i.e., it presents itself through *symptoms* (*sumptoma*)[59] that happen *to* the individual (*pathos*). For the Methodists it will be an imbalance between the constrained and unconstrained. For the Empiricists it will be an imbalance in the tripod, and for the Pneumatists it will be an imbalance in the flow of bodily pneuma

---

[56] For some further discussion see: Anne Roney, *The History of Medicine* (New York: Rosen Publishing, 2012): 121; Pedanius Dioscorides, *De materia medica: Being an Herbal with many other Medicinal Materials,* tr. Tess Anne Osbaldeston, based upon the 1655 translation of John Goodyer (Johannesburg: Ibis Press, 2000)—discusses plants and plant drugs; Susan Francia and Anne Stobart, eds., *Critical Approaches to the History of Western Herbal Medicine* (London: Bloomsbury, 2014): 193; and Vivian Nutton and J. Scarborough, tr. and eds. "The Preface of Dioscorides' *Materia Medica: Introduction, Translation and Commentary*" in *Transactions and Studies of the College of Physicians of Philadelphia* 4.3 (1982): 187–227.

[57] *De differentiis morborum,* VI.836-37K.

[58] In a minority of cases, a person's body and organs may be pre-disposed to operating improperly (via malfunctioning of the condition (διάθεσις)—see *De symptomatum differentiis,* VII.43 K., *Methodus mendendi,* X.63 K., *De Locis affectis,* VIII.25 K. or constitution (κατασκευή)—see: *Methodus mendendi,* X.52 K.

[59] R.J. Hankinson, *Galen on the Therapeutic Method: Books 1 and 2* (Oxford: Oxford University Press, 1990): 152.

via a disturbance in heat—generally toward the hot.[60] But because a principal adherent, Athenaeus, used a background of Hippocratic medicine (Dogmatist) with a sprinkling of Stoicism that was materially based, Galen does not see this sect as a contender to be refuted.

In *Methodus mendendi* Galen addresses the topic of what health is. He begins speaking against Thessalus, a Methodist, who claimed that disease could be conceptualized as unspecified damage to the bodily parts that results in loss of function, Galen claims first that, "Nothing happens without a cause,"[61] This is an important principle that defines what nature$_1$ is, and along with "nature does nothing in vain"[62] form a meta-theoretical approach to confronting nature based upon a materially-based *phusis*$_1$. These passages assume the *coherence* of the anatomy/physiology relation which leads to *completeness*.

In the passage below, Galen uses this meta-theoretical approach to argue against the approach of the Methodists.

> In fact, if health and disease are [contradictory] opposites, in whatever class of things which are in accord with nature health may be, disease will be of that same class of things when they are contrary to nature, so that if health is normal function, it follows that disease is, in all respects, some function which is somehow abnormal. If, however, health is either some *condition* or *constitution* that is normal, disease will also necessarily be some condition or constitution that is abnormal. Thessalus [the Methodist] did not actually attempt to define disease. Rather, we must rely on *divination* as regards the matter to which he applies the name. (*Method of Medicine*, 1.7, after the translation of Johnston and Horsley.)[63]

In this passage Galen begins by a rather logical point: to say that two properties (H [health] and ~H [non-health])[64] that when one of them is possessed by x (say H), then the individual is healthy [Hx], but when the opposite property is possessed (by condition or constitution), then the individual is sick [~Hx]. Such is the position of his opponent. But this would side-step the idea of there being an exact and knowable *cause* for the disease.[65] Further, the notions of *condition* and *constitution* complicate

---

[60] We don't have any full texts of the Pneumatists. The fragments were collected by C.F. Matthaei, *Medicorum veterum et clarorum graecorum varia Opusucla* (Moscow: Royal University, 1808) and M. Wellmann, *Die pneumatische Schule bis auf Archigenes* (Berlin, Weidmann, 1895). Note also, that the writings of Aretaeus and Anonymus Londinensis can add a little to these doctrines.

[61] μηδὲν χωρὶς αἰτίας γίνεσθαι; *Methodus mendendi* I.7.

[62] Οὐδὲν ἡ Φύσις ἐργάζεται μάτην, throughout the writings of Aristotle and Galen—and supports the maker/artifact dynamic.

[63] καὶ τοίνυν εἴπερ ἐναντία ἐστὶν ὑγεία καὶ νόσος, ἐν ᾧπερ ἂν ᾖ τῷ γένει τῶν κατὰ φύσιν ἡ ὑγεία, τούτου τοῦ γένους ἐν τῷ παρὰ φύσιν ἡ νόσος ὑπάρξει, ὥστε εἰ μὲν ἐνέργεια κατὰ φύσιν ἡ ὑγεία, πάντως δή που παρὰ φύσιν ἐνέργειά τις ἡ νόσος ἐστίν, εἰ δ' ἤτοι διάθεσίς τις ἢ κατασκευὴ κατὰ φύσιν ἡ ὑγεία ἐστί, καὶ ἡ νόσος ἐξ ἀνάγκης ἔσται διάθεσίς τις ἢ κατασκευὴ παρὰ φύσιν. ὁ μὲν οὖν Θεσσαλὸς οὐδ' ἐπεχείρησεν ὅλως ἀφορίσασθαι νόσον, ἀλλὰ χρὴ μαντεύεσθαι κατὰ τίνος ἐπιφέρει τοὔνομα πράγματος. (*Methodus mendendi*, 1.7).

[64] Assume "H" is *health* and "~H" is sickness.

[65] In Aristotelian logic (which Galen was familiar with), there are two sorts of opposites: contradictory and contrary. In the first sort of opposite, if "All S is P" and "No S is P" are opposites, then there is no *space* between the first proposition (an "A" proposition in Aristotelian logic) and the second proposition (an "E" proposition in Aristotelian logic). But there *is* a space between the two

the *explanans* so that the dictum that "nature does nothing in vain" (which implies efficient *simplicity*) would also be violated.

In this way, Thessalus (as a stand-in for Methodism) is depicted as rather arbitrary in being able to assess disease and, as such, the treatments for the constricted and unconstructed will be arbitrary, as well. For a Dogmatist, who believes in the rational, material structure of the body (and there being definable, material causes for disease), the *phusis*$_2$ approach of the Methodists just isn't good enough. In its place is the logical, structural materialism (*phusis*$_1$) of the Dogmatists—with Galen at the helm and Hippocrates as the first mate.

The outlines of the Empiricist sect are not as tight as the Methodists. This is because their defining characteristic is two-fold: (a) a reticence to declaring that full knowledge of the body can be known, à la Sextus Empiricus,[66] and (b) a commitment to putting epistemological weight toward that which could be interacted with via the appropriate senses: sight, touch, smell, hearing, and taste (most often the first four).[67] Using this metric, many of the anatomists—such as Erasistratus could be viewed as having a foot in this camp. However, Galen (though he admired anatomical study—as he was at the vanguard of this, himself) did not always approve of the Empiricist tendencies of many of these practitioners.

> But even as regards this doctrine [that nature does everything for a purpose], their agreement is only verbal; in practice Erasistratus makes havoc of it a thousand times over. For, according to him, the spleen was made for no purpose, as also the omentum; similarly, too, the arteries which are inserted into the kidneys[68]— although these are practically the largest of all those that spring from the great artery [aorta]! And to judge by the Erasistratean argument, there must be countless other useless structures; for, if he knows nothing at all about these structures, he has little more anatomical knowledge than a butcher, while, if he is acquainted with them and yet does not state their use, he clearly imagines that they were made for no purpose, like the spleen. Why, however, should I discuss these structures fully, belonging as they do to the treatise "On the Use of Parts," which I am personally about to complete? (*On the Natural Faculties,* II. 4, tr. A.J. Brock)[69]

---

on a continuum in which some are and some are not (the so-called "I" and "O" propositions). This lack of specificity here is troubling because a person can be a "little sick" without being there all the way. To cite concepts of bodily inclination to disease due to certain habits of the body (διάθεσις) or structures of the body (κατασκευή) is to beg the question concerning causation and suggests that nature unnecessarily complicated things—one way of acting "in vain." Thus, Galen's two metatheoretical principles would be violated. However, it should also be noted that there are particularities in different individuals so that habits that are especially helpful for one individual might have only moderate benefit for another.

[66] *Outlines of Pyrrhonism,* VI and XI.

[67] Richard Walzer and Michael Frede, *Three Treatises on the Nature of Science* (Indianapolis: Hackett, 1985): 21–46.

[68] The purpose is set out in II.9, 132.

[69] Ἀλλὰ καὶ αὐτὸ τοῦτο μέχρι λόγου κοινόν, ἔργῳ δὲ μυριάκις Ἐρασίστρατος αὐτὸ διαφθείρει, μάτην μὲν γὰρ ὁ σπλὴν ἐγένετο, μάτην δὲ τὸ ἐπίπλοον, μάτην δ' αἱ εἰς τοὺς νεφροὺς ἀρτηρίαι καταφυόμεναι, σχεδὸν ἁπασῶν τῶν ἀπὸ τῆς μεγάλης ἀρτηρίας ἀποβλαστανυουσῶν οὖσαι μέγισται, μάτην δ' ἄλλα μυρία κατά γε τὸν Ἐρασιστράτειον λόγον, ἅπερ εἰ μὲν οὐδ' ὅλως γιγνώσκει, βραχεῖ μαγείρου σοφώτερός ἐστιν ἐν ταῖς ἀνατομαῖς, εἰδ' εἰδὼς οὐ λέγει τὴν χρείαν αὐτῶν, οἴεται δηλονότι παραπλησίως τῷ σπληνὶ μάτην αὐτὰ γεγονέναι. καίτοι τί ταῦτ'

In this passage, Galen, criticizes Erasistratus, and by extension, the Empirics who use their skepticism (*skepsis₃*) as a reason for making the claim that *just because there are too many possible factors that may influence an outcome, that therefore, no judgment can be made*. This extreme epistemological standard, forces those with this attitude (like the Empirics) to assert that organs that they have seen in dissection (whether of humans or animals) that are not clearly evident in their operation[70] (most organs) are purposeless. This supposedly empirically-based conclusion is flawed in Galen's view because it is factually incorrect and because it violates Galen's and Aristotle's general scientific dictum that *nature does nothing in vain*.

Indeed, when this tenet about the elegance and simplicity of Nature's productive efficiency is accepted in the spirit of the maker/artifact dynamic, then this worldview posit alone, will constitute a prominent argument against the radical skeptics of *phusis₃*.

When engaging in critical science (as the Dogmatists believed they were doing—post Aristotle and the critical Hippocratic writers), the operative meta-method was inductive logic (*epagoge*)[71] which seems to demand that the practitioner develop generalizations (with perhaps some caveats in the case of medicine because of the variation between individuals).

These methodological differences are re-iterated when Galen discusses the case of a wound to an organ, the lung, *Methodus mendendi* V. 10.

> Let us assume again that thick pus has been disseminated by a wound in the lung. Will they now give the green medication or will they direct [the patient] to lick honey? But let them also say from what source they discovered this. They will certainly not say this—that it has a thinning and a cutting potency—these men who deliberately distance themselves from the search for such potencies. It is not because honey was found by the Empirics to be suitable in this sort of syndrome that it is possible for them (i.e., the Methodics) to use it like those men do—first because they despise experience and second because, in such syndromes, the Empiric says he is ignorant as to what the condition in the lung is, although those things that are useful have been observed by him through experience. It is not enough for Thessalus to treat what he does not know at all, but he proceeds from the indication of the affections. Doubtless, if we were to concede to him, as we did before, that he knows everything as we do, he would never shrink from changing the kind of treatment in the different parts. Obviously, to pour melikraton[72] into the uterus because of a filthy wound, to eat some honey, and to wash the wound with a sponge are not the same. But these latter points are minor; those former issues are, however, major. (*Method of Medicine* V.10, tr. Johnston and Horsley).[73]

ἐπεξέρχομαι τῆς περὶ χρείας μορίων πραγματείας ὄντα μελλούσης ἡμῖν ἰδίᾳ περαίνεσθαι;—(*On the Natural Faculties*, II.4, 91–92).

[70] This, of course, is the problem of inferring physiology from anatomy: see Boylan (2015): 79. And vivisection was not the answer because of the trauma caused by the procedure—*ibid*. 83–84.

[71] Cf. Aristotle, *Posterior Analytics*, II.19.

[72] *melikraton* is a fermented or unfermented mixture of honey and water that is generally taken orally. The relative amount of honey and water and its preparation would have been a trade secret of the medical practitioner.

[73] ὑποκείσθω δὲ πάλιν ἐν τῷ πνεύμονι περικεχύσθαι τῷ ἕλκει παχὺ πῦον, ἆρά γε καὶ νῦν τὸ χλωρὸν δώσουσι φάρμακον, ἢ μέλιτος ἐκλείχειν κελεύσουσιν; ἀλλὰ καὶ τοῦτ' αὐτὸ πόθεν

In this passage Galen gets to the essence of why the Dogmatist medical methodology is a better ground for medicine than either the Empirics or Methodists. We begin with the wound to the lung (a victim of a fight or a gladiator?). We then have a discussion of the burden of proof of the physician for knowing *what* was injured and the damage that was caused. Then, we then have a discussion of honey as a possible cure for this problem (as opposed to a herbal remedy, the green medication). But why honey and not something else, like the green medication? This is the issue at stake for Galen. Does honey have some particular properties that make it particularly apt for curing this wound? And if so, then what are they? Does it stop the bleeding? Does it help heal the wound? *What does it do? And why?* These are two very important questions when considering diagnosis, prognosis, and treatment.

Here we are most interested in treatment. But there is more to treatment than knowing whether one should be using honey or the herbal remedy. This is because medications must be applied in a particular way. It is not enough to know that honey is the proper medication. One must know *how* the honey is to be delivered to the patient. Will it be spread on the wound and massaged into the wounded area? Or should it be mixed with water or wine and swallowed by the patient. Obviously, this makes a big difference. One could get the correct medication and the wrong application and therefore fail in the healing effort.

But since the "on-the-ground" practices of both Methodists and Empirics depend upon their own personal experience of trial and error, with no general principles, as such, to guide them, they cannot give the reasons why (*dioti*).[74] This is a very important point. Just because one sees condition x and tries remedy y and it works, does not give the medical practitioner the right to attribute some sort of "causation" between x and y toward patient, z. This is because the medical practitioner does not know the reason *why* it worked.[75] Without the "why" one will fall prey to the logical fallacy, post hoc ergo propter hoc (after the fact, therefore, because of the fact). This is a very common logical fallacy in science—especially to those who practice medicine without a theoretical backdrop of anatomy and physiology and the reasons why things aren't working just right now. Obviously, with a wound to the lung, it is clear why there is a problem, but what sort of problem has been created? This is where

---

εὑρήκασι λεγέτωσαν. οὐ γὰρ δὴ τοῦτό γε φήσουσιν, ὅτι λεπτυντικήν τινα καὶ τμητικὴν ἔχει δύναμιν, αὐτοί γε ἑκόντες ἀποστάντες τοῦ ζητεῖν τὰς τοιαύτας δυνάμεις. οὐ μὴν οὐδ' ὅτι τοῖς Ἐμπειρικοῖς εὕρηται τὸ μέλι κατὰ τοιάνδε συνδρομὴν ἐπιτήδειον, ἔνεστιν αὐτοῖς ὁμοίως ἐκείνοις χρῆσθαι, πρῶτον μὲν ὅτι τῆς ἐμπειρίας καταφρονοῦσιν, ἔπειτα δὲ ὅτι κατὰ τὰς τοιαύτας συνδρομὰς ὁ Ἐμπειρικὸς ἥτις μέν ἐστιν ἡ ἐν τῷ πνεύμονι διάθεσις ἀγνοεῖν φησί τετηρῆσθαι δ' ἐκ πείρας ἑαυτῷ τὰ συμφέροντα. Θεσσαλῷ δὲ οὐκ ἀρκεῖ θεραπεύειν ὃ μηδ' ὅλως οἶδεν, ἀλλ' ἀπὸ τῆς τῶν παθῶν ἐνδείξεως ὁρμᾶται. εἰ δὲ δὴ καὶ πάντ' αὐτῷ συγχωρήσαιμεν, ὥσπερ καὶ πρόσθεν, ὁμοίως ἡμῖν ἐπίστασθαι, τό γ' ἐπὶ τοῖς διαφέπουσι μέρεσιν ἐξαλλάττεσθαι τὴν θεραπείαν κατ' εἶδος οὐκ ἄν ποτε ἐκφύγοι. οὐ γὰρ δήπου ταὐτόν ἐστιν ἢ μελίκρατον εἰς μήτραν ἐγχέαι δι' ἕλκος ῥυπαρόν, ἢ μέλιτος ἐσθίειν, ἢ καταπλύειν σπόγγῳ τὸ ἕλκος, ἀλλὰ ταῦτα μὲν ἔτι σμικρά, μέγιστα δὲ ἐκεῖνα. (*Methodus mendendi* V.10)

[74] The reader is encouraged to go back to Aristotle's four scientific questions in Chap. 3.

[75] I discuss this problem in a general way in Michael Boylan, *The Good, The True, and The Beautiful* (London: Continuum, 2008): ch. 4.

the fact (*hoti*), alone, is not enough. There may be many different sorts of wounds to the lung that require different treatments according to the material nature of the wound. Only one trained first in anatomy and physiology is competent to do this under the *phusis*$_1$ account.

I invite the reader to observe how Galen recommends treatment for fever in *Methodus mendendi* X.11. There he comments on diet (cold juices, gruel, and cold water). He explains that because the fever is a presentation of the hot, and since balance is health, that medication (food and drink) that emphasize the cold are in order. He also talks about the time in the illness in which treatment occurs. Some foods (like honey & water) turn into bile and because bile will increase the hot, and that this treatment should be avoided in raging fevers. Galen, also gives a nod to patient autonomy by suggesting a discussion with the patient about his/her individual nature and how it is affected during times of health in digesting and processing food and drink.

At the end of this passage, Galen discusses *wet and dry* in the stomach and how this also might have an effect upon the pulse.[76] This commitment upon the understanding of medicine as beginning with anatomy and adding to this some interaction with patients using a theory of physiology, made Galen different in a positive way from most of his contemporaries. The result of having these general principles that could be applied in a variety of relevant circumstances means that his theory had wider application (*completeness*) and was less likely to be caught in a sporadic contradiction (*coherence*). Having a limited number of these principles yields both *simplicity* and *elegance*. Thus, in comparison to his predecessors and contemporaries, Galen exhibited superiority in the principal categories of theory evaluation put forward in this book. This was an important step in the history of medicine.

Conclusion

All three of these approaches to medicine are contrary to the Dogmatists. The Pneumatists, Methodists, and Empiricists had their advocates in the ancient world. They viewed nature as *phusis*$_2$ (mostly in the sense of inscrutable causes) with a little *phusis*$_1$ and *phusis*$_3$ (mostly in the sense of the third level of skepticism, *skepsis*$_3$). However, it was Galen who put the most significant stamp on ancient biomedical understandings of nature. He did this by considering the various approaches that came before him along with their different views of nature. Whenever that approach was something different from other opinions by Pneumatists, Methodists, or Empirics, Galen's medical approach argued for his own *phusis*$_1$ standpoint. It offered a comparative advantage because it was built upon anatomical knowledge

---

[76] On the pulse see: Michael Boylan (2015): 133n. 68 and "Galen on the Blood, Pulse, and Arteries" *Journal of the History of Biology* 40.2 (2007): 207–230.

Conclusion

along with a theory of both how the body worked in normal functioning, but also how the body could get out of balance and become un-healthy due to material causes. But all of these understandings were through empirically scrutable, material means. From these he forged his own amalgam, *phusis*$_1$. Thus, it can be properly said that Galen can be seen as a pre-eminent defender of a theoretical position of basing medicine upon a systemic, material understanding of n/Nature, and that this understanding was very influential in the understanding and practice of medicine going forward.

Part II
The Medieval, Modern, and Contemporary Worlds

Chapter 6
The New Paradigm: Changes Leading up to the Seventeenth Century and Beyond

Abstract This chapter begins with brief commentary on the transition of the understanding of nature in the Islamic and European worlds during the Middle Ages transitioning to the European scientific revolution in astronomy during the seventeenth century and links this to mathematical advances. Descartes is an innovator in both mathematics and in critical theory while Newton concentrated more on creating a mathematical expression of motion with several unexamined premises that Kant defended afterwards. In biology/medicine the premier advancement was the circulation of blood model put forth by Harvey.

Keywords Galen · Hippocratic writers · Al Razi (Rhazes) · Al-Zahrawi (Albucasis) · Ibn Sina (Avicenna) · Ibn al-Nafis · Paracelsus · Michael Servetus · Andreas Vesalius · Realdus Columbus · Pope Gregory XIII · Copernicus · Galileo · Descartes · Newton · Kant · Harvey · Circulation of the blood

Introduction

Just as in the first part of this book, in which we balanced our inquiry between basic science and its impact upon developments in medicine, the second part of the book aims to do the same.

It is important to mention that there is a big historical lacuna of more than 1400 years between the first half of the book and the second half. Much went on during this period in the Western World (Europe and the Middle East) to advance the work of the Ancient Western World—most particularly works by Plato, Aristotle, the Hippocratic writers, and Galen.[1]

---

[1] One new book that I'd like to recommend here to help fill in the lacuna is Vivian Nutton, *Renaissance Medicine: A Short History of Medicine in Europe in the sixteenth century* (New York

But starting in the ninth century, things started to change. This change was spearheaded by Islamic and Arab scholars in the so-called *Golden Age*. These are briefly set out followed by the European sixteenth century in which the advances set out in the Golden Age are further expanded via advances in anatomy and pharmacology.

Transition in the Islamic World[2]

Though this subsection is entitled "the Islamic World" there is really quite a bit of cross-pollination of medical writing and practice that pretty much followed existing trade routes between Europe, North Africa (extending south to Timbuktu), and the Middle East (extending to Persia and even to the boarders of India).[3]

Islamic medicine began with the Greek/Roman tradition—especially the Hippocratic writers, Galen (and the other Dogmatists) and the quasi-Empiricist Dioscorides.[4] From the time of al-Razi (865–925/935 CE) to Ibn al-Nafis (1213–1288) Arab philosophers and medical practitioners/thinkers translated and commentated upon their Greek predecessors as well as making discoveries and advancements of their own. For this reason, the time period is often called *The Golden Age in Islamic Medicine.*

This "Golden Age" in Islamic Medicine thrived due to a shared community worldview of adopting a healthy lifestyle that supported religious faith into

---

and London: Routledge, 2022). Other works that discuss this period include: Nancy G. Siraisi, *Medieval and Early Renaissance Medicine: An Introduction to Knowledge and Practice* (Chicago: University of Chicago Press, 1990); Carole Rawcliffe, *Sources for the History of Medicine in Late Medieval England* (Kalamazoo, MI: Medieval Institute Publications, 1996); Peter E. Pormann and Emilie Savage-Smith, *Medieval Islamic Medicine* (Washington, D.C.: Georgetown University Press, 2007); Anne Van Arsdall and Timothy Graham, eds. *Herbs and Healers from the Mediterranean through the Medieval West: Essays to Honor John M. Riddle* (New York and London: Routledge, 2017); David C. Lindberg, *Science in the Middle Ages* (Chicago: University of Chicago Press, 1978); Pieter De Leemans, "Aristotle Transmitted: Reflections on the Transmission of Aristotelian Scientific Thought in the Middle Ages" *International Journal of the Classical Tradition* 17.3 (September, 2010): 325–333; and Peter Distelzweig, Benjamin Goldberg, Evan Ragland, *Know Thyself: Early Modern Medicine and Natural Philosophy* (Cham, Switzerland: Springer, forthcoming).

[2] This could just as easily be called the "Arabic World" as language was a key factor during this era in translations from Greek biomedical writers into Arabic. I have chosen "Islamic World" because during this epoch of history, religion played a big part in regional identity. Note, that for the European World, the "Christianity" achieves a developing parallel identity—especially in the context of the Crusades. See: Jonathan Riley-Smith and Susanna A. Throop, *The Crusades: A History* 4th ed. (London: Bloomsbury Academic, 2022) and Amin Maalouf, *The Crusades through Arab Eyes* (New York: Schocken, 1989).

[3] For a discussion of this and the developing role of physicians in Islamic Society from 650–1500 CE, see: Emillie Savage-Smith and Peter E. Portmann, *Medieval Islamic Medicine* (Washington, D.C.: Georgetown University Press, 2007).

[4] See: Donald Campbell, *Arabian Medicine and its Influence in the Middle Ages* (London: Routledge, 2013): 220.

Introduction

normative posits that were analogous to the Roman poet, Juvenal who declared "Orandum est ut sit mens sana in corpore sano"—"You should pray for a healthy mind in a healthy body"—around the end of the first century AD (Satire 10.356). Thus, medicine and its priests (the physicians—phusis$_2$) facilitate a general normative direction of health that went beyond daily practical comfort, only.

Indeed, in the Hadith, the Prophet is said to have supported a healthy lifestyle— especially regarding the digestive system as the key to general, systemic health.[5] Thus, there was no basic conflict between medicine and religion.

During the various Caliphates (political confederations that sometimes extended as far West as Morocco/Spain and as far East as Persia)[6] intellectual activity was supported by the rulers and it thrived. This ranged from the translation of Greek philosophical and medical texts (particularly the Hippocratic Writers, Galen, and Dioscorides), as well as original research by polymath philosophers and physicians.[7]

This book will highlight four of those best known for their original research: Al-Razi (aka Rhazes)—865–935 CE; Al-Zahrawi (aka Albucasis)—936–1013 CE; Ibn Sina (aka Avicenna)—980–1037 CE; and Ibn al-Nafis—1213–1288 CE.

Our first major figure in Medieval Islamic Medicine is Abu Bakr Mohammad bin Zakariya Al-Razi (aka, Rhazes)—865–935 CE. Rhazes was born in Al Rayy, a town near present-day Tehran in 864 CE. He was interested in music as a young man.[8] Then he began studying philosophy and alchemy. He is credited with the discovery of sulfuric acid and ethanol. His teacher, Ali Ibn Sahl Rabban was a philosopher and physician, who was born a Jew but converted to Islam. This association also promoted spiritual interests in Rhazes.

Soon he surpassed his teacher and his fame as a physician spread widely—even to Caliph Al-Muktafi who made him the head of the largest hospital in Bagdad.

Rhazes chose his place of living in Bagdad by setting out pieces of meat in various locales he was considering. After a few days he checked out each and chose the one that was the least rancid. His reasoning was that he wanted to live in the locale that had the purest air.

After the death of the Caliph Al-Muktafi in 907 CE, Rhazes returned to his hometown, Al Rayy where he took over the hospitals there. Rhazes was known as a

---

[5] See: Ibn Chaldun, *Die Muqaddima Betrachtungen zur Weltgeschichte* (Munich: C.H. Beck, 2011): 391–395.

[6] The various Caliphs (successors of the Prophet Mohamed) include the first: Abu Bakr (632–634), the second: Umar (634–644); the third: Uthman (644–656); and the fourth: Ali (856–661); et al. until the downfall of the Ottoman Caliphte in 1924 at the end of World War I. For more on this see: Daniel Brown, *A New Introduction to Islam* (Hoboken, N.J.: John Wiley, 2011).

[7] One good textual example of this can be found in Glen M. Cooper, *Galen, De diebus decretoriis, From Greek into Arabic: A Critical Edition, with Translation and Commentary, of Hunayn ibn Ishaq, Kitab ayyam al-buhran* (London: Routledge, 2017).

[8] El Gammal SY, "Rhazes Contribution to the Development and Progress of Medical Science" *Bulletin of Indian Institute of the History of Medicine* 25.1/2 (1995): 135–149.

generous man who gave treatment without a fee. In his later years he became blind. He died in Al Rayy in 925 CE at the age of 60.

Rhazes was very prolific in his writings. His key work was a medical encyclopedia: Al-Hawi Fi al-Tibb (known in Latin as Liber Contiens). Rhazes (through his work in alchemy [early chemistry]) was known for his advancements on simple and compound plant medicines. He was also a chronicler of diseases in the tradition of the Hippocratic writers on Epidemics. To this end, he gave some of the earliest accounts of small pox and measles—showing how they were different and could be treated.[9] Rhazes was conservative in his application of medicines. He thought whenever one could treat a condition with a single plant ingredient, then one should choose that over compounds (because the results were easier to predict). This suggested an inclination for physicians to be conservative in treatment except in cases in which death seemed imminent. This inclination persists to the present day in most of Western Medicine.

Our second major figure in Medieval Islamic Medicine is Al-Zahrawi (aka, Abu lcasis)—936–1013 CE. Albucasis was born in Azahara in Northwest Cordoba (the capital of Muslim Spain) around 936 CE. He lived, studied, and practiced medicine in this region most of his life. He served as court physician to Caliph Al-Hakam II. He died 2 years after the sacking of Azahara in 1013 CE at the age of 77.

Around 1000 CE Albucasis wrote his famous work: "Al Tasreef Liman 'Ajaz Aan Al-Taleef'" (The Clearance of Medical Science for those who cannot Compile It) 30 volumes on medical education, training, practice and experience.

In the Al-Taleef three chapters were devoted to surgery including[10]:

- The eye, ear, and throat (including tonsillectomy)
- Instruments for examining the ear
- Removing material from the nasal cavity
- Examining and protecting the throat
- Cauterization to treat skin tumors and abscesses
- Ligatures to stop bleeding using catgut sutures
- Treatment of anal fistulas
- New methods for setting bone fractures
- Using an instrument, he had invented, to inspect the urethra
- Created a new diagnostic procedure for ectopic pregnancy that made treatment more reliable
- Devised a method to create false teeth from animal bones.

In the end, Albucasis is best known for his achievements in surgery and fabrication of medical devices: knives, probes, scalpels and hooks to assist his practice. He invented surgical scissors, grasping forceps and obstetrical forceps.

---

[9] F.H. Garrison, *An Introduction to the History of Medicine* (Philadelphia and London: Saunders, 1929): 129, and Nasim Hasan Naqvi, "A Medical Classic: Al-Razi's Treatise on Smallpox and Measles" *The Historical Medicine Equipment Society* 22 (2009): 1–20.

[10] Ismail A. Nabri, "El Zahrawi (936–1013 AD), The Father of Operative Surgery" *Annals of the Royal College of Surgeons of England* 65 (1983): 132–134.

He has even been credited with using surgery on joints to relieve pain and disfunction.[11]

Our third major figure in Medieval Islamic Medicine is Ibn Sina (Avicenna) (born 980 CE near Bukhara, Persia (now Uzbekistan) and died in 1037 in Hamadan Persia/Iran). He was one of the most prominent philosophers and medical writers in his era. His Book of the Cure/Healing and The Cannon of Medicine were very influential for hundreds of years.[12]

Avicenna was a prodigy. By the age of ten he is said to have memorized the entire Quran. His tutor, Natili, emphasized Aristotelian logic which got Avicenna interested in Aristotle—a lifetime fascination.

By 16, Avicena took a turn toward medicine. He learned quickly and soon was brought into the role of attending physician to the Sultan of Bukhara. Avicenna's treatment was successful, and the Sultan was cured. In gratitude, the Sultan created a library for Avicenna's use. This new library, the Samanid Library, was filled with manuscripts on science and philosophy. Avicenna had access to the best thinkers of the ancient world.

Avicenna was a prolific writer and his Kitab al-Shifa is his 4-part encyclopedia covering logic, physics, mathematics, and metaphysics. As a philosopher, Avicenna (in many ways) followed an Aristotelian critical empiricism. However, unlike Aristotle, his classification of physical objects differs from Aristotle's Categories in which the most real entities were "primary substances"—like Socrates. Instead, Avicenna elevated the qualities, themselves, so that these invisible properties that were "predicable of" primary substances, became the most real.[13] This was rather more Platonic than Aristotelian.

This metaphysical classification that elevated non-sensible properties (phusis$_2$) can be shown through a thought experiment called variously The Floating Man or The Flying Man. In the thought experiment one must imagine that God has created a human out of thin air and this man has had no sensory experience, and thus, has no memories. However, even without any sensory experience or memories, this individual (according to Avicenna) is aware of his own essence as an existing being.[14] Thus, thinking, per se is possible merely through the exercise of self-reflection. Of course, objectors could contend that reflection is only possible through sensory experience by which reflection becomes possible. This would be the position of the naïve Empiricists (phusis$_3$) who ground everything—not just in the

---

[11] K. Markatos, A. Mavrogenis, E. Brilakis, et al. "Albucasis (936–1013): his work and contribution to Orthopedics" *International Orthopedics* 43 (2019): 2199–2203.

[12] Ibn Sina, *The Physics of Healing,* Jon McGinnis, ed. tr. (Provo, Utah: Brigham Young University Press, 2009); and *Al-Qanun fi al-tibb; The Cannon of Medicine,* Oskar Cameron Gruner, ed. tr. (New York: Ams Press, 1973, rpt. 1930 (London)).

[13] One approach along these lines can be found in Ricardo Strobino, "Per se, Inseparability, Containment, and Implication. Bridging the Gap between Avicenna's Theory of Demonstration and Logic of the Predicables" *Oriens* 44.3/4 (2016): 181–266.

[14] Compare this thought experiment to Descartes' thought experience using radical doubt to prove existence—see this chapter.

process of interacting with the world through the five senses, but also by subjecting this experience to searching questions of coherence and completeness (epistemic touchstones discussed earlier in this book).

In medicine, Ibn Sina is known for his work The Cannon of Medicine that is separated into five books.

- Bk 1: Covers the basic principles of medicine—such as anatomy and regimen. It is organized philosophically around Aristotle's four causes and physically (phusis$_1$) around the contraries: hot, cold, wet, and dry.
- Bk 2: Covers simple drugs that can be found in natural herbs. Their effects (respecting the contraries) are set out according to seven rules: (1) The drug must be relatively invariant in its qualities (i.e., not generally changeable); (2) Drugs may have more than one effect on patients; (3) Drugs may even bring out contrary effects, such as making a patient both hot and cold during different intervals; (4) The potency of the drug must be commensurate with the strength of the disease; (5) Some drugs act quickly while others act more slowly—match the time intervals; (6) The drug must generally produce the same effects; (7) Experiments to learn about humans should be carried out on humans and not animals.
- Bk 3: Covers diagnosis and treatments specific to particular regions of the body. It is also the beginning of his discussion of nerves and sets out a very basic beginning of neuroscience.
- Bk 4: Diagnosis and treatment of systemic somatic disorders.
- Bk 5: A formulary of medical compounds (mixtures of herbs to create a new drug. Over 650 compound formulae are set out).

Our fourth major figure in Medieval Islamic Medicine is Ibn al-Nafis—1213–1288 CE. Ibn al-Nafis was born in 1213 in a village near Damascus and died on December 17, 1288 in Cairo. As a young man he studied theology, philosophy, and literature. At 16 he turned his attention to medicine for 10 + years at Nuri Hospital in Damascus. He studied under the physician al-Dakhwar and later went to Egypt to take charge of the Nasiri hospital in Cairo. It was there, perhaps, that he began performing human dissections (in the long tradition from Alexandria to Cairo). There is no evidence that he performed vivisection (as happened in earlier eras in that region of Egypt). This empirical work on the human body allowed Ibn al-Nafis to set out detailed descriptions of anatomy and speculate on the accompanied physiology.

Ibn al-Nafis' most important contribution to the history of medicine in the West, was his speculations on physiology of the circulation of blood (based upon anatomy of the pulmonary circulation of the blood from the right to left ventricle). Galen conjectured that this circulation happened within the heart via small pores in the wall between the right to left ventricle. However, al-Nafis, surmised otherwise (based upon his observations from human cadaver dissections).[15] Thus, Ibn al-Nafis asserted that there were no such pores in the septum between the ventricles. The septum did not exhibit pores (an observation from his work in dissection). Because of this al-Nafis said that if there is communication between the two ventricles, "it must be via the lungs because... the blood after it has been refined in this cavity [right ventricle], must be transmitted to the left cavity [left ventricle] where the [vital] spirit is generated... The penetration of the blood into the left ventricle is from the lung, after it has been heated within the right ventricle and has risen from it, as we have stated before."[16]

Instead, al-Nafis stated that in order to complete this circulatory process there must be small communications or pores (*manafidh* in Arabic) between the pulmonary artery and vein, a prediction that preceded by 400 years the discovery of the pulmonary capillaries by Marcello Malpighi (1628–1694), an Italian physician. Might Malpighi have been influenced by Ibn al-Nafis' work (if he had access to it)? We do not have adequate primary data to make a call.

Ibn al-Nafis also wrote a commentary on Avicenna.[17] In this work, he extends Avicenna's work on anatomy based upon his own work in Cairo on human dissections. Ibn al-Nafis also builds upon Avicenna's work on eye diseases and a theory of nutrition, aka regimen.

In the end, Ibn al-Nafis can best be described as a careful anatomist who exceeded his peers in his questioning of function (physiology) based upon the structure of the components (anatomy). With the help of *vital spirits* (what the Greeks depicted as *pneuma*) Ibn al-Nafis is a fitting backstop on the Golden Age as the Greek Hippocratic/Galenic traditions are extended in positive ways within all four areas of medicine: (a) basic science, (b) diagnosis, (c) prognosis, and (d) treatment.

Transition in the European World

In the European world post-Galen and pre-seventeenth century, there were five individuals who had an effect upon medicine who will be highlighted here: Paracelsus (1493–1541), Michael Servetus (1511–1553), Andreas Vesalius (1514–1564), and Realdus Columbus (1516–1559). These five thinkers will be briefly set out as creating the background for the European exploration of Western Medicine from the seventeenth century, onwards.

---

[15] John B. West, "Ibn al-Nafis, the pulmonary circulation, and the Islamic Golden Age" *Journal of Applied Physiology* 105.6 (2008): 1877–1880.

[16] M. Meyerhof, "La découverte de la circulation pulmonaire par Ibn al-Nafis, Médecin arabe du Caire (XIIIe) siècle" *Bulletin l'Institut d' Egypte (Cairo) 16 (1934): 33–46;* _____. "Ibn an-Nafis and his theory of lesser circulation" *Isis* 23 (1935): 100–120.

[17] Ibn al-Nafis, *A Commentary on Anatomy in Avicenna's 'Cannon of Medicine'* 3 volumes. (Padua: Pierre Manufer, 1477)—Library of Congress, USA, control number: 2021667077.

Our first major figure is Paracelsus. Theophrastus Bombastus Von Hohenheim (**aka,** Paracelsus) was born in the village of Einsiedeln, Switzerland in the latter months of the year 1493.[18] He was the only child of Wilhelm Bombast von Hohenheim (1457–1534), a German chemist, physician and botanist and Elsa Ochsner, a Swiss matron of the local pilgrim hospital. His mother died soon after his birth. Almost a decade later, his father moved to rural Austria near Villach in Carinthia, Austria. In addition to being a physician, his father took on a second job as a manager of mines for the Fugger Family (who also established a hospital). Paracelsus learned something about gasses and chemical reactions within mines from his father during this period.

Around 1507 Paracelsus decided to study medicine and a couple years later enrolled at the University of Vienna to study mathematics, music, and astronomy. In 1515, he took his degree in medicine from the University of Ferrara (Italy).

Then Paracelsus traveled around Europe for 9 years observing the state of medicine wherever he went. He returned to Villach (Austria) to be with his father, who was still alive. Paracelsus tried to make it in Salzburg, but got involved in politics and had to flee to avoid arrest for his participation in the Peasants' Revolt.

Paracelsus then obtained citizenship in Strasburg (Austria) where he began his *Opus Paramirum*.[19] The work covers topics on intellectual history, religion, philosophy, alchemy, and the causes and treatment of diseases.

In 1526 he was summoned to Basel by Johannes Froben (aka Frobenius, 1460–1527), who was suffering from a severe infection that Paracelsus was able to alleviate. Soon Paracelsus received other healing requests beginning with another famous scholar, Erasmus.

This led to an appointment as *town physician* in Basel (which included a professorship in chemistry at the University of Basel). However, in 1528, Paracelsus' abrasive, egotistical manner created enemies who forced him to leave.

Paracelsus then took the opportunity to travel about Europe again. He devoted himself to writing *The Great Surgery Book*. So engrossed was Paracelsus with his writing that he missed hearing of his father's death for 4 years.

In his *Book of Three Principles*[20] Paracelsus argued that diseases should be named after their cures—viz., if a disease were treated with gold dust (like leprosy), it would be named "the gold disease."

---

[18] For further details on the biography of Paracelsus see: Pai-Dhungat, J.V., Parikh, F. "Paracelsus (1493–1541)" *J. Assoc. Physicians India.* 63.28 (2015) 63:28; E. Cockayne, "Theophrastus Phillipus Aureolus Bombastus Von Hohenheim (Paracelsus)—a short biography." *Br. J. Gen. Pract.* 52 (2002):876; G. Feder, "Paradigm lost: a celebration of Paracelsus on his quincentenary." *Lancet.* 341 (1993):1396–1397; Editorial "Remembering Paracelsus (1493–1541) *Indian J. Physiological. Pharmacology.* 37 (1993): 169–170; Bernoulli R. "Paracelsus—Physician, Reformer, Philosophy, Scientist." *Experientia.* 50.4 (1994):334–338; A.H. Fletcher A.H. "The Life and Medicine of Paracelsus (1493–1541)" *Cent. Afr. J. Med.* 4 (1958): 252–256.

[19] A version of this book is available in re-print: Paracelsus, *Opus Paramirum: A bi-lingual edition.* Ed. by Andrew Weeks (Leiden: Brill, 2007)—German/English.

[20] Much of the extant work of Paracelsus can be found in *Paracelsus, Four Treatises of Theophrastus von Hohenheim called Paracelsus,* ed. Henry E. Sigerist (Baltimore: Johns Hopkins University Press, 1996).

Some of the principal ingredients in his toxicology regime included arsenic, sulphur, silver, copper, lead, mercury, and an opium construct called *laudanum* (which he is credited with inventing) and because it is a narcotic, became popular and addictive. These are described in his treatises: *De mineralibus, De Natura Rerum,* and *Archidoxa.* Paracelsus emphasized the importance of carefully measured doses and that particular doses affected particular parts of the body—just as diseases affected particular parts of the body.

Mercury was rarely given alone but was mixed with other metals, such as zinc. However, Paracelsus received some pushback, since many of these ingredients were also *poisons* when given in the wrong mixture or the wrong doses.

It would be fair to parse Paracelsus's trial and error empirical method to be most akin to *phusis₃*—but not entirely so. His skepticism about universal method is more "mitigated" than radical.[21] He puts his faith in his own medical experience treating others. His *Opus Paramirum* is not so much a dogmatic treatise, but an attempt to gain completeness and coherence within the knowledge sphere of the attending physician—especially concerning mixing dosages of herbs and mineral medicines. His doctrine of regions of disease and treatment creates a simplicity that renders an elegance to his account. Especially regarding pharmacology, Paracelsus was very influential during the transition period.

Our second major figure is Michael Servetus. Michael Servetus' birthdate has some uncertainty of its accuracy.[22] Michael's father, Anton Servetus, was a low ranking noble. He was also a notary for the nuns of Sijena. This may have created a religious atmosphere in the life of young Michael.

Michael's father was responsible for his early education.[23] Then, Michael attended the University of Zaragoza that was overseen by the archbishop of Saragossa. He received his B.A. and M.A. in 1523. Then he went to study law in 1528 at the University of Toulouse (where he had access to religious books that discussed the Reformation and, at times, leaned to the extreme).

Around 1529 Servetus left Toulouse to accompany Juan de Quintana (a Franciscan friar who was the confessor to Charles V) to the coronation of Charles V in Bologna. The rich pageantry of the occasion made a negative impression on Servetus as he would later discuss in his *Christianismi Restitutio.*[24]

---

[21] See David Hume's argument on this that supports Paracelsus over and against other more radical *phusis₃* adherants: David Hume, *An Enquiry Concerning Human Understanding* (London: A. Millar, 1748): 18–28.

[22] See: Francisco Echeverria, Maria Chandia, "Miguel Servet o Villanueva documentalmente riavarro de Tudela" *Grupos sociales en la historia de Navarra* 1 (2002): 425–438; Radovan Lovci, *Michael Servetus, Heretic or Saint* (Prague: Prague House, 2008).

[23] Roland H. Bainton, *Hunted Heretic: The Life and Death of Michael Servetus, 1511–1553* (Ashland, Or: Blackstone, 2005).

[24] Peter Hughes, Peter. "The Face of God: The Christology of Michael Servetus." *Journal of Unitarian Universalist History* 40 (2016/2017): 16–53.

In 1531 he spent time in Strasbourg where he published his first book on the Trinity, *De Trinitatis Erroribus, Libri Septem*.[25] His theory sets out that the Christian Trinity (Father, Son, and Holy Spirit) leads one to believe in three gods instead of one. Servetus promoted the monophysit position, instead. Since this position was contrary to the standard dogma of both the Catholic and Protestant Christian Churches, Servetus was in trouble.

Respecting Medicine. In 1536, he returned to the University of Paris to study medicine. He supported himself by lecturing in mathematics, geography and astronomy.

In Michael Servetus' medical education, he excelled in anatomy. However, his religious doctrines on Christology got him in trouble (concerning the Trinity) so he travelled to Montpellier where he obtained his MD in 1539.

Servetus said that he thought about the Holy Spirit in relation to the blood in the body as a creative agent, *phusis*$_2$. This led him to discover that the blood in the heart flows to the lungs for oxidation instead of the Galenic concept of the air in the lungs flowing through a pneumatic tube to the left Ventricle.

What cinched the deal was the bright red color of the blood in the lungs indicating that the change in color was caused by something happening there. Servetus posited that the air or vitalized air mixed with blood in the lungs. He also agreed that the Galenic "septum pores between the ventricles" did not square with his anatomical investigations.

Servetus became a sought-after physician. He was picked to be the personal physician to both the archbishop of Vienna and the Governor of Dauphine.

In 1553 he published another religious work, *Christianismi Restitutio* (Restoring Christianity). The thesis was that in order to understand the Holy Spirit (third person of the Trinity) one had to understand its operation within the human body. It was this principle that he used to support his theory of pulmonary circulation (*phusis*$_2$)..[26]

Much of his work in this area of physiology based upon anatomy overlaps with Ibn al-Nafis; however, no translation was available for Servetus so that this is an independent finding on the heart/lung blood circulation.

In the end, Servetus' renegade religious ideas were used against him in an intolerant era and led to a death sentence that was carried out in the most painful way possible: being killed by being burned at the stake. This speaks to the dangerous environment that the early Renaissance medical writers faced in performing their medical practice and investigating the underlying truths of the human body and how it functioned (a propaedeutic to offering the best possible health care).

Our third major figure is Andreas Vesalius. Andries van Wezel (aka Andreas Vesalius) was born in Flemish Brussels from a family of physicians and pharmacists. He attended the Catholic University at Leuven in 1528–1533. He left for the University of Paris when his father became Valet de Chambre, being promoted from

---

[25] Available from the Michael Servetus Institute: https://www.google.com/search?channel=ftr&client=firefox-b-1-d&q=michael+servetus+institute.

[26] Peter Hughes, "The Early Years of Servetus and the Origin of His Critique of Trinitarian Thought" *Journal of Unitarian Universalist History* 37 (2013/2014): 32–99.

his post of apothecary to Maximillian (later to emperor Charles V). Vesalius toyed with the idea of a military career which would have been helped by his father's political position. To this end, Vesalius went to the University of Paris that had a program that would have prepared him for these skills. Though when he went to the University of Paris, he also continued his interest in anatomy through dissection (often appropriating bodies from the Cemetery of the Innocents.[27]

Vesalius had to leave Paris in 1536 and return to Leuven because of hostilities between France and the Holy Roman Empire. He finished his degree in Leuven writing his thesis on the ninth book of Rhazes. Upon obtaining his degree, he was offered a position at the University of Padua and did some guest lecturing at the University of Bologna.

At Padua, he concentrated upon anatomy and saw himself in the position of correcting some of the work of Galen, who was considered to be authoritative. Since Vesalius had a presence in both Padua and Bologna, he had some freedom to obtain bodies for dissection and not hanging around too long that might get him into trouble.[28]

These endeavors gave rise to his illustrations and commentaries on the human body: *De humani corporis fabrica libri septem (The Seven Books on the Structure of the Human Body,* 1543).[29] In this work, Vesalius strove to present *elegance* (via the beautiful line drawings of the human body) and *completeness* (by not leaving anything important out) and he proved *coherence* by showing by his narrative text how it all worked together. Because of the emphasis upon concrete depiction of the organs and systems, we can classify his approach as *phusis$_1$*.

In 1543 Vesalius left for Mainz to present his book to the Emperor of the Holy Roman Empire, Charles V. The emperor responded by making Vesalius the physician of the Royal Household—all by the age of 28.

In 1544 Vesalius returned to his homeland to marry Anne van Hamme. In 1556, Charles V abdicated and gave Vesalius a lifetime income. Vesalius traveled to Spain to take-up the appointment from Charles' son, Philip, to become physician of the court.

In 1564 Vesalius traveled on a pilgrimage to Jerusalem[30] While in Jerusalem, Vesalius was also said to have been offered another stint at the University of Padua. But the return journey did not go well. His boat was caught in a storm in the Ionian

---

[27] Martin Gumpert, "Vesalius" *Scientific American* 178.5 (1948): 24–31.

[28] For a quick overview on the reluctance to allow dead bodies to be dissected, see the following from the National Library of Medicine, U.S.A. https://www.ncbi.nlm.nih.gov/pmc/articles/PMC4582158/ (accessed 11-3-23).

[29] Andreas Vesalius, *On the Fabric of the Human Body.* 7 vols. (Basel, Switzerland: Joannes Oporinus, 1555–1558). Cf. J.B. de C.M. Saunders and Charles O'Malley, eds. *The Illustrations from the Works of Andreas Vesalius of Brussels* (N.Y.: Dover, 1973).

[30] There is also an unconfirmed story about Vesalius' journey to Jerusalem being a sort of punishment for performing a vivisection. Charles T. Ambrose, "Andreas Vesalius (1514–1564): An Unfinished Life" *Acta Medico-Historica Adriatica* 12/2 (2014): 217–230.

Sea and Vesalius' boat was shipwrecked on the Greek Island of Zakynthos, where he soon died at the age of 49.

At the time of his death Vesalius was in debt and needed a benefactor to pay for his funeral on Zakynthos.

All in all, Vesalius is reputed to have found numerous mistakes that Galen had asserted about the human body—such as the jaw being more than one bone. Other common myths were debunked by Vesalius—such as men having one more rib than women (à la *Genesis*), the sternum having multiple parts (beyond the three Vesalius found to be the case), and the false assertion that men had more teeth than women.

Regarding the nervous system, Vesalius defined nerves via their function: a linear mode of transmitting information and commands throughout the body refuting those who conflagrated nerves with tendons, ligaments, and aponeurosis.[31] Further, his study of the optic nerve demonstrated that the nerve was not hollow (as many promoting *pneuma* as a vital spirit that flowed *through* the nerves contended).

Vesalius' major contribution was his use of careful anatomy as a basis for obtaining knowledge of the body so that these structures might inform on speculations on physiology. Vesalius' depiction of what medicine *is* would be threefold: (1) Drugs (pharmacology); (2) Diet/regimen; and (3) Surgery (which required detailed knowledge of anatomy). This categorization of the field helped define it in ways that helped for the ongoing manner of its development.

Our fourth major figure is Realdo Colombo (aka Colombus) was born in Cremona 1515 in Lombardy, the son of an apothecary named Antonio Colombo. He took his undergraduate education in Milan where he studied philosophy. Then he decided to becoming an apothecary, like his father, but then left to apprentice under a surgeon, Giovanni Antonio Lonigo for 7 years.[32]

In 1538 he enrolled at the University of Padua where he was an excellent student in medicine and anatomy. On the side, he lectured in logic to pay his bills. He became a friend of Andreas Vesalius, who he succeeded as professor of surgery in 1543. In 1546 Columbus became the first professor of anatomy at the University of Pisa. Then, in 1548–1559 he was the surgeon to Pope Julius III, and professor at the Sapienza (Papal University in Rome, 1548–1559).

His only formal work, *De re anatomica* (On things anatomical) 15 volumes (Venice: Nicolai Bevilaequa, 1559) was published in the last year of his life. The work was over 15 volumes and set out what he believed to be the field of medicine:

- Bk 1: on the bones
- Bks 2 & 3: cartilage and ligaments
- Bk 4: the skeleton
- Bk 5: muscles

---

[31] Bruno Splavski, Kresimir Rotin, et al. "Andreas Vesalius, the Predecessor of Neurosurgery: How his Progressive Scientific Achievements Affected his Professional Life and Destiny" *World Neurosurgery* 129(2019): 129–202.

[32] Andrew Cunningham, *The Anatomical Renaissance: The Resurrection of the Anatomical Projects of the Ancients* (Cambridge, MA: Scolar Press, 1997): 143.

- Bk 6: liver and veins
- Bk 7: heart and arteries
- Bk 8: brain and nerves
- Bk 9: glands
- Bk 10: eyes
- Bk 11: viscera
- Bk 12: formation of the fetus
- Bk 13: on the skin
- Bk 14: on performing vivisection
- Bk 15: on rarely seen interior organs

This recitation of the topics of each of the 15 books gives a hint on what he thought the body was all about. His work on the viscera in book 11 was particularly acute for his time period. He continued the work of Michael Servetus on the circulation of the blood from the heart to and from the lungs.

This process begins during ventricular diastole (rest) and is expelled during systole (contraction). He went through the circulation of the venous blood from the right ventricle through the pulmonary artery to the lungs where it transforms to bright red after a mixture with *potent air* found within the lungs. It then returns to the left ventricle via the pulmonary vein. This constitutes the last nail in the coffin of Galen's ventricular pores.

Another thing that made Columbus different is that he not only performed dissection on corpses, but he also performed vivisection in order to get some further information on physiology (though the trauma of vivisection can disrupt normal physiological activity).[33]

One rather singular finding of Columbus was on the female clitoris. Columbus contended that he was the first to connect it to a source of sexual pleasure for females. There is some controversy among contemporary historians of science on whether Columbus thought he was enlightening both males *and* females on this finding, hmmm. . .[34]

Gabriello Fallopio claimed the credit for himself, but did not publish his results until 2 years after Columbus—so the claim of being first belongs to Columbus.[35]

---

[33] One interesting twist on this is the partnership between Columbus and Michelangelo on anatomy for the later's painting and sculpture. See: Duke Pesta, "Resurrection Vivesection: Michelangelo Among the Anatomists" *The Sixteenth Century Journal* 45.4 (2014: 921–950.

[34] Mark D. Stringer and Ignes Berker, "Columbo and the Clitoris" *European Journal of Obstetrics and Gynecology and Reproductive Biology* 151.2 (2010): 130–133; Elizabeth D. Harvey, "Anatomies of Rapture: Clitoral Politics: Medical Blazons" *Signs* 27.2 (2002): 315–346; Cathy McClive, "Masculinity on Trial: Penises, Hermaphrodites and Uncertain Bale Body in Early Modern France" *History Workshops Journal* 68 (2009): 45–68; Michael Stolberg, "A Woman down to her bones: The Anatomy of Sexual Difference in the Sixteenth and Early Seventeenth Centuries" *Isis* 94.2 (2003): 274–299; and Thomas W. Laquer, "Sex in the Flesh" *Isis* 94.2 (2003): 300–306.

[35] Katherine Park, "The Rediscovery of the Clitoris: French Medicine and the Tribate 1570–1620" in David Hillman and Carla Mazzio, eds. *The Body in Parts: Fantasies of Corporeality in Modern Europe* (New York: Routledge, 1997): 171–193.

The European Renaissance medical world had the advantage of exposure to the ancient Greek and Roman physicians as well as the extensions of the same by the Islamic Arab Golden Age. These two complementary eras set the stage for a radical change in medical methodology that rode the path that affected scientific methodology, in general. This would create the foundation of the next three centuries continuing into the present era.

The European Tradition and the Causes for Change

These two historical eras just cited (the Islamic Golden Age, and the European sixteenth century) created the conditions by which major methodological changes in the way basic science and medicine progressed. There are several causes for this change. I will focus upon two: coming from the *genetic order* and from the *logical order*.[36] In the first case the genetic order focusses upon recitation of events as they happen. Aristotle talks of people watching a house being built. One day you see them breaking ground. A month later, maybe the foundation is set, then the house frame, and so on. The genetic order focusses upon the events as they occur in time.

In the logical order one concentrates upon the *reasons why* events occur as they do. In the house example, if the walls aren't plumb, then maybe it's because the wood used is too *green* and should have been dried and seasoned, first. Or if the house is not well-built, perhaps it's because the architect made poor blueprints—architectural plans for the whole. In the case of our examination, I will highlight a few key figures and their ideas that shaped general science and then physiology, in particular, to catch a glimpse of why the seventeenth century constituted a new paradigm. So, let's begin with one popular take on the genetic order of the scientific "revolution" of the seventeenth century.

The Genetic Order for Change

The genetic account that I have chosen to highlight is popular among some.[37] This story starts like this: The Venerable Bede in eighth century England noted that the date of Easter, that is celebrated on the Sunday following the ecclesiastical full-moon on or after 21 March, was moving later—3 days later according to Bede's

---

[36] I am following Aristotle here who in *Parts of Animals* I.1 describes the coming-to-be of a house from the logical order (presumably the builder) and from its material and efficient causes (the point of view of a neighbor watching the house go up)—the genetic order.

[37] Such as Thomas S. Kuhn, *The Copernican Revolution* (Cambridge, MA: Harvard University Press, 1957).

calculation.[38] Roger Bacon at the beginning of the thirteenth century found the error to be 7–8 days.[39] The reason that this is a problem is that the Christian feast of Easter is one of the two principal holidays in the Church year (along with Christmas). Having Easter in the early spring is very important. Easter is about the Resurrection of Jesus from his death on the cross. The metaphorical connection of an early Easter holiday with new flowers and budding trees (which signify *new* life) connects visually with the feast of Resurrection. For adherents, seeing new life in the ground springing forth, parallels the theological message. If Easter were to continue to inch forward year-by-year, then before long, we'd be celebrating Easter in the midsummer when flowers have grown dark and died. The parallel with nature ($phusis_{1+2}$)[40] would no longer work. From a promotional perspective for common folk, it would be a disaster. The "creeping" of the calendar was indeed a serious concern for Christians, in the long run.

There the controversy stops until 1475 onwards when the question is picked-up again, and after some internal debate, was given direction by Pope Gregory XIII (whose reign was 1572–1585). Under Gregory, the project was, itself resurrected, to be completed for all Catholic European countries by 1582. This created a need to enable mathematicians and astronomers to get together and fix this problem.[41] What resulted was more than the Pope had bargained for. Astronomers found new interest in Copernicus' book *De revolutionibus orbium coelestium* that was published on his death bed in 1543. The generating thesis set out in book one of that work was that Aristarchus of Samos' (310–230 BCE) hypothesis was correct: that the earth was round, rotated on its axis, and orbited around the sun. This got rid of the phenomenon of retrogression motion of the planets in the Zodiac along with looped-motion of the planets about the elliptic.[42] In the "old system" based upon refinements of Ptolemy's *Almagest*, artificial solutions such as employing *the epicycle* and *the deferent* were used to "save the phenomena." These two devices *did* save the phenomena (meaning that they offered an account that would account for the movement of the planets). But *the cost* of saving the phenomena was the creation of ad hoc additions to the general account that had no other rationale except that they seemed to "solve the problem." In the language of Aristotle's *Posterior Analytics* II.1, the *dioti* was not organically connected to the *hoti* so that the *ti esti* was totally artificial

---

[38] The vellum manuscript of Bede's calculations, Sp. Coll MS Hunter 85 (T.4.2), is in the Special Collections Department, Library, University of Glasgow, Hillhead Street, G12 8QE, Scotland, United Kingdom.

[39] Lynn Thorndike, "The True Roger Bacon I" *The American Historical Review* 21.2 (1916): 237–257, cf. Ari Ben-Menahem, *Historical Encyclopedia of Natural and Mathematical Sciences* (Cham, Switzerland: Springer, 2009): 863.

[40] In Part Two of this book, I will be using "nature" instead of the Greek "*phusis*."

[41] Two discussions on this of note are: Mark A. Peterson, "Mathematics Old and New" in *Galileo's Muse: Renaissance Mathematics and the Arts* (Cambridge, M.A.: Harvard University Press, 2011): 237–254; and Susan B. Puett & J. David Puett, "Astronomy & Time Reckoning" from *Renaissance Art and Science @ Florence* (University Park: Penn State University Press, 2016): 147–181. In short, the fix was the creation of a "leap year" by adding an extra day every 4 years.

[42] Kuhn, ch. 2.

except under contorted nature$_2$ terms (see Chap. 3). In contradistinction to this *complicated ad hoc* solution, the Copernican hypothesis could describe the same phenomenon by changing the perspective from a geo-centric perspective to a heliocentric perspective,[43] and it could do so in a simpler way. Why would simplicity be relevant?[44] There are many answers here, but they are all nature$_2$ because they are non-empirical. They fall into different realms: (1) Metaphysical, (2) Aesthetic, and (3) Pragmatic.

In the first case, if the structure of nature$_1$ as a material entity is *actually simple*, then to have a theory that is also simple, satisfies the criteria of *correspondence theory of truth*.[45] In this case, since *n/Nature is, itself, simple,* then any account of n/Nature, should mirror that underlying simplicity.

In the second case concerning the aesthetic, we are drawn to the Aristotelian and Galenic claim that "Nature does nothing in vain." This implies an underlying beauty through *functional symmetry*. To observe a functional system whose ability to produce is dependent upon an operation that is constructed in the most efficient (simple) manner possible, evinces a reaction of being confronted by an ingenious, efficient system. This reaction is an experience with symmetry (that can yield *elegance*). This experience (as I have argued in the past) is akin to an aesthetic experience and creates a protective action-response on the part of the observer.[46] This protective reaction inclines the observer to cognitively judge the simple system as

---

[43] Kuhn, ch. 4.

[44] There is quite an extensive literature on the question of simplicity. For an overview of some of the more important contributions see: W.V.O. Quine, "On Simple Theories of a Complex World" *Synthese* 14 (1963): 103–106; Nelson Goodman, "The Logical Simplicity of Predicates" *Journal of Symbolic Logic* 14 (1949):32–41; _____. "New Notes on Simplicity" *Journal of Symbolic Logic* 17 (1952): 188–191; _____. "Axiomatic Measurement of Simplicity" *Journal of Philosophy* 52 (1955): 109–122; Richard Rudner, "An Introduction to Simplicity" *Philosophy of Science* 28 (1961):109–119; S.F. Barker, *Induction and Hypothesis* (Ithaca, New York: Cornel University Press, 1957): 91–105; Karl R. Popper, *The Logic of Scientific Discovery* (New York: Routledge, 1959):6–9; H. Reichenbach, *Experience and Prediction* (Chicago: University of Chicago Press, 1938): sect. 42; F. Attneave and R. Frost, "The Determination of Perceived Tridimensional Orientation by Minimum Criteria" *Perception and Psychophysics* 6.6B (1969): 391–96; S.W. Blackburn, "Goodman's Paradox" in N. Rescher, ed. *Studies in the Philosophy of Science: American Philosophical Quarterly* 3 (1969): 128–142; H. Putnam, "Reductionism and the Nature of Psychology" *Cognition* 2.1 (1973): 131–146; W.C. Salmon, "Statistical Explanation" in W.C. Salmon, ed. *Statistical Explanation and Statistical Relevance* (Pittsburgh: University of Pittsburg Press, 1971): 29–88; Eliot Sober, *Simplicity* (Oxford: Oxford University Press, 1975); M. B. Willard, "Against Simplicity" *Philosophical Studies* (2014): 165–181; James Woodward, "Simplicity in the Best Systems Account of the Laws of Nature" *British Journal for the Philosophy of Science* 65.1 (2014): 191–213; Andrew Brenner, "Simplicity as a Criterion of Choice in Metaphysics" *Philosophical Studies* 174.11 (2017): 2687–2707; Lukas Bielik, "Explanation, H-D Confirmation, and Simplicity" *Erkenntnis* 83.5 (2018): 1085–1104; and Nicholas Maxwell, "Non-Empirical Requirements Scientific Theories Must Satisfy: Simplicity, Unity, Explanation, Beauty" from *Karl Popper, Science, and Enlightenment* (London: UCL Press, 2017): 125–142.

[45] Boylan (2008): chapter 4.

[46] For an example of this see: Michael Boylan, "Worldview and the Value-Duty Link" in Michael Boylan, ed. *Environmental Ethics,* 3rd ed. (Oxford: Wiley-Blackwell, 2022): 114–129.

better because of the positive aesthetic reaction to the symmetry. (This approach also supports the maker/artifact dynamic—see Introduction.)

The third case of pragmatism is that a simpler systems model is easier to understand and to use in the practical world. For Copernicus, this took some time in order to get navigation standards for vessels moving about the Caribbean (and farther) to get to (and then surpass) the established norms of the Ptolemaic norms that existed *before*. But with a little faith in the overall model (based upon simplicity, cum pragmatic), there was a new business model that might incline practitioners to give the new (simpler) model a chance because *if* it turns out to be *just as workable* or even *more workable* than the old Ptolemaic model (that is much more complicated), then the new, simpler model should be chosen because it's easier to teach and to use.

Thus, in the end, the advantage of *simplicity* that promotes a sensibility of symmetry on the part of the researcher, is that sea-traders will more money (and the military-interested Europeans more colonies for expansion).

Each of the three understandings of simplicity generated from the Copernican hypothesis seem to generate positive outcomes. So, what's the problem here? It would seem that if the Roman Catholic Church, via Pope Gregory XIII, who wanted to know the *nature*$_1$ dynamic that was moving Easter forward in the Julian Calendar so that the Calendar might be corrected (in order to keep Easter in the early spring). It would seem that Pope Gregory XIII would welcome any new theories that might give concrete data to set forth a new calendar based upon astronomical *facts*—an *explanans* that was not artificial, but organic.

However, again, theology got in the way. Moving the center of the solar system (understood as the center of the entire universe at the time) to the sun and away from the earth, meant that *we* (the people on earth) might not be the center of *everything*, after all. What might this entail? Well, if the earth was just one of many planets and not the center, then there might be many planets and they might be inhabited by *human-like* creatures. They might be just like earthlings (i.e., sinful and needing of redemption). This might mean that Jesus would have to go to each of these planets and be crucified on each one![47] This was very threatening to the Roman Catholic Church as it struggled for its imprimatur as the single voice that expressed religious truth.[48] *Single Truth*—which is broadly explanatory, is an interesting concept. It possesses symmetry (which I have argued earlier is often used as a *nature*$_2$ justification).

---

[47] For a discussion of this see Michael Boylan, "Henry More and the Spirit of Nature" *Journal of the History of Philosophy* 18.4 (1980): 395–405.

[48] This sort of tension is highlighted in Edward Gibbon, *The Decline and Fall of the Roman Empire* (London: Everyman, 2010 [1776–1789]): volume 6. He's talking here about a fight to be the "mouthpiece" of Catholic Christians between the Latins and the Greeks between 1200 and 1400 BCE. (Note that Martin Luther posted his 95 Theses on the door of the Castle Church in Wittenburg, Germany on October 31, 1517. This was indeed a time in which practical (sociological) effects of actions upon the "numbers of the faithful," from the Roman Catholic point of view, was high in the consciousness of the leaders of the Roman Catholic Church.

The nature$_2$ justification, in this case is truth of religion, is here seen as having sway in the natural realm from the supernatural realm. Whether or not religion has a window to truth in the supernatural realm is separate from whether that asserted supernatural truth has any natural jurisdiction. We found that this was often thought to be the case in Part One of this book. The question is here, whether those on the vanguard of science and medicine in Part Two, also hold this to be the case. However, when Pope Gregory XIII pushed for astronomers to correct the difficulties with the calendar, some cosmological assumptions also came under scrutiny. One such possibility is if the earth is not the center of the universe, but merely one planet (among many) revolving around the sun, then this could lead to the possibility of other planets that revolve around the sun, also being inhabited by rational beings. Cosmological speculations such as these brought forth a re-investigating into truth itself. This is what upset the Roman Catholic Church.

For example, in natural world, there are certainly instances of "single, absolute truth." For example, $2 + 2 = 4$ is a *single, absolute truth*.[49] Following Augustine, if there are single, absolute truths in mathematics, then why not in theology?[50]

If we follow this line of argument, then if Jesus uniquely died on the cross for the sins of humans on earth for his own epoch, and for all time, going forward, then this plurality of possible worlds creates a contradiction in the given account. This is because the conceptual argument (dogma) is structured upon a model in which there is only one planet with living "rational" creatures. What would follow if there were multiple worlds with living "peoples"? Is this a "show stopper"? Maybe or maybe not. But (at the very least) it is an easier "sell" to the regular folk to say that the earth (perhaps the flat earth to most of them) is the center of everything that *is*. Whatever happens *here* is all that *is* under heaven. The rest of the cosmos is created with *us* in mind—because, after all, it's all about **us!**

When this model comes under question, this can be very destabilizing to a religious tradition that seems to require the centrality and uniqueness of the earth as the center of the universe.[51] When the scientists, who were recruited for a very practical task (revising the calendar), move beyond their narrow charge, then this creates a conflict between science and religion. Since the religious entities (Christianity, Judaism, and Islam) are connected to the political power of the countries of the West

---

[49] However, in the nineteenth century, philosophy of mathematics (following set theory) abides by "base" by which the counting system is defined. In the West, 10 has been the norm, but it could easily be another number, but it rarely goes down below "3" so that $2 + 2$ is safe, but Kant's standard reference of "$7 + 5 = 12$" is not so secure.

[50] *De libero arbitrio,* II.17. "1. Mathematical number gives form%2D%2DA. 2. Body and life are nothing without form%2D%2DA. 3. [Form is a 'good' for life and body]%2D%2DA. 4. God is truth%2D%2DF (previous argument). 5. [Truth supersedes number (meaning it gives to number its properties)] %2D%2DF (earlier argument). 6. *God gives form--1, 4, 5.* 7. All body and life are made good by God%2D%2D2, 3, 6."

[51] Not only to the Christian tradition, but in the West (Europe, North Africa, and the Middle East) the Jewish and Muslim traditions were also invested in viewing the earth as the only entity that was populated by rational agents in the universe.

(Europe, North Africa, and the Middle East) they have the potential power to regulate science and its problematic consequences: put the guys in jail or kill them.

But it isn't that easy to put the toothpaste back into the tube once it is removed. You can put Galileo under house arrest, but can you suppress his writings that are secreted out?[52] The Copernican Hypothesis will be addressed and many of Aristotle's conclusions on circular motion (being primary) and the proportionate speed upon which (for Aristotle) objects of different weights return to their *natural* place were being questioned. For institutions that depend upon maintaining the status quo (like the Roman Catholic Church), this is revolutionary stuff—and extends beyond astronomy.[53] Thus, ends the genetic order.

The Logical Order for Change

The logical order best begins with René Descartes, who was born in La Haye en Touraine, France in 1596. His father was a member of Parliament in Brittany. He received his BA in law from the University of Potiers in 1616. He died of pneumonia in the tutelage of Queen Christina of Sweden February 11, 1650. In his work, *Meditations on First Philosophy,* Descartes brings forth a strategy of radical doubt.

Argument 6.1: *Meditation* I (Adams and Tannery VII, 19–22)[54] Minimum epistemological assumptions require what is known be accepted and all else doubted.

1. Madness often deceives—F[act]
2. Dreams often deceive—F
3. Man is often deceived—1, 2
4. A priori knowledge (such as $3 + 2 = 5$ and the sum of the interior angles of a triangle are equal to two rights) is always true—F
5. Sensory knowledge appears to be true of things—F
6. If premise #5 is true, then there is a benevolent God—F
7. If premise #5 is false, then there is a deceiver god—F
8. Premise #7 is a weaker proposition than premise #6—F
9. [Epistemological system building demands starting with the weaker before the stronger]—F
10. Epistemological system building demands we assume a deceiver god—5–9
11. Minimum epistemological assumptions require what is known be accepted and all else doubted—3, 4, 10

He continues with an *a priori* argument that follows this method.

---

[52] Some suggest that in Galileo's case, his daughter may have helped secret out his writings. For a fictive account of what *might have happened* see: Dava Sobel, *Galileo's Daughter: A Historical Memoir of Science, Faith, and Love* (New York: Bloomsbury USA, 2011).
[53] One of the extensions is in physiology, viz., in the circulation of blood—to be discussed shortly.
[54] René Descartes, *Oeuvres*, translated by Charles Adam and Paul Tannery *VII* (Paris: Léopold Cerf, 1897–1913).

Argument Two: *Meditation* II (A&T VII, 25–28) Deceiver or not I exist.

1. An evil deceiver tries to trick me or not—F
2. [I] think or "thinking is"—F
3. [All verbs require subjects and objects]—F
4. *I* exist without a deceiver—2, 3
5. If there is a deceiver, he tries to trick *me*—A[ssertion]
6. I exist with a deceiver—3, 5
7. Deceiver or not I exist—1, 4, 6

In these two arguments Descartes sets out his famous *a priori* epistemological argument that argues that, at the very minimum, he knows that he exists from the very self-reflection act that occasions the question. This argument has often been termed "The Cartesian Circle" because of various assumptions that are set out without justification and, for the claim that the conclusion, follows from a radical skepticism point of view.[55] For example, Descartes claims that he wants to doubt everything (when he means he is doubting some empirical input). In the argument from *Meditation* I, premise #4, he believes that a priori knowledge is beyond doubt. But how does he *know* this? This is especially troublesome since he pretends to doubt everything.

In the second argument from *Meditation* II, the a priori tenet that he accepts without justification is that "all verbs require subjects and objects," premise #3. There is also the acceptance of the fact of his being *aware* of thinking means that this activity is actually occurring. (If an awareness of thinking were "self-referential," then perhaps the illusion of existence might be part of the *created reality*. One might consider "meta-realities" in this context.)[56]

If we give Descartes premise #3, then the fact that the subject is aware of their individual thinking (in this case doubting), means that there must either be a doubter (since the verbal gerund requires a subject) and thus, the subject exists, or if there is an evil genius at play, then the evil genius is tricking something (since the verbal gerund also requires an object). In either case it seems that the subject's own existence has been demonstrated, *a priori*. And since *a priori* knowledge has been asserted to be the gold standard in the first argument, then it would seem that here, in the second argument, Descartes has indeed proven his own existence—even in the face of the radical doubt that was set out as the background condition for the thought experiment.

---

[55] A quick overview of some of the key dimensions of this famous argument can be found in Alan Gewirth, "The Cartesian Circle" *Philosophical Review* 1.4 (1941): 368–395; cf. "The Cartesian Circle Reconsidered" *The Journal of Philosophy* 67.19 (1970): 668–685; cf. Lynn E. Rose, "The Cartesian Circle" Philosophy and Phenomenological *Research*, Sep., 1965, Vol. 26, No. 1 (1965): 80–89; and Gary Hatfield, *"The Cartesian Circle"* in Stephen Gaukroger, ed. The Blackwell Guide to Descartes' Meditations. (Oxford: Blackwell, 2006): 122–141.

[56] One approach along this track is David Chalmers, "Structuralism as a Response to Skepticism" *Journal of Philosophy* 115.12 (2018): 625–660.

The premises that are most problematic in the two arguments are: Argument One, premise 4, and Argument Two, premise 3. These two premises are the lynchpin for the overall inferences. But they could be wrong. Descartes presents no argument for their being true. This is the foundation of the "circularity" claim.

Circular arguments are defective because they assume what they intend to prove. This means, in effect, that a logical inference (the result of a valid deductive argument) is, instead, a mere assertion (the say-so of the speaker, only).[57] This does not qualify as a valid *a priori* deductive argument as Descartes has asserted.

Nonetheless, even if the arguments are circular (as I believe they are), they do intend to advance understanding of nature$_{1\&2}$. And it must also be admitted that philosophy is chocked full of *a priori* arguments[58] without critical evaluation of the premises.[59] Though, Descartes' execution of his design may have been flawed (circular), his *intent* was not corrupt, and its impact on the history of philosophy and to the history of critical philosophy of science (which my book is also an example) is immense. Descartes changed those after him in philosophy to be more self-reflective in the manner of their exposition and their acceptance of new, radical hypotheses—such as Copernicus, Harvey, and Newton. This, alone, catapults Monsieur Cartesian[60] to the role of a major figure in the way we evaluate arguments concerning Nature.[61] Descartes also did original research on anatomy and physiology.[62] Because Descartes advocates skepticism along with critical empiricism when evaluating Nature, his work can be viewed as a combination of Nature 1 and 3, nature$_{1+3}$.[63]

But we cannot stop there when discussing Descartes. Not only was Descartes a pivotal figure in turning philosophy toward critical skepticism on the truth of the various premises presented in arguments about nature, but he also fulfilled an equally important role in the history of mathematics as he executed a reduction of

---

[57] For more information on this see Michael Boylan, *The Process of Argument: An Introduction* (London and New York: Routledge, 2020): Ch. 2.

[58] *Objections and Replies urged against the Meditations on First Philosophy*, reply #1 against the theologians; I believe this influenced another philosopher who advocated for a more mitigated form of skepticism, David Hume, in the last line of his *Enquiry* where he advocates that such books (like the Scholastics who do not engage in critical reasoning of their assertions (especially concerning input of math and empirical input) should be tossed into the fire.

[59] Descartes, *Rules for the Direction of the Mind*, R. 1: The end of study should be to direct the mind towards the enunciation of sound and correct judgments on all matters that come before it (tr. Haldane and Ross).

[60] This was the name that René Descartes was known by in his correspondence, see Boylan (1980).

[61] Descartes also presents arguments for the existence of God in *Meditation* III (cosmological argument) and *Meditation* V (ontological argument)—but in *Meditation* VI, he makes the case for empirical engagement with nature in the critical fashion set out in his overall design to the *Meditations*.

[62] René Descartes, *Treatise of Man*, Thomas Steele Hall, ed. and tr. (Amherst, NY: Prometheus, 2003).

[63] *Phusis*$_3$ (referred to as n/Nature in Part 2) comes from the skepticism while *phusis*$_1$ comes from his original work in anatomy and physiology along with his careful parsing of the roles of empiricism and a priori argument.

geometry to arithmetic. In the ancient Greek World, geometry was primary and arithmetic was derivative.[64] This lent a very empirical bent to the study of arithmetic.[65]

Descartes changed all that. Through his innovation of Analytic Geometry and the Cartesian Coordinate System with x, y, and z axes, arithmetic (through advanced algebra) became the fundamental entity, which more exactly depicted the regular geometric solids, and gave illumination to symmetry via the structure of equations. By changing the way we understand what is primary (algebra over geometry), Descartes landed another laurel, as a "game-changer" pioneer in mathematics as well as in philosophy.[66] Because mathematics, as a representative of Nature falls into the Nature$_2$ category, as setting out a *super-nature* category of numbers and their relations, that can better describe n/Nature than experience alone, Descartes becomes our first highlighted thinker who can be categorized as n/Nature$_{1+2+3}$. This is, indeed, a high accomplishment.

Still, we have carryovers from the ancient world in Descartes investigations into biology.[67] Descartes was engaged in actual physiological investigations, but often this was in the context of teleological posits that framed his accounts of these first-hand empirical studies.[68] As the reader may recall, in Part One, the function of the mind (wherever it was said to have been located—either in the brain as per the Hippocratic Writers or in the region of the heart à la Aristotle) communicated with the parts of the body (to move limbs, etc.) via *pneuma.* Descartes continues along these lines with his use of animal spirits in *Treatise on Man* (unpublished, circa

---

[64] Edward A. Maziarz and Thomas Greenwood, *Greek Mathematical Philosophy* (New York: Frederick Ungar, 1968): Part Two—connecting to Plato.

[65] I am reminded on the way the Montessori method teaches children arithmetic in a very "hands-on" empirical approach. See: Chiara Piroddi and Agnese Baruzzi, *The Montessori Method: Numbers* (N.Y. Union Square Kinds, 2018).

[66] See: Victor J. Katz and Karen Hunger Parshall, "From Analytic Geometry to the Fundamental Theorem of Algebra" in *Taming the Unknown: A History of Algebra from Antiquity to the Early Twentieth Century* (Princeton, N.J.: Princeton University Press): 247–288; and Elena Anne Marchistto, "The Theorem of Pappas: A Bridge between Algebra and Geometry" *The American Mathematical Monthly* 109.6 (2002): 497–516.

[67] A wonderful open-source article on Descartes' biological/medical investigations can be found in Annie Bitbol-Hespéiès, "Medicine, Method, and Metaphysics: Tradition and Innovation in Descartes' Medical Work from the Writing of L'Homme to its Posthumous Publication as *The Treatise on Man,* 1664 Together with *La Description du Corps Humain*" https://www.academia.edu/41769631/*Medicine Method and Metaphysics Tradition and innovation in Descartes medical works from the writing of Homme to its posthumous publication as The Treatise on Man in 1664 together with La Description du corps humain (The Description of the Human Body).*

[68] For an account of the first-hand physiological studies see: Annie Bitbol-Hespériès, "Cartesian Physiology" in Stephen Gaukroger, John Schuster, and John Sutton, *Descartes Natural Philosophy* (New York and London: Routledge, 2000): 349–382. One example of the teleological carry-over was the use of teleology in explaining the mitral valve of the heart and the process of sensation, see: Peter M. Distelweig, "The Use of *Usus* and the Function of *Functio*: Teleology and its Limits in Descartes Physiology" *Journal of the History of Philosophy* 55.3 (2015): 377–399.

1630–1632).[69] This is a vague account of nature$_1$ that asserts the existence of a material account, but giving such a general overview that it borders upon nature$_2$.

However, mitigated skepticism of these former accounts follows from the general skepticism of the geocentric hypothesis and the *Almagest* account of the same, that prompted skepticism of other long-held beliefs—such as the role of *animal spirits*. One example of this is the Englishman, Thomas Willis, who began to speculate further on *how* the animal spirits might physically fulfill the gap between mental intention to raise my arm, for example, and the actual arm raising. This might be characterized as one beginning of nerve-theory.[70]

Thus far, we have seen that the critical examination of Copernicus and his heliocentric hypothesis in *De revolutionibus* (1543)—particularly by Galileo in the *Dialogue* (1632) and *Two New Sciences* (1638)—caused considerable upset, particularly within the Roman Catholic Church.[71] Galileo was put under house arrest. Others, like Descartes, often tried to obscure their authorship because of fear of reprisals.

But in the end, despite the threats of censure, the broad movement toward rethinking previous donées continued. This stimulated an attitude that some of the assumptions of the past—such as Aristotle's doctrine of natural place (to explain why material objects *fall* or flames burned *upward*), should be re-examined. Sometimes this re-examination tried to situate the new account within Aristotelian physics (such as Galileo who tried to save circular motion (Aristotle) from those advocating sub-lunar rectilinear motion (the new post-Copernicus crowd)). But others wished to step outside the existing paradigm all together. Descartes, through his philosophical skepticism (Nature$_3$) and his reliance upon *a priori* deductive logic and analytic geometry (Nature$_2$) was a leader in combining all the senses of nature as important tools moving forward. Along this path was another initiator, Isaac Newton.

The second individual who I wish to bring forward within the province of the *logical order*, falls on the coattails of Descartes, and that is Isaac Newton. Newton was born on Christmas Day 1642 at Woolsthorpe Manor in Woolsthorpe-by-Colsterworth in Lincolnshire, England. His father (of the same name) had died 3 months before. His early education was at The King's School in Grantham. In 1661 he went to Trinity College, Cambridge. He was an undistinguished student, and went back home. By 1667 he returned to Cambridge: having worked on ideas on the calculus, optics, and gravity—all by himself. After receiving his MA, he became Lucasian Professor in 1669. He died March 20, 1727 of complications due to gallstones.

---

[69] See the brief discussion in Dennis L. Sepper, "Animal Spirits" in *The Cambridge Descartes Lexicon*, ed. Lawrence Nolan (Cambridge: Cambridge University Press, 2015): 26–28.

[70] For one account of this see: Louis Caron, "Thomas Wilis, the Restoration and the First Works of Neurology" *Medical History* 59.4 (2015): 525–553.

[71] A general treatment of this can be found in Amos Funkenstein. *Theology and the Scientific Imagination: From the Middle Ages to the Seventeenth Century.* 2nd ed. (Princeton, N.J.: Princeton University Press, 2018).

Now it is true that Descartes revolutionized mathematics by making algebra primary over geometry. This started the direction toward a new Nature$_2$-driven worldview. It was not a driven by a *god-controller*, but instead by another *non-empirical entity*: mathematics. It is often said that among philosophers, they either are drawn to mathematics (like Plato) or are not drawn to mathematics, but to empiricism (like Aristotle). I think this is a reasonable "over simplification" (since this author is drawn to both), but there are contexts in which this is not relevant. However, when considering Nature—especially medicine—it does seem to be a major distinguishing factor.[72]

The application of an underlying mathematical view to Nature can create various competing material visions of how things operate together (what we now call "physics"). Descartes mathematical vision that unites algebra with geometry headed toward his *vortex theory* that has a more dynamic lowest-micro-level built upon *properties*, than that of the ancient atomists which became the model for Newtonians.[73]

Isaac Newton's *The Principia: Mathematical Principles of Natural Philosophy*[74] offers another development along the lines of Descartes' vision that natural philosophy required more development through a non-empirical methodology, viz., mathematics. This Nature$_2$ approach was given a rather broad application in describing the world, at large.[75] Newton's version of the calculus[76] allowed for a quantitative view of motion along with particles that acquired their own definition as "mass" that was acted upon by various forces.[77] This combination created the model of *directed force containing matter* and opened the door for the integration of power and motion. It formed the basis of his theory of gravity that incorporated Euclidean

---

[72] One book that examines some of these dimensions is Eduard Jan Dijksterhuis, *The Mechanization of the World Picture: Pythagoras to Newton*, tr. C. Dikshoorn (Princeton, N.J.: Princeton University Press,1986 [1950]).

[73] For a discussion of this see Dan Garber, *Descartes Embodied: Reading Cartesian Philosophy through Cartesian Science* (Cambridge: Cambridge University Press, 2001).

[74] I will make reference to the edition edited by I. Bernard Cohen and Anne Whitman (Berkeley: University of California Press, 1999 [1687]). I have also been influenced by I. Bernard Cohen's *Introduction to Newton's 'Principia'* (Cambridge, MA: Harvard University Press, 1978).

[75] Some good introductory references on the breadth of these applications can be found in: Rupert A. Hall, *Isaac Newton: Adventurer in Thought* (Oxford: Blackwell, 1992); Rob Illiffe, *Newton: A Very Short Introduction* (Oxford: Oxford University Press, 2007); I.B. Cohen and R.S. Westfall, *Newton: Texts, Backgrounds, and Commentaries, A Norton Critical Edition* (New York: Norton, 1995); I.B. Cohen and G.E. Smith, *The Cambridge Companion to Newton* (Cambridge: Cambridge University Press, 2002).

[76] Of course, there was also Leibniz's version of the calculus, independently created, which was focused on philosophical problems using *infinitesimals* to ground his notion of integration and *differentials* concerned with the problem of finding the rate of change of a function with respect to the variable on which it depends. He used these to develop a theory of material change (*dynamics*) based upon potential energy and kinetic energy. For more on this see: *The Early Mathematical Manuscripts of Leibniz*, ed. J. M. Child (Mineola, NY: Dover, 2005).

[77] Jessica Wilson, "Newtonian Forces" *British Journal for the Philosophy of Science* 58.2 (2007): 173–205.

notions of linear space (which Descartes also used in his Cartesian Coordinate system) within a context of *real,* independent time. Since the writings of Aristotle—space, time, and motion were connected (*Physics* IV, 10–14). Space was exterior to the observer. Motion occurred as bodies moved in space. Time was an interior construct created by the observer (or groups of observers) to measure the locomotion of the bodies.[78] However, without quantitative measuring devices, this meant that observers were forced to rely on intuitive standards of relative speed of motion as "fast" or "slow." (This gave rise to the qualitative messiness that permitted observers such as Zeno of Elea to create his paradoxes—see Chap. 1).

But Newton was interested in quantifiable, universal laws (such as those put forth in mathematics). There is, of course, a question for all *physicists* (using the term in a modern usage), of whether time is *real* (as opposed to conventional) and whether it can be separated from space. The verdict of the post-World War I modern era is that space-time is *real,* but that it is a mistake to try to separate space and time, except in an artificial way.[79]

Since the concentration of the era was the motion of the planets around the sun, a key area of inquiry in this historical moment was why the orbits were not circular (as Aristotle and his Ptolemaic followers had suggested—albeit from their perspective within a geocentric model). Was this because there were minor perturbations within the circular motion model or were the orbits fundamentally different for a concrete, material reason? Bernard Cohen asserts that the *Principia* had two principal goals: (a) the first was to offer the conditions under which Kepler's laws of planetary motion could be shown to be accurately true by using Newton's physical constructs of force and mass along with the mathematical integration that might describe the motion, via his calculus; (b) the second was to show how Kepler's laws had to be modified to show *why* observed planetary observations would fit Newton's own cosmological cast of characters interacting under a new holistic force called *gravity* (which was a force that came from extent of planetary mass (such as observed in Jupiter and Saturn).[80]

Various actors entered this debate putting their bets on some version of the inverse square law. One prominent example of this is Robert Hooke, who was much

---

[78] There are other interpretations on these passages. For an overview, see: Tony Roark, *Aristotle on Time* (Cambridge: Cambridge University Press, 2011): particularly chapters 5, 8, and 9.

[79] Two contemporary accounts that try to compare Newtonian mechanics with some contemporary slants on cosmology are: James Owen Weatherall, "What is a Singularity in Geometrized Newtonian Gravitation?" *Philosophy of Science* 81.5 (2014): 1077–1089; James Owen Weatherall, "Are Newtonian Gravitation and Geometrized Newtonian Gravitation Theoretically Equivalent?" *Erkenntnis* 81.5 (2016): 1075–1091; and Naresh Dadhich, "Einstein as Newton with Space Curved" *Current Science* 109.2 (2015): 260–264.

[80] Isaac Newton, *The Principia,* tr. I. Bernard Cohen and Anne Whitman (Berkeley: University of California Press, 1999): 20–22; cf. I. Bernard Cohen, "Neton's Theory v. Kepler's Theory and Galileo's Theory: An Example of Difference between a Philosophical and a Historical Analysis of Science" in *The Interaction between Science and Philosophy,* ed. Yehuda Elkana (Atlantic Highlands, N.J.: Humanities Press, 1974): 299–388.

interested in the possible effects of the inverse square law.[81] Hooke threw his hat into the ring citing the inverse square law as being the explanation for perturbations and the evidence of the elliptical orbits (under the hypothesis of the helio-centric hypothesis). Newton countered Hooke with a much more mechanical account that was backed-up by the more detailed concepts of mass and force that combined to offer an account of gravity that was supported by a mathematics that could integrate various data points to provide a mathematically deductive account of motion. In the end, this more complicated account resonated well (together that the symmetry of such a complex whole that gave it an aesthetically powerful appeal).[82] However, because gravity was an invisible force that was empirically difficult to measure, as such (in the detailed way that Newton's mathematics predicted), there were skeptics. They claimed that Newton was appealing to vague ideas of electricity or religious forces or powers of the discredited alchemy.[83] This would situate Newton as being accused of using one form of Nature$_2$, mathematics, to give artificial support to another form of Nature$_2$, religion.[84] Both avenues are non-empirical. The former is quantitative while the latter is qualitative. To offer a quantitative account to a non-empirically verifiable phenomenon that is qualitative, creates an illusionary sense of epistemological rigor. However, without rigorous empirical testing, this is not acceptable. The skepticism that arose, nature$_3$ (particularly on the three laws of motion) helped give Newton-supporters impetus to solidify the truth of Newton's project.

Kant was born on April 22, 1724 into a Prussian family in Königsberg, East Prussia. His father was a harness-maker. Kant went to the University of Königsberg. Kant lived in Königsberg for the rest of professional life. He died on February 12, 1804 after a long bout with chronic illness.

When we consider the enthymemes within Kant's opus, there are some unexamined posits in Newton's approach that should be set forth—such as space and time. For a more complete account, Newton should have set these out for scrutiny. Newton

---

[81] For a discussion of the inverse square law and its place in "Hooke's Law" on elasticity which states that the amount of displacement or deformation of an object is directly proportional to the deforming force or load that once removed returns the object to its original state (elasticity). Mathematically, this is expressed as the applied force, F equals a constant, K, times the change in length or displacement or deformity, x: $F = Kx$. When the applied force is very great, the elasticity is often higher so that it is sometimes expressed as $F = \sim Kx$. In this case F is referred to as the equal and oppositely directed restoring elastic force—see: Lisa Jardine, *The Curious Life of Robert Hooke* (New York: Harper, 2004).

[82] Note that here, "complexity" rather than "simplicity" yields the aesthetic symmetry. The fact that the complexity goes together so well and accounts for so much is what gives it the power to win the day, cf. Chaps. 8 and 9 on the micro-level accounts in science and medicine. See also Boylan (2022) on "Worldview and the Value-Duty Link": 114–129.

[83] For an account of this, see: Betty Jo Teeter Dobbs, *The Janus Faces of Genius: The Role of Alchemy in Newton's Thought* (Cambridge: Cambridge University Press, 2022 [1991]): ch. 3.

[84] Another account of this comes from William R. Newman, *Newton the Alchemist: Science, Enigma, and the Quest for Nature's "Secret Fire"* (Princeton, N.J.: Princeton University Press, 2019).

demurred. Kant (a great admirer of Newton) offered, what I am claiming to be, the most philosophically coherent presentation of Newton, in the context of the epistemological and metaphysical principles in his first Critique.[85]

Before presenting the first of four critical arguments that I am suggesting interpret Newton within the radical changes that were started in the sixteenth century and found fruition in the seventeenth century, but were *interpreted* by thinkers like Kant in the eighteenth century, I want to go over what I believe to be the central problems that are at the center of this interpretation.

First, there are four key terms (two groups of two) that are necessary to understand Kant's arguments. The first group constitutes the *analytic* v. *synthetic* distinction.[86] An analytic proposition is "internally" true in virtue of its form. For example, in the proposition "All bachelors are unmarried males" if one understands the subject term "bachelors" then they already understand that they are both unmarried and male. Thus, the predicate term does not extend the province of the subject term. The proposition is internally true and really refers only to itself and its own internal rules of expression. Propositions in mathematics are also analytic. For example, in the equation alpha: $7 + 5 = 12$, if one understands the naming system for numbers and what their names represent (e.g., 7 equal units, 5 equal units, and 12 equal units) along with the combination principle of *addition*, and finally what numeric equality means "=," then one is in a position to say that alpha is true, *necessarily*. The necessity (represented in logic as the operator, $\Box$) follows from meeting the rules of the system that are set out in such a way that they will not vary in any other imaginable, possible world. From this we can say: $\Box$ $(7 + 5 = 12)$, which is read: "necessarily, seven plus five equals twelve."

Analytic propositions speak about themselves and the artificial systems that generate them. If one can prove that *those artificial systems* resemble *n/Nature*,[87] then another dynamic will take place. In that event, *n/Nature* will become an esoteric entity that belies questions about its independent material properties (n/Nature$_2$) or searching skeptical questions about its "truth" (n/Nature$_3$). If only analytic propositions were brought forward in the creation of scientific arguments, then we would

---

[85] There are two versions of the Critique—Kritik der reinen Vernunft, 1. Auflage 1781; und 2. Auflage 1787 (Akademie Textausgabe, rpt. Berlin: Walter de Gruyter & Co., 1968). My translations of the text have been influenced by Norman Kemp Smith, *Critique of Pure Reason* (New York: St. Martin's Press, 1929). I will use Norman Kemp Smith's editing of the "A" and "B" editions side-by-side in my presentation of the four critical arguments.

[86] Since this is an important distinction here, I draw attention to some important work on the subject. D. W. Hamlyn, "Analytic Truths" *Mind* 65 (1956): 359–367; Robert E. Gahringer, "Analytic Propositions and Philosophical Truths" *Journal of Philosophy* 60.17 (1963): 481–502; Henry Ruf, "Transcendental Logic: An Essay on Critical Metaphysics" *Man and World* 2 (1969): 38–64; Harold Brown, "Paradigmatic Propositions" *American Philosophical Quarterly* 12 (1975): 85–90; and Michael Otte, "Mathematics, Logics, and Philosophy: The Analytic/Synthetic Distinction in Kant: Bolzano and Peirce" *Logique et Analyse* 225.57 (2014): 83–112.

[87] Recall, that "Nature" speaks of the systematic features of the objects in the world while "nature" speaks of the individual objects within the system.

148   6 The New Paradigm: Changes Leading up to the Seventeenth Century and Beyond

only have arguments that are justified by *assertion* and *inferences from the assertions.*[88]

What is missing is a connection to the outside world—beyond the subject.[89] The way to get there for Kant (in his defense of Newton) are *synthetic* propositions. There are two important parts to synthetic propositions as set out by Kant: (1) A formal requirement, and (2) A connection to the outside world. Let's address these in order.

First, the formal requirement means in the proposition set out, the predicate term must enlarge the subject term in a significant way. For example, if one were to examine the subject term "force" in Newton's second law of motion, one would not come up with its components purely through logical analysis (characteristic of analytic propositions, n/Nature$_2$).[90] If, indeed, F = ma (Force equals mass times acceleration), then this would have to be proven via externalist epistemology by trying it out in the real world.

Second, because of this "expansionist" role of the predicate in the synthetic proposition, (which must be proven in the real world), synthetic propositions *do apply to the real, experiential world* (the second requirement: a connection to the outside world).

Thus, in the analytic/synthetic distinction, the "analytic" is *interior, compressed focused* on creating an esoteric, nature$_2$ perspective while the "synthetic" is *exterior, expansionist focused*, creating a compatible, material nature$_1$ account that is subject to exterior questions and scrutiny, n/Nature$_3$.

Just like Descartes, who embraced all three aspects of exposition (n/Nature$_{1+2+3}$), the causal accounts of Kant (in his role here as expositor for Newton) will also connects to all three facets (eventually).

A second key distinction that must be made is the *a priori* and *a posteriori* mode of knowing. The key element here is what is *prior* and *posterior* to what? The simple answer is *experience*. This is a difficult question, and in the contemporary era falls into the domain of developmental psychology[91] However, that's not the way Kant saw it:

> There can be no doubt that all our knowledge begins with experience. For how should our faculty of knowledge be awakened into action did not objects affecting our senses partly of

---

[88] For further information on these dynamics see: Michael Boylan, *The Process of Argument: An Introduction* (London and New York: Routledge, 2020): ch. 4.

[89] For a discussion of an epistemological, externalist understanding of nature in a modern context see: Michael Boylan, "What is Nature and Why Should I Care" in Michael Boylan, ed. *Environmental Ethics,* 3rd ed. (Oxford: Wiley-Blackwell, 2022): 15–34.

[90] A good example of this is the ontological argument in which proponents (such as Anselm of Canterbury) believed that one could prove the existence of God, merely from a philosophical analysis of the concept "God." Kant demurs in the First Critique, see: Ideal of Pure Reason, section 4: The Impossibility of the Ontological Proof of the Existence of God: pp. 500–506 (Kemp Smith).

[91] There are, of course, many theories of developmental psychology. For an introductory overview see: Patricia H. Miller, *Theories of Developmental Psychology* 6th ed. (New York: Worth Publishers, 2016).

themselves produce representations, partly arouse activity of our understanding to compare these representations, and, by combining or separating them, work up the raw material of the sensible impressions into that knowledge of objects which is entitled experience? In the order of time, therefore, we have no knowledge antecedent to experience, and with experience all our knowledge begins.

But though all our knowledge begins with experience, it does not follow that it all arises out of experience. For it may well be that even our empirical knowledge is made up of what we received through impressions and of what our own faculty of knowledge (sensible impressions serving merely as the occasion) supplies from itself. If our faculty of knowledge makes any such addition, it may be that we are not in a position to distinguish it from the raw material, until with long practice of attention we have become skilled in separating it (*Critique of Pure Reason,* 1787 ed. B 1–2, tr. Norman Kemp Smith).[92]

Kant is distinguishing the genetic and the logical order. Certainly, in the genetic order we begin as infants. We know little except ourselves and our mother. Gradually, more experience yields more knowledge. In the contemporary realm, this is called developmental psychology (as noted earlier).

The logical order is rather different. In this case, we are looking at what principles are necessarily logically prior before embracing other concepts. These prior concepts are logically required because they are needed to be set forth first in order to generate subsequent inferences (see Chap. 3 on Aristotle's notion of *prior* in the *Posterior Analytics*). Setting out what is logically prior to proceed requires a special sort of argument model. Kant calls this the *Transcendental Argument Form.*[93]

This formula creates an unexpected union between the a priori and the synthetic. Normally, we would expect the *a priori* to be connected to the *analytic*. "7 + 5 = 12" is true in virtue of its form, and that form is given to us without reference to

---

[92] Daß alle unsere Erkenntnis mit der Erfahrung anfange, daran ist gar kein Zweifel; denn wodurch sollte das Erkenntnisvermögen sonst zur Ausübung erweckt werden, geschähe es nicht durch Gegenstände, die unsere Sinne rühren und teils von selbst Vorstellungen bewirken, teils unsere Verstandestätigkeit in Bewegung bringen, diese zu vergleichen, sie zu verknüpfen oder zu trennen, und so den rohen Stoff sinnlicher Eindrücke zu einer Erkenntnis der Gegenstände zu verarbeiten, die Erfahrung heißt? Der Zeit nach geht also keine Erkenntnis in uns vor der Erfahrung vorher, und mit dieser fängt alle an.Wenn aber gleich alle unsere Erkenntnis mit der Erfahrung anhebt, so entspringt sie darum doch nicht eben alle aus der Erfahrung. Denn es könnte wohl sein, daß selbst unsere Erfahrungserkenntnis ein Zusammengesetztes aus dem sei, was wir durch Eindrücke empfangen, und dem, was unser eigenes Erkenntnisvermögen (durch sinnliche Eindrücke bloß veranlaßt) aus sich selbst hergibt, welchen Zusatz wir von jenem Grundstoffe nicht eher unterscheiden, als bis lange Übung uns darauf aufmerksam und zur Absonderung desselben geschickt gemacht hat (*Critique of Pure Reason,* B 1–2, *Akademie Textausgabe* edition, rpt. Walter de Gruyter & Co. Berlin, 1968 [1787]).

[93] There are various versions of how the transcendental argument intends to work (and whether it is successful) see: Stephen Clarke, "Transcendental Realisms in the Philosophy of Science: On Bheskar and Cartwright" *Synthese* 173.3 (2010: 299–315; Lucy Allais, "Kant's Argument for Transcendental Idealism in the Transcendental Aesthetic" *Proceedings of the Aristotelian Society* n.s. 110 (2010): 47–75; Robert Howell, "Kant and Kantian Themes in Recent Analytic Philosophy" *Metaphilosophy* 44.1/2 (2013): 42–47; Maurizio Ferraris, "Transcendental Realism" *The Monist* 98.2 The New Realism (2015): 215–232.

experience.[94] Thus, "7 + 5 = 12" is analytic a priori because if one truly understands "7 + 5" and "=," then "12" as a consequent does not expand our understanding of the left-side of the equation (the subject spot). The proposition would be the same as "All bachelors are unmarried males." To know the subject is to know the predicate. Therefore, because arithmetic is known *a priori* and its form is analytic, it follows the expected model.

Likewise, with the *a posteriori*, since it is known after (and as a result from) experience with the world, it is synthetic and its predicate expands the subject term. "Wood floats in water" is an example of this. One can know the definition of wood without knowing its property to float in water. Only after setting a branch in a lake or stream, does one notice (learn) that in addition to its source from trees, it also floats in water. Because the predicate term expands the subject term in *a posteriori* propositions, they are also synthetic: synthetic *a posteriori*.

These are the normal, expected combinations: analytic *a priori* and synthetic *a posteriori*. But what if there were an unexpected combination: the synthetic a priori?[95] Kant sets this possibility as follows:

1. "In all natural changes the amount of matter is conserved." This sentence is necessary—A[ssertion]/pp. 50 ff. [A9/B13] Kemp-Smith edition (Introduction)
2. "Every action (motion) implies an equal and opposite reaction," This sentence is necessary—A
3. All necessary statements are *a priori*—F[act]
4. #1 & #2 are *a priori*—1–3
5. The concept 'conservation' lies outside the concept of "matter"—A
6. The concept 'reaction' lies outside 'action'—A
7. All statements in which the predicate lies outside the subject are synthetic—F
8. #1 & #2 are synthetic—1, 2, 5–7
9. #1 & #2 are necessary judgments of natural science—F
10. *#1 & #2 are synthetic and* a priori—3, 4, 8
11. Natural science contains synthetic *a priori* judgments as necessary principles—8, 9, 10

---

[94] Note that this was more so the case post-Descartes. Remember, that in the ancient Greek world geometry (that had an empirical base) was the foundation for arithmetic. But after Descartes, as mentioned above, these were reversed. Through analytic geometry, arithmetic (an a priori discipline) was the foundation for geometry.

[95] Two wonderful discussions of this in a different context are: Moltke S. Gram, *Kant, Ontology and the A Priori* (Evanston, IL: Northeastern University Press, 1968) and Albert Casullo, *A Priori Knowledge* (Oxford: Wiley-Blackwell, 2010) and Ralph Wedgwood, "A Priori Bootstrapping" in Albert Casullo and Joshua C. Jhurow, *The A Priori in Philosophy* (Oxford: Oxford University Press, 2013): ch. 10. A good contemporary book on this subject is Albert Casullo and Joshua Thurow, eds., *The A Priori in Philosophy* (Oxford: Oxford University Press, 2013)—see especially Ralph Wedgwood's essay "A Priori Bootstrapping," pp. 226–246.

Argument 6.1: The Synthetic *A Priori* in Nature

Argument 6.1 begins with the analysis of two propositions: the law of conservation of matter (from Lavoisier) and Newton's third law of motion and asserts they are correct (necessarily). To fulfill this condition which would render them *a priori*, they must fit into a complete and coherent mathematical system that dictates this is the case. Kant asserts that they do.

Next, formally, because the predicate expands the understanding of the subject, they are also synthetic. Ergo, we have the *synthetic a priori* proposition.

The next topic to explore is the nature of time and space in nature. These are largely unexamined concepts in Newton. He tacitly accepts Aristotelian time (as a constructed measure of motion) and Euclidean space (the three-dimensional space of naïve visual experience). But these both need further analysis. Kant provides some favorable context for this (in the following argument):

1. Our outer experience presupposes a spatial representation—F/pp. 70–73 [A25/B40] Kemp-Smith edition (Transcendental Doctrine of Elements: Transcendental Aesthetic)
2. Our outer experience is generated by the object—F
3. Our assumptions about the properties of space precede our experience—1, 2
4. Geometry is the science which describes the properties of space—F
5. Geometry describes the properties of space *a priori*—A
6. The character of our outer representations (experience) extends our understanding of spatial representation—F
7. [That which extends our understanding of a concept is synthetic]—F
8. The intuition of space is synthetic, *a priori* and extends our outer experience—3–7
9. All representations are determined in our inner state—F
10. The inner state (the mind) presents (mediated) all appearances—F
11. The organization or sequencing of appearances is time—F
12. Time is necessary for the operation of the inner sense—10, 11
13. Time extends the properties of inner intuition—A
14. Time is synthetic and *a priori*—12, 13
15. *Time is the formal* a priori *condition of all appearances*—9, 14
16. Time is the formal synthetic *a priori* condition of all appearances; while space is the synthetic *a priori* form of our outer intuition—8, 15

Argument 6.2: Time and Space in the Transcendental Aesthetic[96]

Argument 6.2 applies the derivation of the synthetic *a priori* to the concepts of time and space that in the logical order, Kant believes, are presuppositions that make experience possible (a transcendental argument). They do not require separate argument (apart from the contention that without accepting Euclidean Space and Aristotelian Time we will not be able to make any sense out of the world in our day-to-day lives). This argument is a strong support of naïve empiricism. We all view the world this way in our daily lives, therefore; it is the transcendental grounds for all our lived-empirical experience.

Next, we move to how we actually "process" these empirical inputs. The sensory organs present input to the Understanding which must be "processed" to be then forwarded to Reason, itself.

1. The categories of quantity, quality, relation, and modality, constitute an exhaustive analysis of the Understanding—A/pp. 111–119 [A 76/B102] Kemp-Smith edition (Transcendental Analytic: Table of the Categories)
2. These categories are necessary for the Understanding to assimilate appearances—A
3. [That which is necessary is *a priori*]—F
4. The categories are *a priori*—1–3
5. The categories allow us to create a complete wholeness of an object in the Understanding—A
6. The categories are synthetic—5
7. *The categories yield a unity—5*
8. The Understanding represents appearances as existing separately via the categories while combined together in a synthetic *a priori* whole—4, 6, 7

---

[96] Kant's depiction of time and space in the Transcendental Aesthetic has raised much controversy—particularly because of its ties to Newton's conception (which is, itself, tied to Ancient Greek Philosophy). Since, post-Einstein, we have different understandings of these concepts, is this a problem for Kant? For some discussion on time and space in Kant's Transcendental Aesthetic, see: A.C. Ewing, *A Short Commentary on Kant's Critique of Pure Reason* (Chicago: University of Chicago Press, 1938): 28–65; Jonathan Bennett, *Kant's Analytic* (Cambridge: Cambridge University Press, 1966): 15–60; Peter Strawson, *The Bounds of Sense* (London: Routledge, 1966): 39–65; Lorne Falkenstein, *Kant's Intuitionism: A Commentary on the Transcendental Aesthetic* (Toronto: University of Toronto Press, 1995): 3–13; 154–159; 161–174; Michael Friedman, "Kantian Geometry and Spatial Intuition" *Synthese* 186.1 (2012): 231–255; Silvia De Bianchi, J.D. Wells, "Explanation and the Dimensionality of Space, Kant's Argument Revisited" *Synthese* 192.1 (2015): 287–303; Lucy Allais, "Kant's Argument for Transcendental Idealism in the Transcendental Aesthetic" *Proceedings of the Aristotelian Society*, n.s. 110 (2010): 47–75.

Argument 6.3: The Categories of the Understanding

In Argument 6.3 Kant tries to present a logical picture of a genetic, biological process. Since he is interested in the output of the system, he creates a so-called black box that presents input and output but does not give us any details on how this materially comes about.

But Kant *does* recognize that it *must* come about. He does this through the final argument that we will set out from Kant: the schematism. In my present account I am presenting the categories, themselves, (Quality, Quantity, Relation, Modality) along with their application to make both empirical and mathematical concepts presentable to the Understanding and *a fortiori* to Reason, itself. Kant, like Aristotle before him in his rather longer list of categories (see Chap. 3), believes that at least in the logical order, this is the way concepts are formed. The genetic order (the biological account) constitutes just the *details* of how this happens. This part of the mechanical process is attributed to: (a) an unspecified, material process (Kant's general approach, nature$_1$); (b) a semi-specified material process via animal spirits or the developing, vague neurology, nature$_2$; or (c) an unknowable "black box" that cannot, in principle, be known, nature$_3$.

Kant's account of the way the categories work to produce results can be seen in the following argument.

1. All concepts require objects—F (from argument #6.3—on the Categories)/pp. 180–187—[A 137/B176] Kemp-Smith edition (Transcendental Doctrine of Judgments)
2. Sensible images are presented to the Understanding via the reproductive imagination situated in time—A
3. The imagination consists of a reproductive and a pure *a priori* part—A
4. The *a priori* imagination requires the operation of the reproductive imagination to produce a concept (via the intuition of time)—1–3
5. The imagination underlies each of the categories—F
6. The categories, by themselves, require a tie to the external world in order to bring about a *proper representation (schematism)*—A
7. The categories require the schematism in order properly to represent objects—4–6

Argument 6.4: The Schematism of the Categories[97]

In Argument 6.4 Kant logically sets out the concepts that would be necessary for the Understanding to present robust data to Reason, itself. The categories and the parts of the Mind are part of an artificial system that, content-wise (the logical order), are

---

[97] For accounts of the categories and the schematism see: Norman Kemp Smith, *A Commentary to Kant's "Critique of Pure Reason"* 2nd ed. (Atlantic Highlands, N.J.: Humanities Press, 1962 [1923]): 194–201; 334–341; A.C. Ewing: 66–131; Bennett: 93–95; 35–37; Paul Guyer and Allen Wood, *Introducing Kant's Critique of Pure Reason* (Cambridge: Cambridge University Press, 2021): 39–65.

necessary to produce integrated thought. Again, Kant is talking about the systemic components rather than biological/medical descriptions of *how* such interactions must take place. In one way, this creates a sort of methodological goal for future research (teleology). Kant asks the searching questions (nature$_3$) and tries to use the spirit of Newtonian mathematically-based mechanics (nature$_2$) to set out a model for future scientific investigation. But in the Transcendental Dialectic he sets out dilemmas that seem to have no answer—a foray into nature$_3$ (which has significant metaphysical consequences in his metaphysical speculations about freedom and determination, whether the universe has a beginning or not, whether the material universe is reducible to atoms, and whether cosmological speculation into first causes can yield insight into the possibility of there being a God).

As the result of Kant's work in the *Transcendental Analytic* (as summarized above), and through the influence of Hume's more mitigated skepticism in *A Treatise of Human Nature* and *An Enquiry Concerning Human Understanding*[98] philosophers began to consider what other criteria might be brought forward to assist in theory evaluation. As mentioned earlier in the ancient world (particularly regarding axiomatic systems, like Euclid's *Elements*), other formal criteria such as the properties of *completeness, coherence, simplicity,* and *elegance* could be brought forward.[99] These properties will endure as describing nature$_2$, non-empirical properties thought to be essential in guiding nature$_1$ investigations and answering skeptical queries of nature$_3$.

Since Kant is a devoted Newtonian, he believes that down the road nature$_{1\&2}$ will be the way forward—as guided by the useful skepticism of nature$_3$. And like Descartes (and unlike most of the ancient Greek natural philosophers) this embrace of all the senses of nature will provide a more holistic vision of the truth of N/nature. If these *modern philosophers* are correct, our understanding of nature will be the better for it—but that may only be a hope and dream.

Perhaps the scientific advancement most directly focused upon the history of medicine in the seventeenth century, within the logical order of general scientific inquiry, is William Harvey's *De motu cordis* (*On the Circulation of the Blood*).[100] Harvey was born on April 1, 1578 in Folkstone, Kent. His father was a merchant so that Harvey could afford a good education. He went from King's College (Canterbury) to Cambridge, graduating in 1583. Then, he went to one of the most

---

[98] David Hume, *A Treatise of Human Nature,* David Fate Norton, ed. (Oxford: Oxford University Press, 2011[1739]) and David Hume, *An Enquiry Concerning Human Understanding,* Eric Steinberg, ed. (Indianapolis, IN: Hackett, 1977 [1777]).

[99] Cf. My discussion of these properties in all axiomatic theories in Michael Boylan, *A Just Society* (New York and London: Rowman and Littlefield, 2004): chapts. 1 & 2; David Hilbert, *Grundlagen der Geometrie,* ed. P. Bernays (Hamburg: Severus, 2014 [1899]); and Ian Mueller, *Philosophy of Mathematics and Deductive Structure in Euclid's* Elements (Cambridge, MA: M.I.T. Press, 1981).

[100] William Harvey, *De motu cordis & sanguinis in animalibus* (Leiden: Lugduni Batavorum, 1639). Two wonderful accounts of this are Andrew Gregory, *Harvey's Heart: The Discovery of Blood Circulation* (London: Icon Books, 2000) and Thomas Wright, *Circulation: William Harvey's Revolutionary Idea* (London: Vintage Books, 2013).

prestigious universities of the era, Padua. He studied mathematics with Galileo and medicine with Hieronymus Fabricius, who was a pioneering anatomist. Harvey left in 1599.[101]

Padua was a hot spot for change in this era. Such an intellectual environment brought a *questioning atmosphere,* nature₃ that is necessary for change—in this time of theoretical questioning and change. Both Galileo and Fabricius were looking to reconstruct natural philosophy in the context of *a posteriori* experiments—instead of *a priori*, purely logical, speculation. Harvey was attracted to this. It is an interesting coincidence that one of Dr. Harvey's medical patients was Francis Bacon, ardent empiricist, who also suffered from gout.

After Padua, Harvey returned to Britain and Cambridge and obtained his Doctor of Medicine in 1604. He then joined the Royal College of Physicians in 1604 and became the primary physician at St. Bartholomew's Hospital in 1609. It was there that he combined being an attending physician and doing vivisections on animals—particularly dogs.[102]

It should be remembered that at this time in British (and Continental) medicine, many operations were performed in a room that was arranged for observers. Because virtually all operations ended in the death of the patient due to sepsis (see Chap. 7), and because these exercises were severely limited because of the lack of proper anesthetic, operations were viewed as a "last resort."[103]

Because animals were not considered to be given the respect that humans were, they were subject to vivisection. Harvey was inclined to use dogs.

Now one of the issues that bothered Harvey was that according to the existing accounts of blood and how it came-to-be and was used began originally with Aristotle and was refined by natural philosophers until Galen.[104] Blood was said to be produced in either the liver or heart (or both) and it was sent out in different ways through the arterial and venous systems.[105] From there the blood was "used-up" by the body which took this *nourishment* as enabling the body to continue to be healthy and do its normal activities. Let's call this blood-model-A or BM-A.

---

[101] It is possible that at Padua Harvey was also exposed to salvaging vestiges of Aristotelian science—such as the final cause. For a discussion of the final cause in Harvey (e.g., the heart and arteries existing for the sake of circulating the blood) see: Peter Distelzweig, "Meam de motu & usu cordis & circuitu sanguinis sentiam: Teleology in William Harvey's *De Motu Cordis*" *Gesnerus* 71.2 (2014): 258–270.

[102] For a discussion of experimentation on live animals during this period, see: Domenico Bertolani Meli, "Early Modern Experimentation on Live Animals" *Journal of the History of Biology* 46.2 (2013): 199–226.

[103] Thus, just as in the ancient Greek and Roman world, human anatomy was best observed by people who were dying or who just died. Of course, there are stories of grave yard runs by medical students. For an example from the nineteenth century see: https://medicine.yale.edu/news/yale-medicine-magazine/article/anatomy-of-an-insurrection/ (accessed September 1, 2022).

[104] See Boylan (2015).

[105] Michael Boylan, "The Digestive and 'Circulatory Systems' in Aristotle's Biology" *Journal of the History of Biology* 15.1 (1982):89–118.

156   6 The New Paradigm: Changes Leading up to the Seventeenth Century and Beyond

In BM-A, at any given time there is x-amount of blood being made or residing in the heart ventricles (around 70 ml or 1/3 of a cup)[106] and, even as it is made (according to the Galenic model, as later interpreted, some blood was created in the liver and some in the heart),[107] this x-amount is opened up for use by the body. The force that moves the blood along in the blood vessels is *pneuma* (latter called "animal spirits"). So, for example, if the dynamic is that 1/3 of a cup[108] at any given interval is being made: 1/3 of a cup has been made and is ready to be dispersed to the body from the heart (even the blood made in the liver which had to be properly *prepared* for the body in the heart), then the amount of blood at any time in the body is the amount that has been made (1/3 cup)/plus the amount still being made (1/3 cup)/minus the amount being used-up (1/3 cup). The total all together is (1/3 cup, 70 ml). Compare this to our contemporary measurement of 1.5 gallons (5.67 l) of blood in humans (on average).[109]

Of course, performing the application of math at the time was rather sketchy, since all these processes were probably not *in-sync*. However, the 1/3 cup standard, 70 ml, was considered to be a maximum limit when the procedure of bloodletting was done by knife, and 5–10 ml was the standard if done via leeches.[110]

However, there were a couple of sources of logical dissonance in the traditional Galenic-based account, which an observant William Harvey might have seen during ordinary life in London. Butchers, for example, bled their animals before killing them. This was because the customers wanted less-bloody flesh—particularly in pork.[111] It would not have been unusual for a curious customer, such as Harvey, to observe that the amount of blood that came out of these animals was considerably more than 1/3 cup of blood. Now it is true that pigs and cattle have bigger hearts than humans. This might have been the impetus for Harvey to engage in vivisections on dogs (who have a smaller heart than humans). When severing the aorta in vivisection on dogs two things would have been clear: (1) The blood came out in spurts, and (2) The amount of blood from a smaller heart produced much more blood than would have been expected.[112]

---

[106] To be clear, contemporary science does *not* attribute blood creation in the heart (but rather in the marrow of large bones). My comments above are meant to reflect what medical writers at the time, following Galen, perceived to be the case.

[107] Under some interpretations the blood being made in the liver was transported via the venal system and the blood being made in the heart was transported by the arterial system. However, there was not unanimity on this. For some background see: C.R.S.Harris, *The Heart and Vascular System in Ancient Greek Medicine: From Alcmaeon to Galen* (Oxford: Oxford University Press, 1973) and Boylan (2015).

[108] 1/3 of a cup (the English system of measures that was used at the time of Harvey) equals around 70 ml. under the metric system.

[109] The Red Cross says that the average amount of blood in the human body is around 1.5 gallons: https://www.redcrossblood.org/donate-blood/dlp/whole-blood.html (accessed August 1, 2022).

[110] Gerry Greenstone, MD, "The History of Bloodletting" *BCMJ* 52.1 (2010): 12–14. Note, also, that these are approximations. There were no clear, universal measuring protocols at the time.

[111] John Stow, *The Survey of London* (London: Dent & Dent, 1956).

[112] The amount of blood in a dog is a function of its weight. In an "average-sized" dog it is .8 gallons (about half the amount of a typical human).

Argument 6.4: The Schematism of the Categories

These dissonances would have likely put Harvey on his guard about the amount of blood to be found in the body. However, much of this "measuring" was conjectural. This was due to the fact that though the general science of the era was keen on measuring and "mechanization," it was still not entirely clear how this goal could be effected as a proper part of medical procedure.[113] But it is this inclination to *measuring* that gave Harvey the cognitive dissonance that incoherence occasions (see above). If dogs, with smaller hearts had more blood than was predicted in humans and if the blood spurted out rhythmically, then clearly the amount of blood in the human body was greater than previously predicted and its presentation to the blood vessels was not a passive opening of canal gates urged on by animal spirits.

Perhaps, the systolic and diastolic soundings of the heart that had been identified along with the pulse as being of interest,[114] had a greater significance than previously considered. The spurting of blood by the vivisected dog was more consistent with the operation of a pump (something which was being perfected by fire squads in seventeenth century London).[115] As just noted, others have suggested that this technological innovation on fire pumps might have helped Harvey re-think the conventional model with the counter evidence of the powerful, rhythmic spurting of the blood out of the severed arteries of vivisected dogs in the surgical theater.[116]

These inferences suggested a blood model-B, be adopted (BM-B). This was a circulatory model. There was just one critical detail to be worked out: what to make of the veins. Here, we need to go back to Harvey's former teacher, Fabricius who has been anatomically studying the veins in detail (possibly along with his students—Harvey being one of them). The result of this work was a 1603 treatise entitled, *De venarum ostiolis* (the "valves" or "doors" of the veins).[117] Noting the anatomy of the venal valves is an important step toward discovering its physiology. Next, Harvey had to engage in an experiment to help him understand how these *valves* or *doors* worked. Harvey repeated Fabricius' experiment that ran like this.

---

[113] This move toward "mechanization" has been discussed by Peter Distelzweig, "'Mechanics' and Mechanism in William Harvey's Anatomy: Varieties and Limits" in Peter Distelzweig and B. Goldbert, and Evan Ragland, eds. *Early Modern Medicine and Natural Philosophy* (Dordrecht: Springer, 2016): 117–140.

[114] Michael Boylan, "Galen on the Blood, Pulse, and Arteries" *Journal of the History of Biology* 40.2 (2007): 207–230.

[115] For background on pumping devices in London during this period see: G. Basalla, "William Harvey and the Heart as a Pump" *Bulletin of the History of Medicine* 36 (1962): 467–470; K.D. Keele, *William Harvey: The Man, the Physician, and the Scientist* (London: Nelson, 1962); Gweneth Whitteridge, *William Harvey and the Circulation of the Blood* (London: Elsevier, 1971); J. Aubrey, *Aubrey's Brief Lives* ed. by O.L. Dick (London: Harmondsworth, 1972) and R.K. French, *William Harvey's Natural Philosophy* (Cambridge: Cambridge University Press, 1994). This museum in London shows what the seventeenth century fire pumps (which saved the city in the great fire of 1666. https://www.museumoflondon.org.uk/discover/building-17th-century-fire-engine (accessed August 1, 2022).

[116] Wright (2013): ch. 7.

[117] Hieronymus Fabricius, *De venarum ostiolis,* facsimile edition, ed. K.J. Franklin (Springfield, IL: Charles C. Thomas, 1933 [1603]).

Harvey litigated his servant's arm to swell the veins. He confirmed that blood was would not move downward (away from the heart) despite his digital manipulation to bring that about. No matter how much pressure Harvey applied with his fingers, he could not get the venal blood to move downwards (away from the heart). Harvey concluded from this that the valves or doors prevented downward movement of blood. Fabricius had interpreted this result as being a safeguard to keep too much blood from accumulating at the extremities by *delaying* the movement of blood.

Harvey's inclination was different from his teacher.[118] He conjectured that the doors or valves in the veins were in place to promote upward movement of blood: returning to the heart. The valves would stop a backflow of blood.

If this inference was correct, and if the heart were a pump (as suggested by the vivisection with dogs and the rhythmic spurting of blood upon the cutting of arteries [and if dogs were like people]), then the outcome suggested would be that the heart pumped blood in the arterial system away from the heart to the rest of the body— AND the venal system used its valves/doors to encourage via fluid pressure, the blood to return to the heart. If this conjecture were correct, then it would seem that the Aristotelian/Galenic system of blood flow was incorrect. The BM-A system needed to be replaced by the BM-B account.

What would follow from this? First, the role of the heart in blending the humors for the health of the individual and then passively opening up for animal spirits to deliver the blood to the body where it was "used-up" was wrong. Instead, the blood was constantly flowing throughout the body bringing whatever components it delivered to the rest of the body. The BM-B account would allow for a greater amount of blood to be in the body because it was continually being "re-used" and not "used-up." This would better fit with the dog vivisection experiments. As a pump, the rhythmic spurting would also fit the phenomena better. Finally, venous valve experiment (as interpreted by Harvey) solved the last stage of the process—bringing the blood back home to the heart.

Conclusion

Harvey's inference to the most probable explanation[119] turned out to be correct and was one of the most influential discoveries in physiology in the seventeenth century. It followed from a genetic order that promoted questioning from the top of the power structures in Europe (the Vatican's concern about calendars and Easter). This practical questioning stimulated others to reconsider the entire scientific structure.

---

[118] I am influenced in this interpretation of Harvey's experiment by F.R. Jevons, "Harvey's Quantitative Method" *Bulletin of the History of Medicine* 36 (1962): 462–467 and Donald George Bates, "Harvey's Account of His 'Discovery'" *Journal of Medical History* 36 (1992): 361–378.

[119] This is commonly called *abduction* among contemporary philosophers, see Boylan (2020): ch. 6.

In the logical order, Descartes and Newton took on this call and other philosophers (a century or so later) tried to knit together the new science into a coherent whole.

Just as we have seen earlier in this book, questioning in science broadly often has the effect of affecting medicine, in particular. In this case, I have highlighted William Harvey's experimentally-based conjecture that the Aristotelian/Galenic account of the movement of blood was incorrect and that a model that put forth general circulation of blood in the body—both from the heart and back again—was the more probable explanation. It had fewer possible anomalies (and thus was more *complete* and *coherent*). And since blood (and how it behaves in the body) is an essential ingredient for understanding human health, this change made a considerable difference. Though, like Newton, Harvey also held other notions connected to his understanding of Christianity (religion-based $nature_2$) that were inconsistent with the n/$Nature_1$ (or mathematically-based $nature_2$) accounts.[120]

The seventeenth century in the West brought disruption in the accepted paradigms in general science and in biology (a backbone of medicine). This led to Descartes, Newton, and Harvey offering new approaches that seemed to provide a solid, mathematically-supported approach that proposed deductive certainty that led proponents to claim *completeness* and *coherence* in the execution and *simplicity* and *elegance* in the theoretical modeling.

Thus, in our journey to study Western accounts of n/Nature as a picture of the history of medicine, it should be said, that though this was a century of great change and progress, there were still some lingering issues that continue to hang on creating continued complexity and some ambiguity in discourse: onto the nineteenth century!

---

[120] A persuasive argument based upon Harvey's use of the macrocosm–microcosm analogy and of alchemical terminology, was made by Andrew Gregory, "William Harvey, Aristotle, and Astrology" *British Journal for the History of Science* 47.2 (2014): 199–215.

Chapter 7
Evolution, Germ Theory, and Their Consequences

Abstract This chapter explores the general scientific changes in biology of evolutionary theory and germ theory as they were presented in the West during the nineteenth century. These are explored, in themselves, and then applied to general pattern that changed the practice of medicine then and set the stage for the way twentieth century medicine went forward.

Keywords Evolutionary theory · Germ theory · Teleology · Jean Baptiste Lamarck · Alfred Russell Wallace · Charles Darwin · T. R. Malthus · H.S. Kettlewell · G. Mendel · Agostino Bassi · Ignaz Semmelweis · John Snow · Anesthesiology · Epidemiology · Louis Pasteur · Joseph Lister · Antiseptic surgery

Introduction

From the point of view of contemporary medicine, the concepts of *evolution* and of *germ theory* have had an enormous impact the way that medicine is understood and practiced throughout the world. This could obviously be two books all by themselves. However, in our brisk journey I will bring up particular highlights that fit into my general program of highlighting the three aspects of nature along with the reasons that one theory is chosen over another (touched on in Chap. 1 and developed some in Chap. 6, viz., containing the properties of *completeness, coherence, simplicity,* and *elegance*).

Since this chapter is about scientific inquiry in the West during the nineteenth century, we should say a few words on what made the nineteenth century different as an environment in which to work—particular from the point of view of Great Britain.

One macro feature of the century was revolution in transportation. Britain moved quickly (starting around 1825) to create a national railway network that would connect the country with (relatively) high speed connections between major centers (and later between more modest hamlets) for the purpose of economic development and national security.[1]

Then there was the developing fleet of ships (military, merchant marine, and exploratory) that allowed for an attitude described in a poem by James Thomson (1740):

Rule Britannia!

Britannia rule the waves

Britons never, never, never shall be slaves.

Rule Britannia![2]

To most Britons in the Nineteenth-Century these practical opportunities of travel, opportunity to visit and do whatever they liked over much of the world, and communication about what they found, was a comparatively fast development. These, in addition to the colonial empire that had begun in the sixteenth-seventeenth century made a full-blown consolidation as they were able to exploit millions of peoples around the world to gain tremendous wealth that would catapult them to become the richest nation on earth by the end of the century.[3] These factors, together, provided the opportunity for wide-ranging opportunities to explore the *explanandum* (Nature in various venues) which will offer more data for inductive examination that, in turn, will allow for more speculative conjectures that are backed-up by empirical data.

Again, another positive factor was the British postal system, which became one of the most efficient in the world delivering letters two times a day in the major cities, and on a daily basis to many rural areas.[4] And finally, was the proliferation of societies of science that offered interaction and support to scientists both inside and outside the Academy.[5] This was augmented by advances in the telegraph (1836) through most of the country on an ongoing basis. Improvements in communication

---

[1] For a discussion of this see: Michael Cobb, *The Railways of Great Britain: A Historical Atlas* (London: Ian Allen, 2004).

[2] Two academic discussions of note are: Gerald S. Graham, *The Politics of Naval Supremacy: Studies in British Maritime Ascendancy* (Cambridge: Cambridge University Press, 1965) and Richard Harding, *The Emergence of Britain's Naval Supremacy: The War of 1739–1748* (Woodbridge, Suffolk: Boydell Press, 2010).

[3] For an account of this see: P.J. Cain, A.G. Hopkins, *British Imperialism 1688–2015*. 3rd ed. (London: Routledge, 2016).

[4] Catherine J. Golden, *Posting it: The Victorian Revolution in Letter Writing* (Gainesville: U. Press of Florida, 2010).

[5] T.W. Heyck, "From Men of Letters to Intellectuals: The Transformation of Intellectual Life in Nineteenth-Century England" *Journal of British Studies* 20.1 (1980): 158–183.

and feedback from colleagues are crucial for creating an atmosphere for innovation and scientific advancement. It was a time for great progress in scientific discovery.[6]

Another very important advancement in the nineteenth century was in optics with advancements in the compound microscope by Joseph Jackson Lister (1786–1869, father of Joseph Lister, M.D.). The clarity of the optical compound microscope was greatly enhanced, such that with this instrument Joseph Jackson Lister was first able to describe the true form of the red blood cell in mammalian blood.[7]

Evolutionary Theory

Teleology

Certainly, since the time of Aristotle, teleology has been a critical part of biomedicine.[8] Now there are at least two ways to understand teleology in biomedical settings: (a) from the viewpoint of a supernatural creator: God (Nature$_2$, and (b) from the viewpoint of complex evolving material systems, n/Nature$_1$).[9]

In the first instance, we have seen in the ancient Greek and Roman Worldview there was a strong sense of Nature$_2$ (*phusis$_2$*) that was non-empirical in the sense of being controlled by supernatural God figure(s). When Aristotle and Galen say that *Nature does nothing in vain* are they talking about a creative "god figure" (clearly Nature$_2$) or are they describing "systemic features" of an environment that has material feedback input that will reward certain material constructions over others (n/Nature$_1$)? In the case of systemic teleology, the system moves of its own accord (through material factors) toward a particular end. There is no Purposive Being as a controlling entity, but because the system exhibits certain features that act *als ob*

---

[6] Of course, it should be mentioned that much of this "advantageous environment of scientific exploration" was built on the context of colonialism which was inherently unethical and exploitative: Caroline Elkins, *Legacy of Violence: A History of the British Empire* (New York: Knopf, 2022).

[7] This microscope is on display with supporting information at the Welcome Museum, London: https://collection.sciencemuseumgroup.org.uk/objects/co440642/joseph-jackson-listers-microscope-london-england-1826-compound-microscope (accessed May 5, 2022).

[8] For a discussion of Aristotelian teleology and its context compared to evolutionary theory in the contemporary era see Boylan (1983): 102–109.

[9] Further discussion of teleology in the context of biological systems and evolutionary theory can be found in: Javier González de Prado Salas, "Whose Purposes? Biological Teleology and Intentionality" *Synthese* 195.10 (2018): 4507–4524; Bruce H. Weber, "Design and its Discontents" *Synthese* 178.2 (2011): 271–289; Peter Woodford, "Neo-Darwinists and Neo-Aristotelians: How to Talk about Natural Purpose" *History and Philosophy of Life Sciences* 38.4 (2016): 1–22; Lucas Mix, "Nested Explanation in Aristotle and Mayr" *Synthese* 193.6 (2016): 1817–1832; Mariska Leunissen, "Nature as a Good Housekeeper. Secondary Teleology and Material Necessity in Aristotle's Biology" *Apeiron* 43.1 (2010): 117–142; and Monte Ransom Johnson, *Aristotle on Teleology* (Oxford: Oxford University Press, 2008).

there were a controlling entity, the source of the purpose (the system or some supernatural being) is controversial. This is a debatable question in the ancient world (though I believe that both Aristotle and Galen meant the former).[10] However, when it comes to Darwin, I believe that the case is less controversial (meaning a commitment to n/Nature$_1$).

In the ancient Greek World, there were proponents of evolutionary theory. Anaximander suggested that the *eidos* (or form) of an animal type (*genos*) could be altered. Empedocles' account was built upon cycles of Nature coming together (Love) and then it being broken-up again (Strife). This was asserted to occur in progressive cycles that would occur, and then re-occur (see: Chap. 1). Aristotle positioned himself as against this cosmology because he advocated a "steady-state cosmological theory"—meaning that the world/universe has been the way it presently is forever (backwards and forwards). This is one of the three cosmological positions (a-historically) that one might take.[11]

One further position in the ancient Greek World was *devolution*. Under this account, things are beginning to fall apart without the possibility of regaining a past eminence. Hesiod was a prominent advocate of this position.[12] This is a logically possible philosophical stance, but it is not one that is very useful for general science—especially to support medical diagnosis, prognosis, or treatment. More often, the devolutionists were talking about the decline of culture and morality rather than a decline in either scientific knowledge or a decline in the biological functionality of species or particular animals (n/Nature).

In the nineteenth century, the evolutionary hypothesis became a popular question of interest (just as astronomy in the seventeenth century became a common topic of interest). One prominent and influential philosopher, G.W.F. Hegel, set out a philosophy of history that had direction controlled by a God-figure (*Geist*) moving in a positive direction: in the long-run.[13] D.G. Ritchie connected Hegel and Darwin—not only via the role of *Geist* in history, but the connection of this evolutionary

---

[10] Boylan (1983): 225–229.

[11] The three positions are: (a) *The Beginning and End* hypothesis that states the universe had a beginning and will then have an end (possibly Plato—obviously all the Abrahamic religions); (b) The Steady-State theory (Aristotle) that says that the universe is the same as it was and forever will be forever; (c) The Cyclic theory that says that the universe moves in one direction for a long time and then changes direction through catastrophe, and then re-groups (Empedocles—my conjecture that this comes from Hinduism and their cyclic cosmology).

[12] Hesiod, *Works and* Day, Glenn W. Most, ed. and tr. (Cambridge, MA: Harvard University Press, 2006): ll. 109–204. Secondary literature: J.G. Griffiths, "Archaeology and Hesiod's Five Ages" *Journal of the History of Ideas* 17.1 (1956): 109–119; Joseph Fontenrose, "Work, Justice, and Hesiod's Five Ages" *Classical Philology* 69.1 (1974): 1–16; Robert L. Fowler, "Mythos and Logos" *Journal of Hellenic Studies* 131 (2011): 45–66; Andreas T. Zanker, "Decline and Parainesis in Hesiod's Race of Iron" *Rheinisches Museum Für Philologie,* neue folge 156 Bd., H.1 (2013): 1–19. And for a poetic reaction in a contemporary context, see: Michael Boylan, "Hesiod and the Iron Men," from *The Collected Poems of Michael Boylan* (Bethesda, MD: PWI Books, 2022): 205.

[13] G.W.F. Hegel, tr. J. Sibree, *The Philosophy of History* (Mineola, N.Y.: Dover, 2004 [1837]).

thinking to the development of n/Nature, itself, which influenced Darwin.[14] The style of this account is both the genetic order (it happens over time) and the logical order. For example, in the logical order, one could look at most iterations of the Linnaean Classification System (which shows a phylogenetic ordering of animals with mammals and then primates as *higher* (meaning more powerful) positing by such a scale that *human rationality* is the most powerful of all so that it is *obviously* at the top). This logical ordering has been called variously: The Scale of Nature (*scala naturae*) and The Great Chain of Being.[15]

Others who promoted a theory of biological change among species—particularly in the genetic account (with the connotation that such a change was for the better) were Jean Baptiste Lamarck, Alfred Russell, and Charles Darwin.

From our contemporary vantage point, Charles Darwin is attributed as the most *complete* and *coherent* advocate of the theory of biological evolution that has had such a strong historical impact upon how biology and medicine are understood. This hypothesis helps all biomedical writers in its wake to understand the way every species of animal (including humans) exists and changes over time. Because of this, I will attribute Charles Darwin's work as being the most influential (over time) in our contemporary understanding of evolution. However, it is useful to look briefly at others promoting evolution, as well.

Jean Baptiste Lamarck (1744–1829) was born in Bazentin, Picardy in northern France on August 1, 1744 to an impoverished aristocratic family. Lamarck served in the army and fought bravely in the Seven Year War with Prussia. His natural interest was in botany. He became known (as he was self-taught) to be a fine French botanist who created a system for identifying plants meant to be more efficient than Linnaeus' system—particularly for the application by practicing gardeners. The system employed identifying morphological traits via a grid of opposing possibilities in order to make its classifications.[16]

In 1793 he became a professor of the new Museum of Natural History. Because of his emphasis upon morphology and changeable traits of plants and his new interest in invertebrates, he became interested in accounting for how species change their morphological character. For the simplest forms of plant and animal life he adopted the Aristotelian notion of spontaneous generation and a dependence upon environment for an animal/plant's development.

His classification system was built upon complexity—the more complex an organism was in comparison to other sorts of living organisms of the same sort, the *higher* it was (the logical order, cf. *Scala Naturae/Great Chain of Being*). This was in the grand scheme of things. As plants and animals interact with the environment, they change and that change is heritable (a concept that Anaxagoras also held). A

---

[14] D.G. Ritchie, "Darwin and Hegel" *Proceedings of the Aristotelian Society*, old series 1, 4.1 (1891): 55–74.

[15] For Aristotle's uses see: Boylan (1983): 63, 169, 180, 199. On the general idea see the famous: Arthur O. Lovejoy, *The Great Chain of Being* (Cambridge, MA: Harvard University Press, 1936).

[16] Andrés Galera, "The Impact of Lamarck's Theory of Evolution before Darwin's Theory" *Journal of the History of Biology* 50.1 (2017): 53–70.

part of such an analysis leads to the study of the past—particularly through an analysis of fossils, paleontology. Lamarck brought new insights to this study of the past, but his notion of the heritability of acquired traits would have to wait for Darwin and Mendel to refute.

He died on December 18, 1829 (age 85) in Paris.

Before embarking on Darwin, it is useful to examine briefly another advocate of evolution in the nineteenth century, Alfred Russel Wallace. Wallace was born in Monmouthshire, Wales in 1823. He grew up in modest circumstances. He attended a one-room schoolhouse in Hertfordshire, England for 6 years. He became an apprentice carpenter at 14. He self-educated (like many from the working-class). He read and was influenced by Robert Owen and his son Robert Dale Owen, and this inclined him to general skepticism.

In 1834 he became apprenticed in surveying. He performed work in Bedfordshire and then in Wales. He became acquainted with other working-class folk and developed a political sympathy with their lot. As a surveyor he spent a good deal of time outdoors. He read works by William Swainson, Charles Darwin, Alexander von Humboldt, and Thomas Malthus.

Along with his naturalist friend, Henry Walter Bates, he began collecting biological specimens. After a couple of years, the pair went their separate ways, Wallace spent 4 years in the Amazon rainforests. He collected butterflies, insects, and birds. Except for one shipment of his trove back to his agent in England, the rest of his collection was lost at sea in a fire that sank his ship. Fortunately, his journals survived. He regrouped for a trip to the Malay Archipelago.

Wallace spent 8 years there collecting specimens for his research, and for sale back in England. He wrote two articles on the origin of new species. The first was published in 1855: "On the Law which has regulated the Introduction of New Species" *Magazine of Natural History* (1855).[17] The article concludes with this sentence: "every species has come into existence coincident both in time and space with a pre-existing closely allied species."

Wallace proposed that new species arise by progression and then produce off-shoots that outlive their parents in the fight for survival. In 1858 he sent a paper outlining his ideas to Darwin which would eventually lead to a co-authored paper.[18]

Wallace returned to England in 1862 as an established naturalist who had collected more than 125,000 animal specimens. He married and had three children. His narrative of his journey and various conclusions he drew from these were published in 1869 and 1870.[19] Wallace separated from other advocates of evolution at the time declaring that natural selection could not account for human intelligence.

---

[17] Reprinted in *Proceedings of the Linnean Society London* 171.2 (1960): 141–153.

[18] Charles R. Darwin and Alfred R. Wallace, "On the Tendency of Species to Form Varieties; and on the Perpetuation of Varieties and Species by Natural Means of Selection" *Zoological Journal of the Linnean Society* 3.9 (1858): 45–62.

[19] Alfred Russel Wallace, *The Malay Archipelago: The Land of Orang-Utan, and the Bird of Paradise* (London: Macmillan, 1869) and *Contributions to the Theory of Natural Selection* (London: Macmillan, 1870).

Like Lamarck, his idea of a time-line for evolution was very short. Also, like Lamarck, there was no development of the dynamics of morphological traits as working within an environment to create a sense of *fitness* within a given environmental structure. However, what Wallace *did* propose was the concept of common ancestors, which is a critical concept in modern evolutionary systematics. Also, his study of Malthus put *survival competition* to the fore. Darwin used these concepts, as well. How this came about and who is to get the credit is still a topic of some dispute.[20] We can see that the n/Nature depicted is as nature$_1$. There is a sense of completeness with this approach: when food resources diminish, then people will die of starvation. Unfortunately, the British used this idea to allow regular starvations in India and later in Ireland so that wealthy people might not lose their investment income.[21] These evolutionary concepts mix self-interested economics against what ordinary people might legitimately claim as a human right, viz., to be able to eat enough food to survive.[22]

In this way, science is both descriptive (what will happen if x, y, z come forth) and prescriptive (whether the consequences of x, y, and z) are acceptable [morally] and what should follow from these.

Charles Darwin was born on February 12, 1809 in Shrewsbury, Shropshire. His father was a physician who had a *wealthy-patient-practice.* Darwin attended the University of Edinburgh Medical School (very prominent at the time). Darwin, however, didn't much care for medicine. He was more interested in comparative zoology and geology. His father then sent him to Christ's College, Cambridge, but Charles was not keen on pursuing the "typical, prescribed paths." He linked-up with his second cousin, also at Christ's College (who had an interest in insects). This suited Charles very well. He took his degree in 1831.

While at Cambridge, Charles read William Paley's *Evidence of Christianity* (1795) which was a book exhorting God-based-teleology, Nature$_2$ over-against a purely material account, n/Nature$_1$.

The big "turning-point" in Darwin's life was his opportunity (through his brother-in-law, Josiah Wedgwood II), to accept an *opportunity* to be a "gentleman" naturalist on the projected 2-year voyage on the H.M.S. Beagle. Darwin accepted. The voyage began on December 27, 1831 and lasted nearly 5 years (October 2, 1836).[23] The captain of the ship, Robert FitzRoy, was a confident and supporter. Darwin kept a detailed journal and often went ashore to collect a wide-ranging number of dead

---

[20] For some overview on this see: David Bainbridge, *Stripped Bare: The Art of Animal Anatomy* (Princeton, N.J.: Princeton University Press, 2018).

[21] See: Pandit Sunderlal, *British Rule in India* 4 vols. (Thousand Oaks, Ca: Sage Publishers, 2019 [1929]) and Tim Pat Coogan, *The Famine Plot: England's Role in Ireland's Greatest Tragedy* (New York: St. Martin's, 2013).

[22] See Michael Boylan on the moral right for food: Michael Boylan, *Natural Rights: A Theory* (Cambridge: Cambridge University Press, 2014): ch. 6.

[23] Charles Darwin, *The Voyage of the Beagle* (N.Y.: The Natural History Library, Anchor Books, 1962 [1859]).

animal specimens and some fossils. He made numerous observations on geological land formations—perhaps for a future book on geology.

When Darwin reached the Galapagos Islands, he found mocking birds similar to those he had observed in Chile. He noted tortoise shells that varied slightly according to their island of origin. This began to stimulate ideas about the role of the environment in the anatomy of animals from place-to-place. He kept these two examples as important to his developing thought.

Darwin also found instances of species unique to a locale—such as the rat-Kangaroo and Platypus in Australia. Why would there be species unique to a locale instead of being uniform around the world? [The Biblical Creationist Model would suggest a uniform distribution of species throughout the world.]

Captain FitzRoy began collating Darwin's journal notes into a more coherent, holistic manuscript (much like an editor) that would eventually be the basis for the to-be-developed-book: *On the Origin of the Species* (1859).[24]

Other key moments in the voyage included his witnessing of a large earthquake in Chile in 1835 where in a brief period of time the land was raised up. This was contrary to the existing theory of Charles Lyell, who opined that the rising or lowering of land masses occurred only over great periods of time.[25]

When they arrived at Cape Town, South Africa, Darwin and FitzRoy met John Hershel, who was a proponent of Charles Lyell's *uniformtarianism principle*, that assumes that natural laws do not change over time (they are eternal and absolute, Nature$_2$) so that changes in geology or animal species (which are subject to these immutable laws) would occur over a long period of time and not quickly through cataclysms.[26] This position of gradualism has been a feature of Darwinian evolutionary theory that has extended to this very day among Darwinists.[27]

Another point of Lyell that was influential to Darwin was the idea of *replacement species*. When one species went extinct, another would step forward to take its place.[28] Darwin reacted to this in his diary—particularly concerning his observations of mockingbirds, tortoise shells, and the Falkland Islands Fox.[29]

A final point of influence that I'd like to mention at this juncture is the notion of *competition and survival*. Darwin was certainly influenced by Thomas Robert

---

[24] Charles Darwin, *On the Origin of the Species by Means of Natural Selection, or The Preservation of Favored Races in the Struggle for Life* (London: John Murray, 1859).

[25] Charles Lyell, *Principles of Geology* (London: John Murray, 1830).

[26] The opponents here were authors such as James Hutton, *Theory of the Earth* (Edinburgh: Royal Society of Edinburgh, 1788), who contended that several natural catastrophes were what drove major natural changes.

[27] Contemporary examples of the "gradualist group selection approach" include: Edward O. Wilson and William H. Bossert, *A Primer of Population Biology* (Sunderland, MA: Sinauer Publishers, 1971) and Elliott Sober and David Sloan Wilson, *Unto Others: The Evolution and Psychology of Unselfish Behavior* (Cambridge, MA: Harvard University Press, 1998).

[28] See the discussion of John Van Wyhe, "Mind the Gap: Did Darwin avoid Publishing his Theory for Many Years?" *Notes and Records of the Royal Societies* 61.2 (2007): 173–205 (esp. 197).

[29] Darwin (1859): 1.

Malthus.³⁰ Malthus contended that population increases geometrically (doubling every 25 years) while food production increases arithmetically. The consequence of this would be periodic massive famine, unless birth rates would decrease below the food production level. Also, the rise of workers in the population within developing capitalistic countries will mean more workers than jobs—a situation that will result in lower wages. In either case, the result is increasing death and poverty.

Malthus set out his grim predictions within a context of David Hume and Adam Smith, setting out supporting structural concepts that would give a voice to a future in which competition would set out "winners" and "losers" in the game of life.³¹

Obviously, these works set out a context in which competition for survival created a milieu for change. Darwin wrote on this, "In October 1838 ... I happened to read for amusement Malthus on *Population* ... it at once struck me that under these circumstances favorable variations would tend to be preserved, and unfavorable ones to be destroyed. The result of this would be the formation of new species."³²

Alfred Russel Wallace also stated, "But perhaps the most important book I read was Malthus's *Principles of Population* ... It was the first great work I had yet read treating of any of the problems of philosophical biology, and its main principles remained with me as a permanent possession, and twenty years later gave me the long-sought clue to the effective agent in the evolution of organic species."³³

The Origin of the Species

We will now examine four key passages that are essential to Darwin's theory.

A. But my tables further show that, in any limited country, the individuals, and the species which are most widely diffused within their own country (and this is a different consideration from wide range, and to a certain extent from commonness), oftenest give rise to varieties sufficiently well-marked to have been recorded in botanical works. Hence, it is most flourishing, or, as they may be called, the dominant species, those which range widely, are the most diffused in their own country, and are the most numerous in individuals—which oftenest produce well-marked varieties, or, as I consider them, incipient species. And this perhaps, might have been anticipated; for, as varieties, in order to become in any degree permanent, necessarily have to struggle with the other inhabitants of the country, the species which are already dominant will be the most likely to

---

³⁰ Thomas Robert Malthus, *An Essay on the Principle of Population* (London: J. Johnson, 1798). (Note, the work was originally published anonymously.)

³¹ David Hume, "On the Populousness of Ancient Nations" rpt. in *Population and Development Review* 3.3 (1977 [1752]) and Adam Smith, *An Inquiry into the Nature and Causes of the Wealth of Nations* (London: W. Strahan and T. Cadell, 1776).

³² Charles Darwin, *The Autobiography of Charles Darwin*, ed. Nora Barlow (London: Collins, 1958 [1887]): 128.

³³ Alfred Russel Wallace, *My Life: A Record of Events and Opinions* (London: Chapman and Hull, 1908): 38.

yield offspring, which, though in some slight degree modified, still inherit those advantages that enabled their parents to become dominant over their compatriots.[34]

In this passage Darwin is concerned about variation within a species. Variation within a breeding population is a key concept for Darwin's theory. For example, among wheat plants some individuals within the species have heritable traits (α and β, that will make them more or less susceptible to particular insects that, in turn, can impede that plant's ability to grow to maturity and reproduce.[35] Depending upon what we now call the environmental fitness pressure, this could have an effect upon the population containing traits α and β.

Darwin contends here that within populations of a given species, there is more variation within large and common populations—even those with a dominant type. The traits that made the dominant group the most successful at reproducing to the next filial generation are inherited traits that can be passed on.[36] But the ability to be dominant may be transitory so that those variants within the population are called incipient—since they, themselves, may become dominant under new environmental circumstances.

B. As the species of the same genus usually have, though by no means invariably, much similarity in habits and constitution, and always in structure, the struggle will generally be more severe between them, if they come into competition with each other, than between the species of distinct genera. We see this in the recent extension over parts of the United States of one species of swallow having caused the decrease of another species. The recent increase of the missel-thrush in parts of Scotland has cause the decrease of the song-thrush. How frequently we hear of one species of rat taking the place of another species under the most different climates! In Russia the small Asiatic cockroach has everywhere driven before its great congener. In Australia the imported hive-bee is rapidly exterminating the small, singles native bee. One species of charlock has been known to supplant another species; and so, in other cases. We can dimly see why the competition should be most severe between allied forms, which fill nearly the same place in the economy of nature; but probably in no one case could we precisely say why one species has been victorious over another in the great battle of life.[37]

In this passage, Darwin is highlighting the struggle for existence. It is probable that he was influenced by Malthus in this regard.[38] The struggle for life is

---

[34] Charles Darwin, *The Origin of Species* (New York: New American Library, 1958 [1859]): 68.

[35] This is based upon an actual contemporary study of wheat plants: Xiang-Shun Hu, et al. "Multigenerational Effects of Different Resistant Wheat Varieties on Fitness of *Sitobion avenae* (Hemiptera: Aphididae)" *Journal of Insect Science* 21.5 (2021): ch. 9.

[36] Note that they are *not* acquired traits during the lifetime of those individuals. Inherited traits are given at birth. More on the mechanics of inheritance will come from Mendel.

[37] Darwin (1958 [1859]): 84.

[38] See: Heather Remoff, "Malthus, Darwin, and the Descent of Economics" *The American Journal of Economics and Sociology* 75.4 (2016): 862–903.

most severe between individuals and between varieties within the same species. As we saw in the "A" passage, it is a given that there is some variation of physical, inherited traits (nature$_1$) even within a dominant cadre within that species. This creates the basis of the competition for food and habitat that allow individuals within a species to survive. Such a "battle" is not like military battles that may bring a war to a conclusion through the signing of a peace treaty. Even if the sub-group with β (inherited morphological traits) out-reproduces the sub-group with α (inherited morphological traits) at time$_1$, this does not preclude another sub-group with γ (inherited morphological traits) at time$_2$ from becoming dominant in numbers within a given environmental niche. This can be so extreme as to lead to extinction of some sub-group populations.

C. We shall best understand the probable course of natural selection by taking the case of a country undergoing some slight physical change, for instance, of climate. The proportional numbers of its inhabitants will almost immediately undergo a change, and some species will probably become extinct. We may conclude, from what we have seen in intimate and complex manner in which the inhabitants of each country are bound together, that any change in the numerical proportions of the inhabitants, independently of the change of climate itself, would seriously affect the others. If the country were open on its borders, new forms would certainly immigrate, and this would likewise seriously disturb the relations of some of the former inhabitants. Let it be remembered how powerful the influence of a single introduced tree or mammal has been shown to be. But in the case of an island, or of a country partly surrounded by barriers, into which new and better adapted forms could freely enter, we should then have places in the economy of nature which would assuredly be better filled up, if some of the original inhabitants were in some manner modified; for, had the area been open to immigration, these same places would have been seized on by intruders. In such cases, slight modifications, which in any way favored the individuals of any species, by better adapting them to their altered conditions, would tend to be preserved; and natural selection would have free scope for the work of improvement.[39]

In this passage, the focus is upon *natural selection*. The "nature" in natural selection is clearly (n/Nature$_1$). We begin with climate change as a factor introduced to create survival pressure that will reveal which sub-groups within the population will reproduce and thrive. Now climate change has always been a controversial subject—even in Victorian England.[40] But Darwin sets this out as a given and has two parts to his model of *survival of the fittest*. (A) within a given geographical space that is an island with little intercourse with other regions and (B) within a given geographical space that is (in contemporary language) an ecosystem (small space) to a biome (large space) there is movement

---

[39] Darwin (1958 [1859]): 88–89.

[40] For an overview of the Victorian climate change deniers see: Allen MacDuffie, "Charles Darwin and the Victorian Pre-History of Climate Denial" *Victorian Studies* 60.4 (2018): 543–564.

between the geographical unit in question and other contiguous areas of land or sea (C).

Both (A) and (B) can be subject to climate changes. In (A) because of its relatively isolated situation the challenge to fitness will be greater than in (B) in which individuals, which make-up the sub-species and have heritable diversity, can travel geographically, the challenge to fitness may be less.[41]

In the history of the world, climate change has been a pressure that has caused changes in the species that could survive and those that become extinct. Global temperature, such as the various ice ages, put considerable pressure on the ability of species to continue. Darwin, a student of geology and fossils, saw this clearly (C).

D. It is very difficult to decide how far changed conditions, such as of climate, food, etc., have acted in a definite manner. There is reason to believe that in the course of time the effects have been greater than can be proved by clear evidence. But we may safely conclude that the innumerable complete co-adaptation of structure, which we see throughout nature between various organic beings, cannot be attributed simply to such action. In the following cases the conditions seen to have produced some slight definite effect: E. Forbes[42] asserts that shells at their southern limit, and when living in shallow water, are more brightly colored than those of the same species from further north or from a greater depth; but this certainly does not always hold good. Mr. Gould[43] believes that birds of the same species are more brightly colored under a clear atmosphere, than when living near the coast or on islands, and Wollaston[44] is convinced that residence near the sea affects the colors of insects. Moquin-Tandon[45] gives a list of plants which, when growing near the sea-shore, have their leaves in some degree fleshy, though not elsewhere fleshy. These slightly varying organism are interesting in as far as they present characters analogous to those possessed by the species which are confined to similar condition.[46]

---

[41] For some discussion of climate change to biological fitness see: Martha Munoz and Craig Moritz, "Adaptation to a changing World: Evolutionary Resilience to Climate Change" from *How Evolution Shapes our Lives: Essays on Biology and Society* (Princeton, N.J.: Princeton University Press, 2016): 238–252; and J. David Archibald, "Private Musings Then Shared Sketches" in *Origins of Darwin's Evolution: Solving the Species Puzzle Through Time and Place* (New York: Columbia University Press, 2017): 127–146.

[42] Edward Forbes work on sea shells and mollusks can be found in: Edward Forbes, Sylvanus Charles Thjorp Handly, *A History of British Mollusca and their Shells* (London: John Van Voorst, 1853).

[43] Gould's work on birds can be found in: John Gould, *The Birds of Great Britain* (London: Taylor and Francis, 1873).

[44] Thomas Vernon Wollaston's work on insects can be found in: Thomas Vernon Wollaston, *On the Variation of Species with Especial Reference to the Insecta* (London: John Van Voorst, 1856).

[45] Alfred Moquin-Tandon's work on plants can be seen in: Alfred Moquin-Tandon, *Éléments de botanique médicale* (Paris: J. B. Baillière et fils, 1866).

[46] Darwin (1958 [1859]): 131–132.

Laws of variation: use and disuse, correlated variation, organism parts that are developed in an unusual manner, secondary sexual characteristics, reversions to long-ago ancestors are the focus of Chap. 5. Much of these cited observations (noted below) were about coloration within certain environments that gave predators and prey certain advantages within those contexts. A contemporary example of this can be seen in the 1950s experiment by Henry Bernard Davis Kettlewell. Kettlewell took two subspecies of moths—one which was light colored and the other which was dark colored. He released the dark colored moths into the variegated, clean woods of Dorset and the light-colored moths into the industrialized sooty woods around Birmingham. In each case the moths were being put into environments in which their predators, birds, could see them more plainly against the contrasting background. This caused their populations to decrease markedly whereas when the moths were blended in to the environment that matched their coloring, their populations were steady and increasing because of the camouflage.[47] This is a good example of the principle that Darwin is putting forth on the definition of natural selection as a fitness measure of one soma-type within a particular environment. It isn't that the dark or light soma-types of the moths are "superior." Neither characteristic, by itself, has higher or lower fitness associated with it, *per se*. It is only when placed in a dark or light variegated environment that the coloring makes a difference. The combination of heritable morphological traits within a certain context creates positive or negative fitness coefficients that will "select" which sub-species will survive.

The "selecting" is a mechanical, material, nature$_1$ process that is teleological in the systemic sense discussed earlier. This will be involved in the process of inheritance from the parents to offspring from even more remote filial ancestors. In a broad sense, even Aristotle recognized this in the *Generation of Animals* (715a 20 ff). Since *completeness* is one of the criteria for evaluating a scientific theory, and since Darwin's account in this regard is lacking the mechanism for the manner that the *how* takes place, we must await the work of Mendel to take the next step on this journey.

The Darwin model can give an account for how species change: *completeness*. However, there may be cases in which it seems that one cannot properly predict which somatic changes will yield positive inheritance outcomes—*coherence*. The general rule is *simple*, but it is of some debate whether such a combative environment is really beautiful or otherwise chaotic: ~ *elegant*.

Gregor Mendel (floruit for experimentation: 1856–1863)[48] was born on July 20, 1822 into a German speaking family in Heinzendorf bei Odrau (now Czechia). The family lived and worked on a farm. As a young man he attended a gymnasium in Troppau. From 1840–1843 he studied philosophy at the University of Olomouc. He was always fighting sickness. He became a Moravian monk (partially because of poverty) He entered the Augustian St. Thomas Abbey to train to be a priest. Then he

---

[47] H.B.D. Kettlewell, "Recognition of Appropriate Backgrounds by the Pale and Black Phases of Lepidoptera" *Nature* 175.4465 (1955): 943–944.

[48] See Edward Edelson, *Gregor Mendel and the Root of Genetics* (Oxford: Oxford University Press, 1999).

attended the University of Vienna to train to be a high school teacher. However, he returned to the abbey and experimented with hybridization of pea plants in the monastery garden. He planted and recorded his trials of more than 5000 plants looking at empirically visible traits in the plants themselves (phenotypes). The four major categories he observed were: (1) Seed color/form (gray versus white/ and round versus wrinkled); (2) Flower color (white versus violet); (3) Pod form/color (full versus constricted/ and yellow versus green); and (4) Stem shaping/size (bushy axial pods versus straight, thin pods/and long versus short). Using these categories Mendel performed Mono-hybrid crosses (altering one phenotypic trait) and Dihybrid crosses (altering two phenotypic traits).

From this "upper-level" observation, Mendel hypothesized the existence of a "lower-level" (genotype) that was responsible for these phenotypic expressions. For Mendel the path from the genotype to the phenotype was reductionistic and simple.[49] Therefore, he could posit a genetic level that was clearly and directly responsible for a phenotypic trait. The concept of a "gene" was set out to be a lower-level biological phenomenon that was responsible for a particular phenotypic trait. There was a one-to-one association between "gene" and "expressed phenotypic trait." Mendel assumed that "parent-1" and "parent-2" (oftentimes called "male" and "female") presented genes for phenotypic traits that were either the same (homozygous) or different (heterozygous). In the latter case, a "battle" between the two would exist and the dominant (for that trait) would determine the character of the zygote.[50]

This simple reductionistic relationship was the basis of the Mendelian Laws of Genetics:

1. The Law of Dominance and Uniformity. Alleles are dominant or recessive within different contexts (i.e., the creation of the gamete from the two parents presents a particular trait X which comes from one parent dominating the other at the allele in the gamete formation—however the dominance might be different if the partner parent were different, a recessive gene in one context could become dominant in another context (with another parent partner)).
2. The Law of Segregation. During gamete formation each gene segregates from other genes so that each genetic interaction occurs separately. Thus, whatever a gene is at the lower-level is for phenotypic trait X will separate itself from all other genes so that each genetic interaction for each phenotypic trait occurs sep-

---

[49] In the contemporary era, from 1970s onward the so-called unit of selection has called into question this simple notion of reductionism. Philosophers of biology have suggested that there are many complications to this simple relationship such as "feedback-loops" and intermediate causal steps in the developmental process. For a brief discussion of this see: Elliott Sober and Richard Lewontin, "Artifact, Cause, and Genetic Selection" *Philosophy of Science* 49 (1982): 157–180 and Kim Sterelny and Philip Kitcher, "The Return of the Gene" *Journal of Philosophy* 85.7 (1988): 339–361.

[50] Aristotle talked about such "battles" as well. See: Michael Boylan, "The Galenic and Hippocratic Challenges to Aristotle's Conception Theory" *Journal of the History of Biology* 17.1 (1984): 83–112.

arately. The "father" may dominate with regard to trait A, but be recessive to the "mother" in trait B.[51]

3. The Law of Independent Assortment. Genes of different traits can segregate independently during the formation of gametes.

The significance of Mendel's work in the context of evolutionary theory is that it provided a conceptionally material account, nature$_1$, for the mechanics of how species might change: those with key phenotypic genetic traits that were successful in a particular environment, could pass those traits forward by reproduction with a better chance for success for their offspring to move forward: filial generation (f_1 could pass on environmentally favorable traits that would give the offspring a better chance of reproducing successfully in f_2). The traits, themselves, were not *better* or *worse* except in the context of particular environmental conditions.

In 1865 Mendel presented his findings to the Natural History Museum of Brno (later published in 1866).[52] His phenotype-outcomes-based model was captivating—especially as it was mathematically interpreted by Mendel (cf. Newton in Chap. 6). And though the model posited an entity that could not be observed, the gene, the concept (however it might be materially constructed at the lower-level) was simple in its descriptions and was able to make predictions that were backed-up by lots of empirical data.

Using our criteria set out in the beginning of this book: *completeness, coherence, simplicity,* and *elegance* to judge scientific theories, Mendel's presentation is certainly big on coherence, simplicity, and elegance. This certainly gave the theory great appeal to his peers.[53]

When combined with Darwin's and other zoologists and botanists working in the nineteenth century, who were inclined toward natural selection, the basic science of how to understand life on earth took a positive step forward that affects our understanding of Aristotle's critical questions: that [the specimen in question] is, the cause of the fact [it's somatic characteristics], what this tells about the essence [of the specimen before us], and "if it is" meaning the necessity or permanence of that essence. It is this last category that has changed the most with Darwin and natural selection—all is very contingent upon changing conditions that will favor some inherited traits over others with environmental change.[54] The big question mark that

---

[51] Of course, in the contemporary—at least suggested from the middle of the twentieth century onwards, the existence of *crossover* genetic expression can link more than one trait and seem as if it is an exception to Mendel's second law—see: Carl Veller, Nancy Kleckner, Martin A. Novak, "A Rigorous Measure of Genome-Wide Shuffling that Takes into Account Crossover Positions and Mendel's Second Law" *Proceedings of the National academy of Science of the United States of America* 116.5 (2019): 1659–1668.

[52] J.G. Mendel, *Versuche Über Pflanzenhybriden* (Leipzig: Wilhelm Engelmann, 1866 and 1870).

[53] Because the number of cases in animal and human genetics are extremely complicated, and because Mendel concentrated on a small number of traits in pea plants only, he could not claim completeness. This would be a goal for his successors until the present.

[54] With significant environmental change today, there will likely be extinctions of many species and the survival of others. For some insights on this see Michael Boylan, ed. *Environmental Ethics* 3rd

possibly challenges completeness and coherence is the inherent changeability that occurs within species *all the time* under evolutionary theory. We will see further repercussions of this in Chap. 9.

He died on January 6, 1884 of chronic nephritis at the age of 61.

Germ Theory

Introduction

As we transition from general theory about life on earth to particular application in humans (and the role of medicine to maintain their health), it is important to examine a transition from the Hippocratic and Galenic humor theory and the Hippocratic understanding of the roles of the hot, cold, wet, and dry within environmental factors—such as the air we breathe regarding these variables to another model in which microbes living outside of the human organism might enter an individual and become a cause of disease. This is a huge change in the way we think about individual and public health (on par with some of the revolutionary changes described in Chap. 6 on physics in the seventeenth century). We move from *imbalance* (in the outside air and within our bodies) as the source of disease to *being attacked by outside agents that enter our bodies and do harm*. We will call this approach of there being an outside agent (of microscopic size) that causes disease in animals (including humans) as *germ theory*.[55]

We will examine some of the important figures in this movement that changed the practice of medicine. Before we begin, it is useful to note the developments in inductive logic that occurred during the nineteenth century. Though Aristotle wrote generally about inductive logic in *Posterior Analytics* II.19; and Francis Bacon advocated generally on *careful empirical scrutiny,* it was still a rather vague concept that was meant to be a contrast to deductive logic.[56] However, in the writings of

---

ed. (Oxford: Wiley-Blackwell, 2022).

[55] The term "microscopic size" implies the existence of the microscope, which was the result of many who studied light and lenses from Galileo to Hans and Zaccharis Jansen (a father-son team who created the compound microscope that improved both magnification and resolution—sixteenth century) to Anton van Leeuwenhoek (who made improvements in grinding and polishing lenses that also improved both magnification and resolution—sixteenth century) to Chester Moore Hall (who created achromatic lenses that greatly improved resolution by minimizing refracted light—1729) to August Köhler and Ernst Leitz who separately made changes that allowed the possibility of photography and the use of multiple lenses on a moveable turret with controllable light source (late nineteenth century) and Joseph Jackson Lister (who increased clarity by using a combination of weak and strong lenses together). For more on this see: Catherine Wilson, *The Invisible World: Early Modern Philosophy and the Invention of the Microscope* (Princeton, N.J.: Princeton University Press, 1997).

[56] Francis Bacon, *The Advancement of Learning* (N.Y.: O. Smeaton, 1962 [1605]); and _____. *Novum Oeganum Scientiarum,* ed. Joseph Devey (N.Y.: Collier, 1902 [1620]).

Blaise Pascal—especially in his letter exchange with Pierre de Fermat on "the problem of points" a mathematically-based theory of probability took over as the way induction was forever-after to be understood.[57]

In the nineteenth century, Pierre de Laplace extended the mathematical understanding of inductive logic as probability. His *central limit theorem* asserts that independent events, with randomly distributed variables will converge rapidly according to an expanding sample space to represent "normal distribution."[58] So, for example, if one were to toss a two-sided coin (heads and tails) into the air, it is possible for a small sample space to produce 10 heads and 0 tails. Since the toss is random and independent, when the sample space approaches n (a very large number of trials), the recorded value of heads v. tails should approach 50% for each (the expected normal distribution).

The description of the process of induction as a logic that helps us understand through empirical trials what is most reasonable to believe (unlike deductive logic which tells us what we *must* hold to be true)[59] sets the stage for the advocates of germ theory (though it can be used to support Mendelian interpretations of inheritance and evolution, as well).

One philosopher who simplified this process into three strategies of approaching empirical data from separate sources is John Stuart Mill.[60] Mill sets out a way to choose how to reason an outcome from these separate sources via their *agreement, difference,* or a *joint method of agreement and difference.*[61] In the *agreement* method, for example if you were trying to determine what was causing your child's asthma attacks, you could list all the exterior factors that confronted your child when they suffered an attack. If there were a common element, then that would be *agreement* and indicate the cause.

If for the same child you took away one of the possible causal factors and asthma occurred (because this factor was no longer there), then that would be an example of *difference.*

Working with both, at the same time to support each other, would be the *joint method.*

---

[57] Blaise Pascal, *Pensées*, ed. P. Sellier (Paris: Bords, 1991 [1670]). His letters to Fermat on the "problem of points" is a thought experiment in which there are two people gambling against each other. In each round each player has an equal chance of winning the pot. The first player to win x number of rounds is the overall winner and will take the entire pot of money (for all the rounds). But what happens if there is an external event that interrupts the game before either player has won x number of rounds? This gets to the understanding of what "fair" means: is it a proportion on the number of rounds played "so-far"? Or should another rule be used? For a further discussion of this in terms of the letters between Pascal and Fermat on this thought experiment, see: Keith Devlin, *The Unfinished Game: Pascal, Fermat, and the Seventeenth-Century Letter that Made the World Modern* (N.Y.: Basic Books, 2010).

[58] Pierre de Laplace, *Théorie Analytique des Probabilités* (Paris: Courcier, 1812).

[59] I set this and other key distinctions down on the epistemic claims that can be made via non-deductive logics in *The Process of Argument: An Introduction* (N.Y. and London: Routledge, 2020).

[60] John Stuart Mill, *A System of Logic* (London: John W. Parker, 1843).

[61] Boylan (2020): 88–91.

Now, let's turn to some of the prominent authors who advocated for germ theory and some of the outcomes that resulted.

Edward Jenner was born on May 17, 1749 in Berkeley Gloucestershire. His father was a clergyman. At 14 he was apprenticed to Daniel Ludlow, a surgeon at Chipping Sodbury, South Gloucestershire. At age 21 he was then apprenticed in surgery and anatomy to John Hunter at St. George's Hospital, London. Jenner later received his M.D. from the University of St. Andrews, Scotland.

Though medical inoculation was already being used in parts of Asia and Africa, in the UK, there were serious concerns about getting infections in the process (which were not fully resolved until first, Pasteur's methods that became *le rigueur*, along with the hygienics of Lister).

In 1768 the English physician John Fewster became aware that the exposure to cowpox caused a person to become immune from smallpox.[62] A similar observation was made in France by Jacques Antoine Rabart-Pommier in 1780.[63]

Thus, the connection between cowpox and smallpox was in the general medical discussion at the time. Jenner advanced a delivery device to connect these *post hoc ergo propter hoc* observations. Jenner inoculated James Phipps, the 8-year-old son of his gardener, with the pus from a cowpox blister.[64] Like the proponents in ancient Greek Medicine who set forth the nature$_3$ position of finding con-commitment correlation between antecedent input and their consequences, Jenner did not comprehensively speculate about *the causation* for his project (*dioti*, Chap. 3). It was enough for him to set out general words like "properties" and "constitution." In effect, this skirts germ theory, but it *does act* as an invitation to others to speculate about the mechanics of the causation involved.

Agostino Basi was born on September 25, 1773 near Lodi, Lombardy, Italy. He died on February 8, 1856 in Lodi. He was the son of a wealthy farmer and lawyer. Though Agostino had a passion for biology (especially insects), his father was more practically oriented and put-up barriers to such esoteric interests.

In 1807 Bassi, began investigating the silkworm disease, *mal de segno* (aka *muscardine*). He paid especial attention to the white powder that could be seen on the silk worms before their deaths. Using the notion of *post hoc ergo propter hoc,* Bassi devoted his attention to the white powder. After 25 years of research, he contended that the white powder was the sign of a disease that was contagious and was caused by a microscopic parasitic fungus. He published his results in 1836.[65]

---

[62] George Pearson, ed. *An Inquiry Concerning the History of Cowpox, Principally with a View to Supersede and Extinguish the Smallpox* (London: J. Johnson, 1798): 102–104; and L. Thurston and G. Williams, "An Examination of John Fewster's Role in the Discovery of Smallpox Vaccination" *Journal of the Royal College of Physicians of Edinburgh* 45 (2015): 173–179.

[63] P.C. Plett, "Peter Plett and Other Discoverers of Cowpox Vaccination before Edward Jenner" *Sudhoffs Archim* 90.2 (2006): 219–232.

[64] Edward Jenner, *An Inquiry into the Causes and Effects of the Variolae Vaccinae* (London: D.N. Shury, 1801).

[65] Agostino Bassi, "Del mal del segno, calcinaccio o moscardino" *Societa Italiana di Biochimica e Biologia Molecolare*. Rpt. Jacques Barbo, tr. (Paris: Bonbée, 1836).

Bassi saved the Italian and French silk industry by recommending disinfectants to kill the white fungal spores on the silk worms. Bassi later generalized from his work with silkworms to disease, in general (affecting all animals and humans). These, also, might be caused by microorganisms. He was one of the first in the nineteenth century in Europe to assert a germ theory of disease and a means of containing it.

Ignaz Semmelweis was born on July 1, 1818 in the Kingdom of Hungry (now a part of Budapest) and died on August 13, 1865 in Oberdöbling in the Austrian Empire (now a part of Vienna). He was from a family of prosperous grocers (who were ethnically German). He began studying the law at the University of Vienna in 1837, but changed after a year to medicine. He took his M.D. in 1844. Because he was unable to get a job in internal medicine, he took an appointment in obstetrics at Allgemeines Krankenhaus, a large teaching hospital in Vienna in 1846. He observed that there was a higher mortality rate of post-delivery deaths (aka childbed fever) among women delivered by male physicians and medical students than those delivered by midwives (13–18% vs. 2%).[66] Some physicians thought that childbed fever (aka puerperal infection) was caused by poor ventilation that gave rise to polluted air, miasma. However, a particular incident turned Semmelweis in a different direction. A friend of Semmelweis, Jakob Kolletschka, was poked by a medical student's scalpel while Kolletschka was performing a post mortem examination. The resulting infection killed his friend. In the autopsy, the cadaver resembled those who had died of childbed fever. This led Semmelweis to make a connection between the two and to explore what was different in the midwife group as opposed to the physicians who delivered babies. The answer turned out to be that the physicians and medical students came in for their shift and performed autopsies on dead babies before going over to deliver women in labor. Semmelweis reasoned that there might be some sort of pollutant in the cadavers that was lethal (just as it had killed his friend, Kolletschka). The midwives came in from their normal clean and proper lives (for the most part) and their death rates were much lower. The key variable, process of elimination (*method of difference*), is contact with the cadavers. Ergo, make all caregivers clean-up before delivering babies. This was Semmelweis' solution and it worked. Infant mortality declined 90%.

What is behind this policy is the idea that there was a polluting factor in the cadavers: germs. These germs could decline significantly with simple hand washing. This was a big step forward in providing empirical support for germ theory.

John Snow was born on March 15, 1813 in York, UK. He died on June 16, 1858 in London. His father was working-class, who spent some time employed at a coal yard, but later become a farmer in a rural village, north of York. The conditions of his upbringing (as the first of nine children) were rather unsanitary.

---

[66] Ignaz Semmelweis, tr. K. Codell, *Etiology, Concept, and Prophylaxis of Childbed Fever*, vol. 2 (Madison, WI: University of Wisconsin Press, 1983); Theodore G. Obenchain, *Genius Belabored: Childbed Fever and the Tragic Life of Ignaz Semmelweis* (Tuscaloosa, AL: University of Alabama Press, 2016); and K. Codell Carder and Barbara R. Carter, *Childbed Fever: A Scientific Biography of Ignaz Semmelweis* (New York and London: Routledge, 2005).

Snow had a gift for math. When he was 14, he went into a medical apprenticeship with William Hardcastle in Newcastle-upon-Tyne. By 1836 he began formal medical education at the University of London. In 1849 he became a licentiate of the Royal Physicians of London. He lived and practiced medicine in the Soho neighborhood of London's West End. Snow's contributions to Western medicine are twofold: anesthesiology and epidemiology.

Anesthesiology

By 1846 Snow learned about the use of *ether* in America to relieve pain during surgery. Within a year he became the anesthesiologist at St. George's Hospital. He also began using *chloroform* and created a safe delivery method for the drug.[67] This represented the first of two critical steps for surgery to move forward (anesthesia and hygiene). Snow's work focused upon the first. The second was put forward by Joseph Lister.

Epidemiology

The second major area of medicine that Snow made advancements is in epidemiology and its relation to public health. In 1831–1832 a cholera epidemic hit London. A second cholera epidemic struck in 1848–1849. Snow was among a group of physicians who formed the London Epidemiological Society which wished to study these epidemics as a public health problem so that future outbreaks could be mitigated. Snow's approach was that cholera was caused by deadly microbes (germ theory). This was counter to the bad air model (miasma) that had been around since the Hippocratic work *Airs, Waters, and Places*.

Snow initiated two classic studies during the third cholera epidemic of 1853–1855.[68] The first study centered around the Broad Street Pump. Snow observed and documented the number of cholera cases on a London Map containing the various districts of the city using dots to represent cholera cases. He found an unusual concentration around the Broad Street Pump region. Using Mill's method of inductive agreement, he decided to inactivate the use of the pump by having the handle removed.[69]

After the pump had been inactivated, the number of cholera cases in that district dropped precipitously. This led Snow to conclude that the water from the Broad Street Pump had been the cause of the cholera in that district (Mill's *method of difference*). However, it still left open the cause of the *germs* that had invaded what had been a safe water source.

To answer this second question, he examined the sources of the water to all of central London (including the Broad Street Pump). He found that there were two water companies: Lambeth Waterworks and Southward & Vauxhall Waterworks. The former company sourced its water from the upper Thames which was free from urban pollution. The latter company sourced its water from the Thames around

---

[67] John Snow, *On Chloroform and other Anesthetics* (London: John Churchill, 1858).

[68] John Snow, *On the Communication of Cholera* (London: John Churchill, 1855).

[69] Obviously, he had to obtain permission from the local council, which handles such matters.

central London and was subject to the dumping of human sewage and other pollutants. The areas serviced by Southward and Vauxhall Works had cholera rates that were 14 times higher than Lambeth. Therefore, by Mill's *method of inductive difference,* the sewage and other dumping was deemed to be the culprit for fostering disease via the germs created by the pollution *method of agreement* (those who received water from the source that was subject to sewage got sick at a much higher rate from something within the sewage: germs).

Snow's two applications of inductive logic (agreement and difference) helped launch epidemiological public health and give another empirical support to germ theory.

Louis Pasteur was born on December 27, 1822 in Dole, Jura, France and died in St. Cloud, France on September 28, 1895. He was from a working-class family of tanners. Pasteur was a poor student early-on, but entered the Collège Royal and earned a general degree in 1840. He continued his studies and was admitted to the École Normale Supérieure in 1842. He stayed for only a year and then returned 3 years later as an assistant to faculty, but left for a better position at a lesser university. Then, in 1848, he became professor of chemistry at the University of Strasbourg, where he met the woman he would marry in 1849. In 1852 he became chair of chemistry at Strasbourg. Then in 1854 he became dean of faculty at the University of Lille where he began his work on fermentation when he was asked to solve some problems with alcohol production at a local distillery. This began his work on fermentation. The central question concerned the cause of fermentation (cf. the reasoned fact, Chap. 3). In 1857 he returned to Paris to become director of scientific studies at École Normale Supérieure. About the same time, one of Pasteur's students wanted to create alcohol from beetroot. But instead of alcohol the fermentation produced lactic acid. At that time, fermentation was thought to be purely a chemical reaction transforming sugar into alcohol.

This changed when Pasteur first proved that a microscopic plant was involved in the souring of milk (lactic acid fermentation). Pasteur proved that yeast (a eukaryotic single-celled microorganism—a fungus) acted on the sugar to form alcohol. He also showed that microorganisms in the air were responsible for contaminating the process and this could be avoided by heating the mixture: a process later called "pasteurization."

Pasteur believed that if microorganisms created and contaminated the fermentation process, they might also be the cause for disease in plants and animals. This was his "ah-ha moment." He addressed such creative insights by saying, "In the field of observation, chance only favors the prepared mind."[70]

This was an important boost to the germ theory of fermentation and *a fortiori* to germ theory, in general.

---

[70] "Dans les champs de l'observation, le hazard ne favorise que les éspirits prepares" in L. Pasteur, "Discours prononcé à Douai, le 7 décembre 1854 à l'occasion de l'installation solemnelle de la Faculté des Lettres de Douai et la Faculté des Sciences de Lille" rpt. in Vallery-Radot, ed., *Oeuvres de Pasteur*, vol. 7 (Paris: Masson & Co., 1939): 131.

One of the clinical results of this was the advent of immunization via inoculation. This began in the 1870s. Pasteur turned his attention to farmers who were losing chickens due to cholera. The disease could wipe out all a farmer's chickens in 3 days. Pasteur and his research team found that if they created a small, weaker quantity of the microbe[71] and injected it into the unaffected chickens and then waited for several days before re-injecting the birds with the full-strength microbe, that the previously injected birds were unaffected by the full-strength cholera exposure.

This created the so-called "Pasteur vaccine approach."[72] The medical team weakens the full-strength virulent sample and uses the weakened form to build-up the animal's/human's body's immune system[73] in order to withstand a full-strength attack. This approach proved successful with chicken cholera, anthrax, and rabies. In the case of small pox, a slightly different approach was used than Jenner's method. This improved strategy of inoculation to prevent some infectious diseases, is an important step forward in public health,[74] and much of this progress is in the application of *germ theory* to population-wide medical prognosis in order to prevent the disease from taking place.[75] This will prove to be an important step towards minimizing deaths within recognizable populations (epidemics) and across various national boundaries (pandemics).[76]

---

[71] The original sample became "weaker" by two methods: the first was to heat the virulent sample partially (i.e., a ways toward "pasteurizing" but stopping short, and secondly by taking an infected animal who has died and hanging up the corpse for a few days before taking fluid samples from the blood. Of course, in some of these diseases that Pasteur worked he confronted viruses instead of bacteria (like rabies), but the procedure still worked. For further discussion see: https://www.vbi-vaccines.com/evlp-platform/louis-pasteur-attenuated-vaccine/ (accessed June 2, 2022).

[72] For some further context on this on how Pasteur developed on the work of Jenner see: Robert A. Weiss and José Esparza, "The Prevention and Eradication of Smallpox: A Commentary on Sloane (1755) 'An Account of Inoculation'" *Philosophical Transactions: Biological Sciences* 370.1666 (2015): 1–11; Jack H. Botting, "Smallpox and After: An Early History of the Treatment and Prevention of Infections" in *Animals and Medicine: The Contribution of Animal Experiments to the Control of Disease* (Cambridge: Open Book Publishers, 2015): 3–16; and Frank M. Snowden, "The Germ Theory of Disease" in *Epidemics and Society: From the Black Death to the Present* (New Haven: Yale University Press, 2019): 204–232.

[73] It should be noted that the "immune system" was only a vague, general concept at this time.

[74] There were, of course, others who also were on the front line of inoculation, e.g., Edward Jenner and Daniel Sutton. For an account of their contribution see: Gavin Weightman, "Sutton and Jenner: The Legacy" in *The Great Inoculator: The Untold Story of Daniel Sutton and his Medical Revolution* (New Haven: Yale, 2020): 156–161.

[75] See my historically-based essay: Michael Boylan, "The Context and Foundation of Public Health Policy" in *Ethical Public Health Policy Within Pandemics* (Cham, Switzerland: Springer: 2022).

[76] Of course, the vaccination process would later bifurcate from the Pasteur process of weakening a "live" (now attenuated element (bacteria/virus)) to using an inactivated version of a bacteria/virus that would *awaken* the body's immune system, but could not cause the disease, as such. The latter is safer, but depends upon the body's immune system recognizing the inactivated input as sufficient for creating effective antibodies.

Joseph Lister.[77] was born on April 15, 1827 to a Quaker family in Upton, West Ham, Essex and died on February 10, 1912 in Walmer, Deal, Kent. His father made significant advancements in the design of achromatic object lenses for the compound microscope. Joseph attended a Quaker school for his primary education in Hitchin, Hertfordshire, and then another Quaker school for secondary education, Grove House School in Tottenham, where he studied mathematics, natural science, and languages. For tertiary education he attended the non-sectarian University College London, Medical School. In 1847 he took a degree in classics and botany. He became bachelor of Medicine at the University of London in 1852 followed by a fellowship from the Royal College of Surgeons. In 1853 James Syme, a professor of clinical surgery at the University of Edinburgh—one of top surgeons of his day—invited Lister to come aboard, and the partnership was fruitful. In 1860 Lister was elected as a fellow to the Royal Society.

Lister's connection to James Syme developed professionally and personally (as Lister married Syme's daughter, a non-Quaker) on April 23, 1856.[78] Lister spent much of his honeymoon visiting leading hospitals in France, Germany, Switzerland, and Italy.

When Lister returned, he began to approach a problem that had troubled him for some time: why so many patients died post-operation (post-op). There seemed to be two logical causes: (a) either the surgery was poorly performed and there were later consequences to this sloppy work; or (b) another health dynamic entered into the picture, post-op. The first option seemed to be rarely the case through Lister's examination of operations in Edinburgh and London. It should be noted, that in the mid-nineteenth century not all operations were performed in hospital. Some were performed in the physician's surgery (clinic) and others at the surgeon's house—even on the dining room table just after the surgeon had been cleaning up garbage from yesterday's dinner![79]

If (a) is eliminated, then the next option is (b). This became Lister's research plan.

In 1859, Lister wrote a letter describing his research on *inflammation*—using the powerful microscope developed by his father. What was key was his material/mechanical procedure. He believed that inflammation and ultimately disease (what is generally called *sepsis*)[80] was caused by microbes that he could examine and understand. In the healing of wounds (the center of his investigation) he identified microbes that might be the cause of infection. He set out his preliminary results in the Croonian Lecture at the Royal Society in 1863. In this lecture, Lister specified the results of his observations and experiments on both mechanical and electrical nerve stimulation, and the manner in which blood clotted. These experiments led to

---

[77] For a good general treatment of Lister's life and work, see: Lindsey Firzharris, *The Butchering Art: Joseph Lister's Quest to transform the Grisly World of Victorian Medicine* (New York: Farrar, Straus, and Giroux, 2017).

[78] Lister was soon to move away from the Society of Friends and become an Episcopalian (Church of England).

[79] See Fitzharris for examples of this.

[80] Sometimes this presented as *gangrene*, instead.

the material explanation of nature that allowed Lister to employ a germ theory to account for wound sepsis. If his account was correct, then a possible solution to the problem would be to kill the germs once the wound was opened—a prime example being before, during, and after surgery.

Lister's "germ killer" was carbolic acid to be applied to surgical instruments, the surgical table, and the wound post-op. The regimen worked by moving the post-op mortality percentages from 45% to 15%.[81]

Though it was slow to catch on, Lister's antiseptic approach to wounds and surgery transformed the surgery as a procedure (a treatment) from a "last-ditch *hail Mary*" to a reasonably safe option to address accidents and tumors that might arise. This antiseptic approach set out a standard for future improvement (that exists until this very day)[82] towards an antiseptic mind-set to make hospital surgeries and other stays more viable.

Conclusion

In many ways, the practice of medicine before the nineteenth century (in the Western Tradition) was concerned with putting forth explanations of Nature via the metaphysical directions of nature$_1$, nature$_2$, and nature$_3$ with some attention given to each as a primary direction for study. The goal was to create a model that would explain general principles (such as the fundamental elements that were the groundwork of the experiential environment) in order to provide the physician with tools to apply to the human body as it moves from its normal, "healthy" state, to one that has been injured or diseased. These abnormalities, whether they are short-termed or chronic, will seriously affect the well-being of the patient and are lumped together as *health* (see Appendix).

For the most part the developments in science before the nineteenth century focused upon diagnosis and prognosis that would lead to humor-based treatments

---

[81] Statistics here are controversial: Ulrich Tröhler, "Statistics and the British Controversy about the effects of Joseph Lister's System of Antisepsis for surgery: 1867–1890" *Journal of the Royal Society of Medicine*. 108.7 (2015): 280–287; see also: B. Jessrey, "Joseph Lister (1827–1912): A Pioneer of Antiseptic Surgery Remembered, A Century after His Death" *Journal of Medical Biography* 20.3 (2012): 107–110; and Edward R. Howard, "Loseph Lister: His Contribution to Early Experimental Physiology" *Notes and Records of the Royal Society* 67 (2013): 191–198.

[82] "What do we know about Early Management of Sepsis and Septic Shock in Polish Hospitals?" *Healthcare Basel*, 9.2 (2021): 140; Jeremy Kahor, Bille Davis, et al. "Association Between State-Mandated Probilized Sepsis and In-Hospital Mortality Among Adults with Sepsis" *Journal of the American Medical Association* 322.3 (2019): 240; Ubed-Iglesias, Fernandez-Burgos, et al. "Letter to the Editor: Admission Characteristics Predictive of In-Hospital Death from Hospital-Acquired Sepsis: A Comparison of Community-Acquired Sepsis" *Journal of Critical Care, Philadelphia* 56 (2020): 318; and Faheem Guirgis, Teresa Padro, et al. "Response to Editor Letter 'Admission Characteristics Predictive of In-Hospital Death from Hospital-Acquired Sepsis'" *Journal of Critical Care, Philadelphia* 56 (2020): 319–320.

Conclusion 185

that sought *balance (symmetry)*. But these treatments turned out to be rather broad and led to poor outcomes more often than not (cf. *Epidemics III,* and the description of what follows the crisis—it seemed rather in the realm of nature$_2$ and out of the control of the attending physician).

It has been the methodology of this volume to extend the general science and the advancements in anatomy and physiology to the additional scrutiny of philosophy of science by the principal practitioners of the times in which these discoveries were put forth. The purpose of this level of scrutiny is in the tradition of nature$_3$ which exhorts us all to use mitigated skepticism whenever novel changes are put forth against the status quo. This is not to suppress change, but to ensure that novel substitutions to the given scientific paradigm are worth accepting.

During the nineteenth century (in the Western Tradition), there were certainly advancements in the theoretical realm—such as evolutionary theory and germ theory.

The meta-understanding of evolution and its mechanical explanation (via inheritance theory) which will stand as a research paradigm for the twentieth century in order to more fully to investigate (vis-à-vis how inherited diseases can be combatted in the clinical realm and how general strategies can be employed in the public health realm).

In the case of germ theory, the development of *disease inoculation* (from germ theory), and *the reshaping of medical surgery* towards including antiseptic cleaning (also from germ theory) and anesthesia transformed medical practice at the clinical level. Using our evaluative terms of *completeness, coherence, simplicity,* and *elegance*, germ theory offered a conceptually *simple* account that brought about an explanatory apparatus that made strides in *completeness* and *coherence.* Though the bacterial world was various, the concept of specific "germs" being responsible for specific diseases is a rather *simpler* approach than having to gather a large variety of environmental sources coming from air, water, and places. From this *simplicity*, *elegance* is achieved.

The nineteenth century (in the Western Tradition) was certainly a blend of theoretical advancement (that had immediate applications) and practical implementation of theory that affected patient outcomes. This trend continued. For more this, we will turn to the final chapters: Chaps. 8 and 9.

Chapter 8
Nature Split: Big Nature and Little Nature

Abstract As in Chaps. 6 and 7, this chapter highlights large advances in general science and in their application to medicine, in particular. There always is a lag between the advances in general science and how they might be applied in medicine. There is also a philosophical response to what the structure of the scientific/medical paradigms *mean* and whether they should be permitted to go ahead without oversight. Two explanatory devices that sought to characterize this new science in the most positive way possible are also explored concerning their claims.

Keywords Albert Einstein · Relativity · Zeno · Immanuel Kant · Space · Time · Spacetime · Alfred North Whitehead · James Clerk Maxwell · Marie Curie · Niels Bohr · Werner Heisenberg · Max Planck · Ernest Rutherford · Erwin Schrödinger · Wolfgang Pauli · Rudolf Carnap · Nature · Reductionism · Logical empiricism · General relativity · Quantum mechanics · Internal medicine · Surgery · Public health · Causation in biology · Cancer · Heart disease · Infectious disease

Introduction

The bifurcation of n/Nature has, in one way, always been with us. The Greek philosophers Democritus and Leucippus put forth a reductionistic strategy that centered around the organization of the smallest units gathered together, and this composition created a material entity at the experiential level—that level at which normal human vision could see a completed entity.[1] The flip side of this process is to determine just *how small* the tiniest entities could be. Since there were no microscopes in ancient Greece, this was largely a thought experiment. Might there be a

---

[1] See Chap. 1.

point in which the sharpest, thinnest knife could no longer cut something into something smaller? If so, then we would have reached the uncuttable, *atoma*. This was later termed (in other languages) as "atom." Though the term was coined in ancient Greece, the word and concept continued on. It was the building block of physics and chemistry from the end of the nineteenth century when Niels Bohr and Ernest Rutherford set out their influential model of the atom and its structure.[2] This set the stage for some crucial re-examination by Max Planck and others on the details of this account.

In the realm of physics-writ-large, we will begin with questions about the nature of time and space and how traditional Euclidean notions of space (such as those that Newton used and Kant defended) may not have been correct and that this created a problem that needed radical reformulation. The same holds true with the re-examination of Aristotelian time.

Just as in Chaps. 6 and 7, there are some fundamental changes in foundational science that affect bio-medicine and the trickles down to patient care (the general mission of medicine).

Before beginning on our quest, it is important to set out two background conditions that underlie these two examinations. They concern field theory of electromagnetism (including the theory of light) and the field theory of radiation. These were in minds of scientists going forward. We will give a brief glimpse of these through James Clerk Maxwell and Marie Curie.

James Clerk Maxwell was born on June 13, 1851 in Edinburgh, Scotland.[3] His father was brother to a Baron and had "old money." James attended the Edinburgh Academy for primary and secondary education and then went to the University of Edinburgh from 1847–1850—studying logic, metaphysics, and mathematics. From there he went to Cambridge from 1850–1856. After that he became a chair of natural philosophy at Marischal College at Aberdeen for four years before going to Kings College, London (1860–1865) where he was made chair of natural philosophy. After a break between 1865–1871, he returned to Cambridge until his death in 1879.

What I want to emphasize here is his work on electromagnetism which began with a paper on Michael Faraday's understanding of "force."[4] Soon, Maxwell set out that electromagnetic fields moved at a speed equal to that of light (which was

---

[2] Ernest Rutherford was a New Zealand physicist who directed the Geiger-Marsden experiment of 1909 that indicated that atoms had a positively charged nucleus that was surrounded by a negatively charged cloud. Rutherford pictured this as a model similar to the solar system with the nucleus as the sun and the electron cloud as the planets. his gave way to Niels Bohr to further definition and structure to this concept of the atom: Niels Bohr, "On the Constitution of Atoms and Molecules" *The London, Edinburgh, and Dublin Philosophical Magazine and Journal of Science"* 26.151 (1913): 1–25 (followed shortly by three other articles in the same journal). Then, a summary occurred in *Nature:* Niels Bohr, "Atomic Structure" *Nature* 107.2682 (1921): 104–107.

[3] For a critical biography of Maxwell's life and work, see: P.M. Harman, *The Natural Philosophy of James Clerk Maxwell* (Cambridge: Cambridge University Press, 2001).

[4] James Clerk Maxwell, "On Faraday's Lines of Force" *Transactions of the Cambridge Philosophical Society* 10.1 (1855): 3.

measured by Michelson in 1879).[5] This led him to conclude that light is an electromagnetic disturbance propagated through a field according to electromagnetic laws.[6] Maxwell's notion of light as electromagnetic radiation traveling through space, was very influential going forward as a background condition of Nature.

A second pivotal background condition scientist was Marie Curie.

Maria Solomea Slodowska-Curie was born on November 7, 1867 in Warsaw.[7] Her parents were both teachers who schooled Marie early on. At 10 years old she attended J. Sikorska's boarding school, then a school for girls from which she graduated with top honors in 1883. She could not attend an established university because she was a woman. Instead, she attended a clandestine university for women.

In 1891 she left Poland for Paris (where her married sister lived). She enrolled at the University of Paris and in 1893 she was awarded a degree in physics and then a year later a second degree. She began studying magnetism-field theory.

In 1885 Wilhelm Röntgen discovered x-rays.[8] In 1896 Henri Becqueri discovered that uranium salts emitted rays that resembled x-rays.[9]

Curie entered this milieu trying to measure intensity with a device that her husband and his brother constructed. She used this device to help her distinguish various sorts of radiation along with her conjectures about the nature$_1$ causal mechanisms.

In her work, Curie hypothesized that the radiation from uranium came from the atoms themselves. This set her apart from her contemporaries. In 1902 she isolated radium as radium chloride and determined its atomic weight. She distinguished radium from the new element that had been brought forth and named after her native Poland, *polonium*.

She went on to discover other sources of radiation from natural ore and (in what turned out historically to be an important medical achievement) she observed that radium exposure killed diseased tumor-forming cells.

Curie expanded the medical uses of radiation to be applied to surgical patients in WWI, and in sterilizing tissue with radon. This work opened the vista of radiation sources and how they might be used in medical practice.

The focus of this group that I am characterizing as "background to the rest of the discussion" is that they are concerned with the way *nature really is, materially,* n/Nature$_1$. On the four key areas of evaluation (completeness, coherence, simplicity, and elegance), they are most interested in completeness and coherence. By getting more of n/Nature$_1$ into the picture, viz., radiation, they aspire to include factors that

---

[5] John C.H. Spence, *Lightspeed* (Oxford: Oxford University Press, 2019).
[6] James Clerk Maxwell, "A Dynamical Theory of Electromagnetic Field" *Philosophical Transactions of the Royal Society of London"* 155 (1865): 459–512; Pierre Maurice Marie Duhem, *The Electric Theory of J. Clerk Maxwell* (Cham, Switzerland: Springer, 2015).
[7] Further details can be obtained from the biography written by her daughter: Eve Curie, *Madame Curie: A Biography,* tr. Vincent Sheean (New York: Doubleday, 2013).
[8] Wilhelm Röntgen, George Sarton, "The Discovery of X-Rays" *Isis* 26.2 (1937): 349–369.
[9] Further information including his Nobel Prize address can be found at: https://www.nobelprize.org/prizes/physics/1903/becquerel/facts/ (accessed July 23, 2022).

had not been taken into account. This also might make a difference in possible contradictions, i.e., coherence. With this background, we move to large changes in the way general science is conceived.

Big Nature

Since the days of the ancient Greeks and the axiomatic geometry of Euclid, the parallel postulate has been a problem in the foundations of the system. Euclid set out a system of postulates that are laid down without justification along with primitive rules of inference. From the postulates and the rules of inference, theorems were derived. They followed necessarily given the foundational postulates and rules of inference. In principle, all theorems derived in this manner could be fit into a rather long deductive argument that went back to the original postulates and primitive rules of inference. This gave power to the tree-structured system which produced outcomes that, on the principle of heritability, possessed conclusions that were equally *necessary*, i.e., could yield the same level necessity, $\Box$,[10] to subsequent theorem demonstrations at levels remote from the origins of the axiomatic system.

Historically, one postulate in Euclid's system was considered by some not to be a postulate, but instead to be a theorem. This is the *parallel postulate* that contends that on a given plane defined by three points, P_{1-3}, one can construct a line, L_1, through two of the points and through the third point, P_3, there is unique line, L_2, that can be constructed that will never intersect with L_1. Euclid considered this to be a postulate.[11] Others, throughout history, have thought that it was a theorem and could be proven with other postulates through the construction of angles and planar figures.[12] None of the attempts to derive the parallel postulate as a theorem have proved successful. Thus, it has historically remained in Euclidean axiomatics as a postulate: an unproven posit that just has to be accepted, intuitively, as true. The reader may be reminded that in Chap. 6 and our discussion of Kant's Transcendental Aesthetic, this was the way space and time were put forward and this acceptance was required to construct and make applications of Newtonian physics.[13]

---

[10] The box quantifier, $\Box$, denotes absolute necessity in quantificational logic.

[11] Mueller (1981):30 ff.

[12] For a discussion of this in the context of the upcoming non-Euclidean geometries see: Harold E. Wolfe, *Introduction to Non-Euclidean Geometry* (Mineola, N.Y.: Dover, 2012 [1945]): ch. 2.

[13] It should be noted that Kant based his notions of time and space in the Transcendental Aesthetic upon an individual's experience in the world. This argument is not disproven by the possibility of positively or negatively curved space that can only be proved through astrophysics using very large distances far beyond ordinary experience.

Big Nature   191

However, in the nineteenth century Nikolai Lobachevsky set out a possible challenge to the assumption of the parallel postulate.[14] Instead of there being a unique line, L_2 that could be constructed through P_3, there are instead an indefinite number of lines that can be constructed if the special understanding of the initial plane is a negatively curves space that can be rotated to allow even more lines to be constructed. This negatively curved space is generally depicted as being like a horse saddle that allows lines to go down to the lowest point and then come up at curved angles that can also be rotated.[15] If space can possess negative curvature, then Euclid's parallel postulate fails because there are numerous lines that can be constructed through P_3 that never intersect with L_1. Thus, one possible challenge to Euclidean space is that it may possess negative curvature and this would alter the deductive inferences that can be brought forward.

Another possible refutation of the parallel postulate occurs if space can have a positive curvature as set out by Bernhard Riemann.[16] What if space were positively curved, like a globe? If one were to attempt to construct a line through P_3, they would find that it would be impossible to construct *any* lines that did not intersect with L_1. To visually picture this, the reader can imagine the globe of the earth at the equator. If one were to pick two particular points, $P_{1\ \&\ 2}$ (one at the equator and one just north of the equator, and draw a line, L_1 through the two points the line would go north until it reached the north pole and then continue until it ended in the south pole. Now, it would also be the case that any other point chosen on the equator, P_n and another connecting point chosen, P_x that was just north of the equator, and a line was drawn between these two points, this line, L_m would intersect with L_1 at both the north and south poles. Since this is true of any given point on the equator that was not P_1, it is the case that (given the hypothesis of positively curved space) there are *no* parallel lines that can be constructed. Thus, a second possible challenge to Euclidean space is that it may possess positive curvature and this would alter deductive inferences can be brought forward.

The possibility of space being non-Euclidean—even though Euclidean space is the day-to-day space that we all experience, opens possibilities for big Nature. The traditional model of Euclidean space formed the basis of Newtonian physics and the way it described and predicted events occurring within it. Kant set this out in the Transcendental Aesthetic (see Chap. 6). Newtonian mechanics was very useful for a few hundred years and is still very accurate for the "medium Nature" in which each of us lives our normal lives. But within the astrophysics realm, big Nature, these minute differences at medium Nature are enlarged so that potential problems arise.

---

[14] Nikolai Lobachevsky, ed. and tr. Athanase Papadopoulos, *Pangeometry* (Zuürich: European Mathematical Society, 2010 [1855]).

[15] For a pictorial description of negative space as being like a horse saddle see: Rudolf Carnap, *An Introduction to the Philosophy of Science*, ed. Martin Garner (New York: Basic Books, 1966):137–139.

[16] Bernhard Riemann, "Über Die Hypothesen welche der Geometrie zuGrunde liegen" Abhandlugen der Königlichen Gesellschaft der Wissenschaften zu Gõtingen" 13 (1867): 1–15.

Then there is the issue of time. Newton and his apologist, Kant, used another ancient Greek thinker, Aristotle, to set out time as an absolute measure of motion that all could universally agree upon. Again, for the most part, this fits most of our everyday notions of time. However, in Chap. 1 we examined the work of Zeno of Elea, who created two frames of relativity that skewed unreflective notions of time. One frame, through *The Achilles* sought to create a problem by asserting that on a line segment AB, Achilles starting at A would have to go half-way to B *before* reaching B, and then half-way again *before* reaching the mid-point. This added "half-way" requirement could be continued ad infinitum. Therefore, motion was impossible (and Parmenides' doctrine that all is one/static [$Nature_2$] was upheld.

The second frame included space in its analysis of time.[17] This included the Arrow and the Stadium. In the Arrow, the arrow was fixed where it was because there was no coherent account of how one gets from one established place to another. In the stadium, the pace of motion was not absolute but relative to whether one was an observer in the stands (seeing two columns which were initially together, marching in opposite directions) or whether one was a member of one of the columns. From the point of view of the observer in the stands, both columns were marching at the same speed, whereas from the standpoint of the column marchers who measured their speed by looking at the other column, they were going twice as fast.

Thus, even from the context of the ancient world, time and (time with space) were conceptualized as relative. However, this takes a big step forward within the realm of quantification through mathematics when we get to Einstein.

Albert Einstein was born in Ulm, which was in the Kingdom of Wüttemberg, a part of the German Empire, on March 14, 1879.[18] His father was a salesman and engineer. Albert attended a Catholic elementary school in Munich from ages 5–8. (Einstein's family were secular Jews.) He transferred to Luitpold-Gymnasium (also in Munich) where he finished primary and secondary school. At 16 Einstein went to the Swiss Federal Polytechnic School in Zürich. He transferred to Argovian Cantonal School to complete his secondary education.

In 1900 he looked in vain for a teaching post and then, instead, took a job at the Swiss (his new country of citizenship) patent office. On the side, he completed a dissertation on molecular dimensions and became a Ph.D. from the University of Zürich in 1906.[19] In 1908 he became a lecturer at the University of Bern, and in 1911 he became a full professor at the University of Prague (when he also changed his citizenship to Austrian—since Prague was a part of the Austro-Hungarian Empire). Thereafter, he moved about regularly since he loved travel.

---

[17] This was prescient as one of the alterations made by Einstein and the writers on big nature in the early twentieth century was to fuse space and time together as spacetime—see the discussion below on Special Relativity.

[18] A wonder biography of Einstein is: Walter Isaacson, *Einstein: His Life and Universe* (New York: Simon and Schuster, 2007).

[19] His dissertation was: *Eine neue Bestimmung der Moleküldimensionen*, 1905 (A New determination of molecular dimensions).

Special and General Relativity

Einstein had a breakout year in 1905.[20] Besides his dissertation (mentioned earlier) he published four papers[21] that were original and proved to be influential: a. "On a Heuristic Viewpoint Concerning the Production and Transformation of Light" (a discussion of the photoelectric effect on how energy is released in bursts (quanta))—this paper was an early influence of quantum theory; b. "On the Motion Required by the Molecular Kinetic Theory of Heat—of Small Particles Suspended in a Stationary Liquid" (a discussion of Brownian Motion that gives empirical support for the burgeoning atomic theory); c. "On the Electrodynamics of Moving Bodies" (this sets out his theory of special relativity by reconciling Maxwell's equations for electricity and magnetism by showing that the speed of light is independent from the motion of the observer); d. "Does the Inertia of a Body Depend upon its Energy Content?" (this essay asserts that matter and energy are equivalent, $E = MC^2$, one of Einstein's most famous equations that sets out that energy equals the mass of converted matter times the speed of light squared. This concept also undergirds that gravity can bend light and space—which is essential to Einstein's new understanding of gravity [contra Newton]). The way that Einstein presents this concept comes from examining a body at rest. If that body spontaneously emits two identical pulses of light in opposite directions (and the object doesn't move), then the exiting light pulses will lessen the energy in that body. Now, what would this event look like to a *moving* observer? From this perspective, the object, itself, it would appear to move in a straight line as the pulses exited the body. The energy level of the pulses would be different: the energy of the pulse moving forward (as perceived by the stable observer) would be greater than the energy of the pulse moving backward. This differential means that the object not only lost energy, but it lost mass as well. Thus, mass and energy (under certain conditions) are interchangeable according to this thought experiment.

Two critical concepts that begin here and continue to be developed are *special relativity* and *general relativity*.

In *special relativity*, that begins in (c), above, Einstein starts with the nature of light as *oscillating electric and magnetic fields* rippling at 186,000 miles per second. Now, if a person could move at that speed, would the oscillating fields seem stationary to him/her? In *one way* "no" (call this α) since Maxwell's equations set out that these oscillating fields could never be actually stationary.

---

[20] There is some evidence that Einstein's wife was an unacknowledged collaborator with him, see Martinez (2011).

[21] (a) Albert Einstein, "Über Einendie Erzeugung und Verwandlung des Lichtes betreffenden heuristischen Gesichtspunkt" *Annalen der Physik* 322.6 (2006 [1905]): 132–148; (b) _____. "Über die von der molekularkinetischen Theorie der Wärme geforderte Bewegung von in ruhenden Flüssigkeiten suspendierten Teilchen" *Annalen der Physik* 322.8 (2006 [1905]: 549–560; (c) _____. "Zur Elektrodynamik bewegler Körpers" *Annalen der Physik* 322.10 (2006 [1905]): 891–921; (d) _____. "Ist die Trägheit eines Körpers von seinem Energieinhalt abängig" *Annalen der Physik* 323.13 (2006 [1905]): 639–641.

But in another way (call this β), the vantage point of someone traveling at the speed of light ought to be visible also to a stationary observer or to an observer moving in a fast train (say at 100 miles per hour) and the velocity of the light beam should be measurable at 186,000 miles per second whether you were in a fast train or not. But (à la Zeno of Elea), it might seem that the viewpoint of measuring a beam of light moving in the opposite direction would turn out to be 186,000 miles per second minus the speed of the train.

These two different answers to the problem of various observers' measurements (α and β) seem to create a conundrum. However, in 1905 Einstein came up with a thought experiment that undergirded his notion of special relativity that would give an explanation.

Imagine a train moving from point A to point B on a straight rail track. Also, imagine that at a given moment one lightning bolt struck the front of the train and another struck the rear. Now, to a stationary observer on a nearby hill adjacent to the tracks and sitting at mid-train when the lightning bolts hit the front and back—seemingly to this observer *at the same moment, i.e., simultaneously.* From this observer's perspective the lightning bolts hit the train simultaneously.

Now, from the perspective of a passenger in the train sitting mid-train, the light from the rear of the train would have to travel farther than the light from the front due to the directional movement of the train. This would cause the mid-train observer to report that the bolt that hit the front of the train struck first.

This thought experiment emphasized that the context of measuring an event depends upon the context of the people doing the measuring. The stationary subject on the hill would present a different story than the observer mid-train.

Special relativity, in a quick epitome, is about the nature of spacetime emphasizing an examination of the relativity of "simultaneity," time dilation,[22] and that objects are measured with respect to their directional movement from the perspective of particular observers and their situation in space time (given that the speed of light is constant).

---

[22] This feature of special relativity asserts that clocks on a moving object record time differently than clocks on a stationary object. The faster the moving object the greater the disparity. The most famous exemplification of this tenet is the so-called "Twin's Paradox" in which one of two twins travels near the speed of light and returns after his trip to find that he is younger than his twin brother who stayed stationary on earth. For a discussion of this paradox see: Cord Friebe, "Twins' Paradox and Closed Time Line Curves: The Role of Proper Time and the Presentist View of Spacetime" *Zeitschrift für allgemeine Wissenschaftstheorie*, 43.2 (2012): 313–326; Charles L. Adler, "Interstellar Travel and Relativity" in *Wizards, Aliens, and Starships: Physics and Math in Fantasy and Scientific Fiction* (Princeton, N.J.: Princeton University Press, 2014): 176–187; Gergely Székely, "A Geometrical Characterization of the Twin Paradox and its Variants" *Studia Logica* 95.1/2 (2010): 161–182; and R. Shankar, "Special Relativity II: Some Consequences" in *Fundamentals of Physics I: Mechanics, Relativity and Thermodynamics, Expanded Edition* (New Haven, CT: Yale University Press, 2019): 209–226.

General Relativity assumes the truth of special relativity, but sets its emphasis upon gravity as it affects spacetime.[23] What special relativity left out was the relationship between spacetime and gravity. A key way to see the significance of this is to compare it to Newtonian Mechanics. For Newton, Gravity was all about a force that was created between the mass of two bodies and their distance apart.[24] Newton presented a flat, Euclidean view of space and a separate Aristotelian notion of time as the simple measure of motion. Instead of this model, Einstein set forth positively curved (Reiman) space. Newton asserted his three laws of motion based upon the fact that: 1 Bodies keep their state of inertia unless acted upon; 2. Force equals mass times acceleration; and 3. For every action there is an equal and opposite reaction.

In contrast to this, Einstein offers the principle of equivalence (asserting the equivalence of gravitational and inertial mass). According to this principle, in its initial rendering (sometimes called the "weak form"), all objects in free fall accelerate identically (sometimes also referred to as the Galileo principle).[25] They do not fall because of the attraction of a large mass, such as the earth, but rather they act as if they were *weightless*. The trajectory of a point mass is determined by its initial position (at measurement) and its velocity. The motion occurs in curved spacetime (gravitation).

In the strong version of equivalence, it is added that the context of the body in free fall makes no difference, *"The outcome of any local experiment (gravitational or not) in a freely falling laboratory is independent of the velocity of the laboratory and its location in spacetime."*[26] Thus, what becomes front and foremost is the geometric structure of curved space and bodies moving within that special structure.

---

[23] Albert Einstein, "Die Feldgleichungen der Gravitation" *Sitzungsberichte der Preussischen Akademie der Wissenschaften zu Berlin* (1915): 844–847.

[24] A famous empirical proof intended to prove this approach is Cavendish's experiment (and many that came after along the same basic lines). In Cavendish's experiment, as originally set out, there were two wooden rods that were suspended by a thin wire. On each end of the rods were metal weights. On one rod the metal weights were 1.6 pounds each (.73 kg). On the other rod they were much heavier, 348 pounds (158 kg). The heavier weights created a measurable twisting upon the lighter wrights. This twisting was within the predicted hypothesis of the attractive force, F, being equal to G times the product of their masses (m_1, m_2) divided by the square of the distance between them (r^2), $F = G (m_1 * m_2)/r^2$. The experiment was performed by Henry Cavendish in 1797–1798 and has been repeated in many forms—sometimes using gold foil instead of wire and much lighter weights. The point of the experiment is to show the attraction of masses to each other (the basis of Newton's theory of gravitation). For more see: Russell McCormmach, "John Mitchell and Henry Cavendish: Weighing the Stars" *British Journal for the Philosophy of Science"* 4.2 (1968): 126–155.

[25] Named after Galileo's famous experiment dropping two weights from the Leaning Tower of Pisa in order to disprove the so-called Aristotelian Weight Principle which suggested that heavier objects would fall faster. See: Ernest A. Moody, "Galileo and Avempace: The Dynamics of the Leaning Tower Experiment" *Journal of the History of Ideas* 12.2 (1951): 163–193; and Alberto A. Martinez, "Galileo and the Leaning Tower of Pisa" in *Science Secrets* (Pittsburgh: University of Pittsburgh Press, 2011): 1–12.

[26] Mark P. Haugen and Claus Lämmerzahl "Principles of Equivalence: Their Role in Gravitation Physics and Experiments that Test Them" *Gyros* 562 (2001): 195–212.

This geometric structure and the activity within it (as influenced by this structure), are what gravity is all about.

In short, the overall differences from Newton amount to these: (a) there is gravitational time dilation; (b) the model of the orbiting planets is no longer dependent upon centrifugal force working against the attraction of large masses through angular motive-dynamics, but is rather due to precession (changes in the Euler angles that are torque-induced due to the effects of rotating masses); (c) because space is curved, every object (including waves such as light) will bend in the curved spacetime (gravity); (d) rotating masses will drag spacetime along with it (like a spinning object in a jar of honey will spin the honey in its vicinity along with its angular motion); (e) spacetime is expanding and will continue to expand; (f) gravitational lensing will lead to the red-shift of light; and the suggestion of (g) the occurrence of singularities and the production of black holes from dwarf stars. These are major changes from the "big Nature" theory of Newton and are part a new vision within the twentieth century and going forward.[27]

Logicist Thesis

Before moving on to "little Nature" it is instructive to examine the philosophical context of science (in the Western Tradition) during the late nineteenth century and into the twentieth century. This context will aspire to be relevant to both the big Nature and the little Nature perspectives. Our story here begins with the logicist thesis. Since much of the work of theoretical physics is done via mathematical formulae in *a priori* speculation (nature$_2$),[28] it is important to understand the relationship between mathematics and the logic it is meant to express.

This relationship was significantly developed by German philosopher/logician/mathematician Gottlob Frege (1848–1925) in his foundational work, *The Foundations of Arithmetic: A Logical-Mathematical Inquiry into the Concept of Number*[29] that set out arithmetic as largely about *numbers* (what they are and how they act to form foundational concepts). For Frege, numbers are not merely psychological constructs of measurement (as Aristotle's understanding of time suggests, *Physics* IV, 10–19), but rather are real entities that are understood in their particularity, e.g., (2 + 2 = 4); or in general form, e.g., (a + b = b + a). In contemporary terms

---

[27] Hanoch Gutfreund and Jûrgen Renn, *Einstein on Einstein: Autobiographical and Scientific Reflections* (Princeton, N.J.: Princeton University Press, 2020).

[28] This is not to suggest that there haven't been empirical testing of Einstein's *special relativity* and/or *general relativity* (nature$_1$). For a discussion on the former see: Abraham Pais, *Subtle is the Lord: The Science and Life of Albert Einstein* (Oxford: Oxford University Press, 2005 [1982]). On general relativity see: Sean M. Carroll, *Spacetime and Geometry: An Introduction to General Relativity* (Cambridge: Cambridge University Press, 2019).

[29] Gottlob Frege, *Die Grundlagen der Arithmetik: eine logisch-mathematische untersuchung über der Begriff der Zahl* (Breslau: W. Koebner, 1884).

this position supports mathematical realism.[30] Numbers speak about the real world. Sometimes they can be thought of as *adjectives*: "There are 4 chairs in the room," cf. "There are blue chairs in the room." As real *properties*, numbers are assertions about concepts and point back to their explanatory origins, viz., logical truths. As mentioned in Chap. 6, real, non-empirical truths are characterized as nature$_2$.

This notion that logic is primary as compared to mathematics (which is explanatory of the real world for speculative scientists, such as Einstein), puts a rather high role for logic (it becomes something like Aristotle's first philosophy). However, Frege goes farther in asserting that logic also is about relations between concepts which are captured in his theory of quantification (the universal quantifier (x) "for all x such that" *and* the existential quantifier ($\exists$x) "there exists an x such that") which connect the logical statements set out in what we now call a Z-F context[31]— often through material implications, addition, or disjunction [or some combination therein within a deductive argument form].[32]

The reductionistic strategy behind this approach harkens back to axiomatic theory structure described earlier in terms of geometry.[33] What axiomatic systems aspire to in setting out truth in a formal setting are: *completeness* (every well-formed formula can be brought back to theorems and rules of inference that can be traced to the postulates and primitive rules of inference of the initial situation) and *coherence* (one cannot produce a well-formed formula that is committed to espousing both ϕ (a truth-functional proposition) and $\sim\phi$).[34]

The positive spin on the logicist approach is that if one can produce a system of logic that is at the foundation of everything and is complete and coherent, and if this system gives birth to mathematics that *inherits* these properties from its progenitor, and if mathematics can then be used to describe n/Nature, then the advocates of nature$_1$ will be put in a secure place to contradict the skeptics who reside in the worldview of nature$_3$. This line of argument dangles *certainty* in front of its audience. This is the same alluring certainty that Kant used to burnish Newton to

---

[30] For a discussion of some of the issues concerning mathematical realism see: Shahid Rahman, Giuseppe Primero and Mathieu Marion, eds. *The Realism-Antirealism Debate in the Age of Alternate Logics* (Dordrecht: Springer, 2012).

[31] For a discussion of Z-F set theory see: Heinrich Vieler, *Untersuchungen über Unabhängigkeit und Tragweite der Axiome der Mengenlehre in der Axiomatik Zermelos und Fraenkels* (Göttingen: W. Fr. Kaestner, 1926).

[32] For more on this see: G. Boolos, *Logic, Logic, Logic* (Cambridge, MA: Harvard University Press, 1998) and M. Dummett, *The Interpretation of Frege's Philosophy* (Cambridge, MA: Harvard University Press, 1981).

[33] The two paradigmatic axiomatic geometric systems are: Euclid, *The Thirteen Books of the Elements*, 2nd revised edition, ed. and tr. Thomas L. Heath (Mineola, N.Y.: Dover, 1956 [4th century B.C.]) and David Hilbert, *Grundlagen der Geometrie* (Leipzig: Teubner, 1903).

[34] The reader can note that two of the four criteria for measuring the adequacy of a medical/scientific system in this book (completeness, coherence, simplicity, and elegance) are registered in the goals of axiomatic systems.

philosophers of his time. It cannot be refuted, that the possibility of such certainty is attractive to anyone inclined to deductive logic (*elegance*). These outcomes would satisfy *coherence* (ensuring that there are no internal contradictions). They would also satisfy *simplicity* because through the tree-structured system one can create many subsequent theorems. These are connected to a generating system that is *simple* and easily, mechanically reduced to the generating process. If it really *worked*, then it was also *elegant* because a large, complex system could be reduced to a small number of symmetrical, interacting postulates and primitive rules of inferences that, in turn, generate an indefinite number of theorems, moving downward in their applications.[35]

Such a system, properly executed, is also *elegant* in its adherence to the rules of inference and their extended applications. Thus, the logicist thesis would satisfy all four of our criteria of assessment: completeness, coherence, simplicity, and elegance. If it were true, it would go a long way in justifying developing scientific theories.

It was this to this end that Alfred North Whitehead and Bertrand Russell explored some of these possibilities in their three-volume work, *Principia Mathematica*.[36] Both of these authors sought to put forth another support of the logicist thesis, but in a different way than Frege, whose system, they contended fell prey to what is commonly called "Russell's Paradox:" ([37] This paradox is central to deductive systems that attempt to be self-referential: for example, here are two famous self-referential paradoxes:

1. Epimenides Paradox: "Epimenides, a person from Crete, says, "All Cretans are liars."[38]
2. Barber Paradox: "A certain barber shaves all the men in the town who do not shave themselves. Does the barber shave himself?"[39]

Whenever one tries to assert membership criteria (in which the set is asserted to be a member of itself), there is a problem concerning both completeness and coherence. In the first case we are presented with a well-formed formula that the system cannot answer, and in the second case any given answer appears to involve a contradiction.

---

[35] Properly justified theorems may be used to justify new, potential theorems.

[36] Alfred North Whitehead and Bertrand Russell, *Principia Mathematica* 3 vols. (Cambridge: Cambridge University Press, 1910, 1912, 1913).

[37] If R is the set of all sets that are non-members of themselves, then if R is not a member of itself, then its definition, x, implies that *it is a member of itself*. In other words, if it *not* a member of itself, then self-referentially, it *is* also a member of itself: this creates a contradiction. For more on Russell's paradox see: Janet Farrell Smith, "The Russell-Meinong Debate" *Philosophy and Phenomenological Research* 46.3 (1985): 305–350; and Bernard Linsky, "The Resolution of Russell's Paradox in 'Principia Mathematica'" *Language and Mind* 16 (2002): 395–417.

[38] Obviously if Epimenides is telling the truth, then he is lying, and Cretans are truth tellers, but that would mean he's lying, etc. If he is lying, then he is a truth teller, but that means he's lying, etc.

[39] If the barber shaves himself, then he does not shave himself, which means he shaves himself, etc.

Russell used this paradox to attack Frege's version of logicism, but then created a *theory of types* to solve his own version of logicism.[40] This attempt to solve the problem tried to use various strategies to separate the *application process* from *types or classes* which are being examined. Other strategies to do this formally, took away *reductionism* and imposed a version of *truth tables* to the process.

Self-referential paradoxes, however, still stuck around and were the foundation of Gödel's Proof, which applied itself to mathematics, itself, with a self-referential paradox (Richard's Paradox) that ended up showing that if mathematics is complete, then it is not coherent and if coherent, then not complete.[41]

Together, these two self-referential paradoxes slowed down the strategy of reductionism behind the logicist thesis, but not completely. In its stead, arose logical empiricism which employed a different sort of reductionism that followed the move towards "little Nature" as quantum mechanics took center stage.

Little Nature

Little Nature begins with the atom. As mentioned earlier, the notion of a smallest particle that was no longer "cuttable" into a smaller entity has been with us since the ancient Greeks. However, in the time period of this chapter, the examination of this. This sub-section will examine the work of five writers on little nature and sketch out how physicists struggled to set out a coherent account. Then we will transition to the results of big and little nature in medical research and patient care.

Max Planck was born in 1858 in Kiel, Holstein. He came from a large, blended family. When Max was ready for school, his family moved to Munich, where he enrolled in Maximiliansgymnasium School where he excelled in mathematics and astronomy. He was also keenly interested in music. In 1874 he studied physics at the University of Munich and then in 1877 he went to study at Friedrich Wilhelms University in Berlin. He defended his dissertation on the second law of thermodynamics (on the entropy of isolated systems always increasing) in 1879 receiving his Ph.D. in theoretical physics.

Planck moved his principal research in 1894 to black body radiation—concerning the intensity of electromagnetic radiation emitted by a black body (also called a "high absorber"). He created a radiation principle called the "Planck Black Body Radiation Law" which was meant to describe the experimental black body spectrum of observational data. This was influential because it satisfied the *coherence* criterion of our model of theory evaluation.

---

[40] Bertrand Russell, "Mathematical Logic as Based Upon a Theory of Types" *American Journal of Mathematics* 30.3 (1908): 222–262.

[41] A plain language account of this is given by Ernest Nagel and James R. Newman, *Gödel's Proof* (New York: NYU Press, 1958). For a connection to Russell's paradox see: Alasdair Urquhart, "Russell and Gödel" *Bulletin of Symbolic Logic* 22.4 (2016): 504–520.

Planck continued to revise his "law" until 1900 when he claimed that electromagnetic energy was emitted in *quantized form*. The elemental unit could be described as E= hv (energy equals h (to be called Planck's Constant) times v (the frequency of radiation waves). "hv" came to be known as quanta photons which possessed their own energy level.

By moving from a continuous stream to discrete units, Planck was among those who were responsible for developing quantum theory.[42]

Ernest Rutherford was born on August 30, 1871 in Nelson, New Zealand. His father was a Scottish Wheelwright who immigrated to New Zealand in 1842. His mother was a schoolteacher who immigrated in 1855. Ernest's primary education was at government schools. At 16 he attended Nelson Collegiate School. In 1889 he attended the University of New Zealand, Wellington—on scholarship. He graduated with an M.A. in 1893 with double firsts in math and physics. In 1894 he received a scholarship to attend Trinity College, Cambridge under Joseph John (J.J.) Thomson, whose work on atomic structure intrigued him—especially the plum-pudding model" (in which negatively charged electrons are surrounded by a volume of positive- charged material (the pudding)).[43]

In 1898 Rutherford left for Canada to take-up the Macdonald Chair of Physics at McGill University, Montreal. He returned to England in 1907 to become the Langworthy Professor of Physics at the University of Manchester, and in 1919 he became Professor of Physics at Cambridge.[44]

Rutherford's research began with an examination of the magnetic properties of iron and then moved to examining other high frequency oscillations at the atomic level—particularly originating from the nucleus. His doctoral thesis was "Magnetization of Iron by High-Frequency Discharges." He built upon the "plumb-pudding" model of his teacher, J.J. Thomson, using the idea of electrons and positively charged atomic material interacting, but set out a more detailed account of the nucleus (his long-lasting contribution to the structure of the atom).

---

[42] It should be remembered from Chaps. 1 and 3, that in the Ancient Greek World, there was a similar dispute with Aristotle arguing for the *continuous* and Zeno and the atomists arguing for the discrete units.

[43] For account of the plum pudding model and the interaction of various the various atomic models that various atomic physicists were putting forth can be found in Charles A. Perkins, "Experiments of J.J. Thomson on the Structure of the Atom" *Science* n.s. 12.297 (1900): 368–370; Alberto A. Martinez, "Thomson, Plum-Pudding and Electrons" in *Science Secrets: The Truth about Darwin's Finches, Einstein's Wife and other Myths* (Pittsburgh: University of Pittsburgh Press, 2011): 147–163; and Maria A. Rodrigues, Mansoor Niaz, "A Reconstruction of the Structure of the Atom and its Implications for General Physics Textbooks: A History and Philosophy of Science Perspective" *Journal of Science Education and Technology* 13.3 (2004): 409–424.

[44] Thaddeus J. Trenn, Hans Geiger, Ernest Marsden, E. Rutherford, "The Geiger-Marsden Scattering Results of Rutherford's Atom, July 1912-July 1913: The Shifting Significance of Scientific Evidence" *Isis* 65.1 (1974): 74–82; and Ernest Rutherford, F. W. Aston, J. Chadwick, C.D. Ellis, G. Gamow, R.H. Fowler, W.W. Richardson, D.R. Hartee, "Discussion of the Structure of the Atomic Nuclei" *Proceedings of the Royal Society of London* 123.792 (1929): 373–390.

Niels Bohr was born on October 7, 1885 in Copenhagen. His father was a professor of physiology and his mother was from a wealthy banking family. Bohr attended primary school at Gammelholm Latin School. In 1903 he enrolled at Copenhagen University majoring in physics. Bohr defended his thesis in 1911 on electron behavior in metals. From there, he travelled to the Cavendish Laboratory at Trinity, Cambridge and J.J. Thomson (who worked on experimental investigations on the conductivity of electricity within gasses). The following year he went to Manchester to work with Ernest Rutherford's laboratory that was working on the disintegration of the elements, and chemistry of radioactive substances.

Bohr returned to Copenhagen in 1912 for his wedding and on his honeymoon, he travelled to British Laboratories working on atomic physics. His post-wedding period began in 1913–1914 when he held a lectureship in physics at Copenhagen, and in 1914–1916 at Victoria University in Manchester. He returned to Copenhagen in 1916, first to the faculty and then to head a research institute that he held until his death in 1962. During his post-wedding period, he adapted Rutherford's nuclear structure to Planck's quantum theory to form his own theory of atomic structure—for which he is best known.[45] In this theory a small, dense nucleus is surrounded by energy levels or "shells" in which electrons rotate around the nucleus. In some ways this can be visualized like the planets orbiting around the sun.

Bohr also contributed to the Copenhagen Interpretation of Quantum Mechanics along with Werner Heisenberg (to be discussed in the Heisenberg section).

Erwin Rudolf Alexander Schrödinger was born on August 12, 1887 in Erdberg, Vienna, Austria. His father came from old Bavarian family money. He later became a botanist. His mother was a professor of chemistry.

Erwin spent his primary school at the local gymnasium where he studied general science, classical grammar, and German poetry. From 1906–1910 he studied at the University of Vienna and he received his doctorate in 1910. During this time, he studied under Fritz Hasenöhrl who in 1904 published on the equivalence of matter and energy from the perspective of black body radiation in a moving body and its equivalence to radiation energy as expressed in the equation: $E = 3/8mc^2$.[46]

During World War I he served as an artillery officer for Austria. From 1920 onwards he spent time teaching at Breslau, Berlin, Magdalen College, Oxford, Princeton, and finally to Ireland at the Institute for Advanced Studies. He became a naturalized Irish citizen in 1948 where he retired in 1955.

In 1925 Schrödinger read Louis de Broglie on wave mechanics. This prompted Schrödinger to propose that electrons moved as waves. Using Bohr's model of the atom, with its various electron levels, Schrödinger proposed that the electrons moved as waves (though they could also be conceptualized as particles). To this end, two of Schrödinger's key contributions concern these issues: (a) his wave equation

---

[45] For an overview of this see: Helge Kragh, *Niels Bohr and the Quantum Atom: The Bohr Model of Atomic Structure, 1913-1925* (Oxford: Oxford University Press, 2012).

[46] For a description of this in the context of Einstein's work see: Stephen Boughn, "Fritz Hasenöhrl and E = mc²" *The European Physical Journal H* 38 (2013): 261–278. If the dates are as they seem, then Hasenöhrl preceded Einstein.

and (b) his cat thought experiment. In the first case, his wave equation is fundamental to understanding how standing waves interact within the state of the system and its energy. One form of the equation is:[47]

$$H(\Psi) = E(\Psi)$$

The variable Ψ describes the behavior of the electron but is silent about its nature. When the time component is added to the equation, then the probability of an electron's wave movement is given more specification, but what the electron "is" remains problematic (wave or particle). And because the vector's position is probabilistic, one cannot state with certainty *where* some given electron will end up. This is partially due to the dual nature of the electron as a wave and as a particle (called complementarity)—which seems to violate the logical laws of non-contradiction ~ (p & ~p) and of the excluded middle (p v ~p).[48]

The wave equation raises fundamental questions on the electron's vector in a way that has been linked by some to Newton's second law (Force equals mass times acceleration, $F = ma$).

The cat thought experiment has fascinated many as it tries to match randomness with our preconceptions of certainty.[49] In Schrödinger's thought experiment, a cat is put into a box with a small amount of a radioactive substance. If that radioactive substance decays, it will trigger a Geiger counter that is wired to a device that will cause *either* the release of poisonous gas or an explosion—in either instance the cat would die.

It is important to note that the decay of the radioactive substance is governed by the laws of quantum mechanics. This means that the atoms of the radioactive substance are in a combined state of "not going to decay" and "going to decay." Since the cat is in a sealed box, there is no observer. This eliminates the "observer-driven" interpretation of quantum mechanics from the problem. Without the observer, the radioactive material is in a state of "going and not going to decay at the same time." This means that the cat is in the state of being both dead and alive at the same time, which is absurd. None of this depends upon an observer so that we should not characterize wave function decay in terms of an observer-driven interpretation. What has kept this thought experiment in the fore among many is its metaphorical suggestion that the cat may be understood to be both alive and dead at the same time within the box. This would violate the Logical Law of Non-Contradiction and Excluded Middle which are necessary for deductive logic—and in a subsidiary fashion to mathematics, whose structure is fundamental to much of

---

[47] Where H is a "Hamilton Operator," which is a set of operations setting out the state of the system, and E which is total energy of the particle. Ψ, that is the state of the vector of the quantum system.

[48] These are read as "not the conjunction of p and its opposite (not-p)" and "p or not-p." These are essential to the creation of any necessary deductive, logical system. See: Michael Boylan, *The Process of Argument: An Introduction* (London and New York: Routledge, 2020): 47.

[49] Paul Halpern, *Einstein's Dice and Schrödinger's Cat: How Two Great Minds Battled Quantum Randomness to Create a Unified Theory of Physics* (New York: Basic Books, 2016).

shared-body-of-knowledge that affects communities around the world.[50] Some have tied the extremes of interpretation (on the demise of the Law of Non-Contradiction) to post-modernism that plays with some of the same ambiguities).[51]

A second interpretation by David Lewis that has been brought forward is that when a quantum event occurs, there is a superposition of outcome-branches at various levels creating a growing plurality of outcome possibilities that expand in an exponential pattern. This might create various *supervision* outcomes that spawn other supervisions (making the cat live on and on in each evolving branch). This obviously creates an unacceptable confusion.[52]

Another, more plausible, interpretation of this thought experiment (that eliminates these extreme consequences) is to introduce probability theory into the mixture. At the instant that the experimental apparatus is set-up (in theory) at time, t_1 there is a percentage chance that the cat will die in the interval t_n (where n is > than 1). This mortality percentage will grow as "n" increases. At some point in time (unless the cat is fed), it will die of starvation. And if the cat is fed, there is another point in time that the cat will die of disease or internal system failure. No living organism is immortal so that the cat *will die*, it is a question of when and due to what causal mechanism. Each of these *descriptions* (in the sense of Donald Davidson) will dictate their own probability assessment.[53] This is not unlike the way we treat patients in the emergency department of the hospital. They arrive due to sickness, organ failure, or accident. If there are few openings in *intensive care*, for example, a triage formula is used based upon probability of survival (the most emergent get care *first*).

If we take this "rather literal approach to Schrödinger's Cat" it expresses a probabilistic expression of a nature₁ event and does not imply the loss of the Law of Non-Contradiction, or the Law of Excluded Middle, or the adoption of epistemologies that see accommodation to the loss of these principles—such as post-modernism.

Wolfgang Pauli was born on April 25, 1900 in Vienna. His father was a chemist. His primary and secondary education was at Döblinger-Gymnasum in Vienna where he graduated in 1918. He then attended Ludwig Maximillans University in Munich

---

[50] For a discussion of the "shared body of knowledge" see Michael Boylan, *The Process of Argument: An Introduction* (London and New York: Routledge): 36–38.

[51] For one account that is sympathetic to this approach see: Hugh Tomlinson, *After Truth: Post-Modernism and the Rhetoric of Science in Dismantling Truth: Reality in the Post-Modern World* (New York: St. Martins, 1989).

[52] David Lewis, "How many Lives has Schrödinger's Cat: The Jack Smart Lecture Canberra, 27 June 2001" *Australasian Journal of Philosophy* 82.1 (2004): 3–22; cf. David Papineau, "David Lewis and Schrödinger's Cat" *Australasian Journal of Philosophy* 82.1 (2004): 153–169; Paul Tappenden, "The Ins and Outs of Schrödinger's Cat Box: A Response to Papineau" *Analysis* 64.2 (2004): 157–164; Christian J.Laventhal Vercler, Naomi Tricot, "Schrödinger's Cat and the Ethically Untenable Act of Not Looking" *American Journal of Bioethics* 20.6 (2020): 40–42; Bartiomiej A. Lenart, "Why We Shouldn't Pity Schrödinger's Kitty: Revisiting David Lewis' Worry about Quantum Immortality in a Branching Multiverse" *Metaphysics* 20.1 (2019): 117–136.

[53] See Donald Davidson, "Davidson Responds to Adverbs of Action" in *Essays on Davidson: Action and Events,* 2nd ed. (Oxford: Oxford University Press, 2001): 163–180.

where he received a Ph.D. in 1921 (working under Arnold Sommerfeld) for a thesis on the quantum theory of ionized diatomic hydrogen. He then took various posts from the University of Göttingen, to Copenhagen to Hamburg, to the Federal Institute of Technology in Zürich. Then, he went to the University of Michigan and to the Institute for Advanced Study at Princeton before creating his own center of research in Zürich until his death in 1958.

Pauli's most significant contributions were in quantum mechanics. He set out that in an atom no two electrons could occupy the same quantum state at the same time—Pauli's exclusion principle. He hypothesized the existence of neutrally charged particles, neutrinos and his introduction of the quantum number (later emerged as "spin") created a norm for quantum mechanics that does not concern itself with "visibility" which created its own issues down the road, cf. Schrödinger's cat.

Werner Heisenberg was born on December 5, 1901 in Würzburg, Germany. His father taught middle and modern Greek at the secondary-school-level. Werner received his primary and secondary education at the Maximilian School in Munich, graduating in 1919 when he went to the University of Munich where he received his Ph.D. in 1923. He then became assistant to Max Born at the University of Göttingen. In 1926 he was Lecturer in Theoretical Physics at the University of Copenhagen under Niels Bohr and then the next year he became Professor of Theoretical Physics at the University of Leipzig. After some international lecture tours and the Second World War, he returned to lead the Max Planck Institute for Physics and Astrophysics.

The work for which is he is best remembered came from an essay published in 1925 when he was only 23 years old.[54] Heisenberg's version of quantum mechanics was grounded in empirical observation of how radiation was emitted from source radioactive material. This inspired him to come up with an explanation that was consistent with the irregular patterns that were noted by measuring devices. Heisenberg wanted to capture some of this irregularity as he returned to the Bohr model of the atom and explained (at first) the movements of electrons. He believed that we cannot always assign a place for an electron in space at a given time (this led to the concept of electron clouds). We also cannot have both space and time together, nor can we follow its orbit in the Bohr-atom-electron levels. This doctrine (that we cannot know both position (at a given time) and the speed of the particle (photon or electron) with accuracy (in principle)) is called The Uncertainty Principle. This principle demarcates the sub-atomic world as different from the visual, empirical world (cf. Kant's Transcendental Aesthetic concerning space and time). In our everyday world we can determine speed and location all the time—just ask an American Baseball hitter who gets a base hit.

Heisenberg's Uncertainty Principle makes the physics of quantum mechanics seem odd to those of us living at the level of human experience. Following on the work of Bohr, Heisenberg's work contributed to a fuller understanding of the

---

[54] Über quantentheoretische Undeutung kinematischer und mechanischer Beziehungen" *Zeitschrift für Physik* 33.1 (1925): 879–893.

Copenhagen Interpretation of Quantum Mechanics.[55] Under this account, there is an epistemic differentiation between the observer and the explanandum being observed. This encompasses an application of the Uncertainty Principle such that the idea of *complementarity* arises, which suggests that certain sub-atomic particles cannot be both observed and measured simultaneously. On the face of it, this interpretation would seem to violate our evaluative terms of *completeness, coherence,* and *simplicity*.[56] Because of this we should classify the principle as nature$_2$ —which also spawned the thought experiment of Schrödinger's cat, and other speculation about whether this approach saves the principle of non-contradiction, or not.

Logical Empiricism and Reductionism

As mentioned earlier, the new concepts of nature as *big* and as *little* were received by much of the Western Philosophical tradition with an inclination towards a mechanistic approach that suggested that where "the real action" was (what was *primary*) in the *explanans* was at the lowest physical levels, which then became expressed upwards to the objects and their behavior that we observe in life. For example, consider a social event like war, marriage, or a baseball game, might be reducible to sociology, which is really applied psychology, which is really applied biology, which is really applied chemistry, which is really applied physics at the atomic level. So that, just as in axiomatics, all well-formed formulae can be listed backwards to all the applied theorems and primitive inference statements. It is mechanically constructed to be perfectly "deconstructable."[57] If the empirical world were just like the perfect axiomatic mathematical system ideal, then pure reductionism would be correct and science and biology would be directed that way in order to determine causation. If correct, this approach might regain *completeness, coherence, simplicity,* and *elegance* which may have come under question with big Nature's relativity and little Nature's uncertainty principle.

---

[55] For more details see: Roland Omnès, "The Copenhagen Interpretation" in *Understanding Quantum Mechanics* (Princeton: Princeton University Press, 1999): 41–54.

[56] One way to make this firmer is to move towards statistical probability instead of Newtonian certainty. This was set out via Max Born, "Zur Quantenmechanik der Stossvorgänge" *Zietschrift für Physik* 37 (1926) 863–867. For further discussion of this see: Asher Peres, Daniel R. Terno, "Quantum Information and Relativity Theory" *Review of Modern Physics* 76.1 (2004): 93–123.

[57] Of course, as we saw earlier with the logicist thesis, some logical and mathematical paradigms are not as perfectly complete and coherent as they plan themselves out to be. On the issue of "deconstruction" see Michael Boylan, *The Good, The True, and The Beautiful* (London: Continuum, 2008): chapter 8.

The Unity of Science and Reductionism

One of the premier proponents of logical empiricism (aka logical positivism) was Rudolf Carnap. Carnap believed in the unity of science (that all the sciences could ultimately be reduced to physics). "The question of the unity of science is meant here as a problem of the logic of science, not of ontology. We do not ask: 'Is the world one?' 'Are all events fundamentally of one kind?' 'Are the so-called mental processes really physical processes or not?'" ... "When we ask whether there is a unity in science, we mean this as a question of logic, concerning the logical relationships between the terms and the laws of the various branches of science. Since it belongs to the logic of science, the question concerns scientists and logicians alike."[58]

In this quotation Carnap shows that his assertion about reductionism (the grounding of the unity of science) is based upon logic [and mathematics], aka nature$_2$. This process of moving between the sciences depends upon the creation of a transitional language that moves us between the practice of empirical science and its expression through a systematic structure of laws and their applications. This transitional structure Carnap thought could be some sort of language of induction—such as Ramsey Sentences.[59]

Another powerful advocate of reductionism and the unity of science was the team of Hilary Putnam and Paul Oppenheim.[60] "The unity of science ... is attained to the extent to which the laws of science become reduced to the laws of one discipline... The exact meaning of 'unity of laws' depends, again, on the concept of 'reduction' employed." One way the authors show this is by a listing of six levels of complexity:

6 ... Social Groups
5 ... (Multicellular) living things
4 ... Cells
3 ... Molecules
2 ... Atoms
1 ... Elementary particles.

These six levels move upward carrying all of the properties that they possess up with them: $(x)\ (x \text{ is transparent} \equiv (y)\ (y \text{ is an atom of } x \rightarrow \text{transparent}(y)))$.[61] This logical expression of how reductionism works is at the heart of the unity of science

---

[58] See Rudolf Carnap, "Logical Foundations of the Unity of Science" from O. Neurath, R. Carnap, and C. Morris, eds. *International Encyclopedia of Unified Science: Volume 1* (Chicago: University of Chicago Press, 1938–1955): 42–62.

[59] For a brief view of Carnap's view of Ramsey Sentences, see: Rudolf Carnap, *An Introduction to the Philosophy of Science* (New York: Basic Books, 1966): Ch. 26.

[60] Paul Oppenheim and Hilary Putnam, "Unity of Science as a Working Hypothesis" *Minnesota Studies in the Philosophy of Science: Volume II*, H. Feigl, M. Scriven, and G. Maxwell, eds. (Minneapolis: University of Minnesota Press, 1958): 3–36.

[61] Read as: for all x, x is transparent if and only if for all y, y is an atom of x and all y is transparent.

hypothesis—which was at the heart of both philosophers of science and the theoretical practitioners of both the big nature and the little nature. They become connected under logical empiricism with little nature becoming causally more primary.

Linguistic Expressions and Meaning

Another takeaway of logical empiricism was its emphasis on careful linguistic expression of empirical data and what they suggest or do not suggest. Again, we return to Carnap and his chapter on empiricism, semantics, and ontology.[62] Carnap sets out different sorts of entities (ontology) and how we know what the things are (empiricism) and how we can talk about these entities (semantics). The real emphasis in the chapter is upon semantics. Beginning with numbers and what they signify, Carnap moves to *kinds of things* to adjectives and then to adverbs. This follows in the tradition of Russell's denotation to A.J. Ayer's linguistic analysis to Davidson and many others. The goal was an exact *description* of what was being communicated and whether its basis was justified for action (cf. Quine on the application to action[63]).

Behind it all, the logical empiricists took the new big and little Nature as a way of *doing philosophy*. All the basic questions that could or should be asked should begin with Nature in one of its forms. In this way, these philosophers from the end of the nineteenth century until the early 1960s were pre-eminent in the Western Tradition. And they set the stage for one application of their theories into the focus of this book: medicine.

Big and Little Nature in Medicine

At the beginning of the twentieth century medicine was making a clean break from the ancient Greek World. My grandfather, who was in medical school in the 1890s, still studied Galen as part of his studies, but those days were rapidly ending. In its place was a new medical delivery system that in many ways mirrored the advances in physics and chemistry: big and little Nature. And as they did so, there was also an evolving understanding of what it means to be healthy (see Appendix). In a quick assessment in this context, we can look at the sources of mortality in the United States and in Europe in the pre-World War 1 era.[64] The rough percentages of mortality in the United States were 35% infectious disease (pneumonia, influenza,

---

[62] Rudolf Carnap, *Meaning and Necessity* (Chicago: University of Chicago Press, 1956): 205–221.
[63] The best presentation this side of Quine can be seen in *Word and Object* (Cambridge, MA: M.I.T. Press, 1960) and *The Web of Belief, 2nd ed.* (New York: Random House, 1970).
[64] Here are the official records from the United States in 1910: https://www.cdc.gov/nchs/data/vsushistorical/mortstatbl_1910.pdf (accessed August 4, 2022).

tuberculosis, diphtheria, et al.); "Other" were 35% (suicide, accidents, unexplained); heart attack, coronary diseases = 20%; cancer 5%; gastro-intestinal disease 5%. In Europe this skewed even more toward infectious disease as the leading killer in 1910.[65]

20th Century Medicine

The scope of this sub-section is the first sixty years of the twentieth century. Since the leading killer in the Western World (the focus of this book), was infectious disease, it is logical that the initial response of the medical systems in this part of the world would be toward creating vaccines (for prevention) and for medicines (for treatment). Since, in both cases one is involved in pharmacological research and application, the social development of an internal research and development structure to create these drugs was crucial to making infectious disease a lesser cause of death by 1960 (the endpoint of analysis for this chapter).[66]

Medicine followed the trends towards science being considered in both the big Nature sense (via public health for populations) and in the clinical sense by treating the phenotype via surgery. It also took direction toward little Nature in its work toward creating vaccines and drug treatments.

The major scientific achievements within medicine were mostly involving little Nature. And within this sphere the goal from the beginning was to complete what Mendel had started in the last century: to provide a n/Nature$_1$ account of who we *are* from the bottom-up (using the reductionistic strategy of the big Nature. This moved basic research to finding out the most primitive descriptions going down to the atomic level, if necessary.

This timeline began in 1869 when Swiss physiological chemist Friedrich Miescher identified a unique atom-type that he called "nuclein" inside the nuclei of human white blood cells (later re-named nucleic acid).[67] In subsequent work Miescher showed that nuclein was a characteristic component of all nuclei and hypothesized that it would prove to be inextricably linked to the function of this organelle. He suggested that this compound existed prior to subsequent cell division—thus making it part of the operating machinery of the cell.

---

[65] Here is a summary from UK and Europe: https://www.ons.gov.uk/peoplepopulationandcommunity/birthsdeathsandmarriages/deaths/articles/causesofdeathover100years/2017-09-18 (accessed August 4, 2022).

[66] For a historical account of the rise of the pharmaceutical companies in the twentieth century see: Alfred D. Chandler, *Shaping the Industrial Century: Pharmaceutical Industries* (Harvard Studies in Business History) (Cambridge, MA: Harvard University Press, 2004).

[67] For more details on this see: Ralf Dahm, "Discovering DNA: Friedrich Miescher and the Early Years of Nucleic Acid Research" *Human Genetics* 122.6 (2008): 565–81. Copies of Miescher's original letters setting this out can be found at: http://www.ub.unibas.ch (accessed August 5, 2022).

The next stage of the process, can be attributed to the Russian biochemist, Phoebus Levene. He was the first to set out a polynucleotide model containing a sugar molecule and a phosphate group.[68] This was followed-up by Erwin Chargaff who extended Levene's work by analyzing whether there was a commonality among living organisms (animals and plants) or whether there were entirely separate systems for each. Chargaff concluded that they were common and that the amount of adenine (A) is usually similar to the amount of thymine (T), and the amount of guanine (G) usually approximates the amount of cytosine (C). In other words, the total amount of purines (A + G) and the total amount of pyrimidines (C + T) are usually nearly equal. (This second major conclusion is now known as "Chargaff's rule.").

Erwin Chargaff expanded Levene's work. In 1944 he read Oswald Avery's essay on bacteria that cause pneumonia.[69] At the time, it was observed that some of these bacteria had a *cap* or outer layer. They were called "S-type." Other of these bacteria did not have this cap and were called "R-type." Avery set up a series of experiments that showed that only DNA could change an *R-type* into and *S-type*. This indicated that DNA could communicate between cell-types. This was not true of other intracellular substances. Thus, by the argument from remainders (Mill's method of difference), DNA was held to be the causal agent.[70]

This was ground breaking work that was confirmed 8 years later by Alfred Hershey and Martha Chase.[71] They found that bacteriophages (viruses that infect bacteria) are composed of two substances: protein and DNA, but not both. They were able to show this by radioactive labels that demonstrated that DNA alone transferred directly from bacteriophages into bacteria when infected by one of those viruses. But the mechanics of the process were still unclear.

What the Hershey-Chase experiment(s) showed was that it was DNA which was the key actor in hereditary interaction—rather than proteins, as some had been promoting (such as Linus Pauling). (Bacteriophages are useful for exploring heredity because they insert their genetic material into their host cell's genetic material. And also, because they multiply quickly and are easy to isolate and retrieve.)What was essentially left in the exploration of DNA was setting out its double-helix structure and how it acted to create proteins. This important work was begun by James Watson, Francis Crick, Rosalind Franklin, and Maurice Wilkins.[72]

---

[68] Phoebus Aaron Theodor Levene, "The Structure of Yeast Nucleic Acid IV. Ammonia Hydrolysic" *Journal of Biological Chemistry* 40 (1919): 415–434.

[69] Oswald Avery, "Studies on the Chemical Nature of the Substance indicating Transformation of Pneumococcal Types" *Journal of Experimental Medicines* 79.2 (1944): 137–158.

[70] For a discussion of "argument from remainders" see Michael Boylan, *The Process of Argument: An Introduction* (London and New York: Routledge, 2020):79–80.

[71] Alfred Hershey and Martha Chase, "Independent Functions of Viral Protein and Nucleic Acid in Growth of Bacteriophage" *Journal of Genetic Physiology* 36.1 (1952): 39–56.

[72] Rosalind Franklin (the only chemist in the group) and Maurice Wilkins (a physicist who created a crystallographic device that helped evaluate their experimental hypotheses), are frequently left out of the story. For a brief account of their contribution to the project, see: https://www.science-

It is my contention that identifying DNA as the key structure for inheritance, its chemistry, structure, and operation, constituted the key little Nature discovery relating to biomedicine in the first half of the twentieth century.

Treating Patients

One way to understand big Nature versus little Nature within the clinical setting (one patient and one health care provider) is by looking at some of the major cases of death from 1900 to 1960.

Little Nature in Medicine: Infectious disease

As mentioned above, vaccination and treatment moved to the creation of medications (internal medicine). This delivery system existed in private practice through "general practitioners" (the only doctor many ever saw—especially in rural settings, small town settings, and segregated ghettos within cities). This became especially effective after the discovery of penicillin which was made available to larger populations after World War II (1946 and thereafter) to treat diseases caused by virulent bacteria.[73] Other, more specific drugs for prevention included the Salk and Sabin polio vaccines that neutered this vicious virus-killer/de-habilitator. But even this life-saving measure was not without controversy.[74] However, the emerging pharmacological industry was here to stay as it both saved millions of lives and made millions in profits.[75] This is an example of the role of internal medicine as a means of distribution of pharmaceuticals as a way to control disease at the clinical level.

Big/Little Nature in Medicine: Accident, Cancer, Heart Attack (Surgery)

From earliest times, simple fractures could often be treated. As time advanced and as surgical practice became sanitary (see Chap. 7), general surgery began taking on more complicated cases. Injuries (such as car accidents) that caused spinal and

---

history.org/historical-profile/james-watson-francis-crick-maurice-wilkins-and-rosalind-franklin (accessed August 8, 2022). For a more complete account see: Howard Markel, *The Secret of Life: Rosalind Franklin, James Watson, Francis Crick and the Discovery of DNA's Double Helix* (New York: W. W. Norton, 2021).

[73] Though penicillin saved countless lives, it became overused. It was prescribed for viral infections for which is was useless, and was even used to fatten cattle for profit. This caused bacteria to mutate and become immune to the anti-biotic. For a discussion of this scenario see: Robert Bud, *Penicillin: Triumph and Tragedy* (Oxford: Oxford University Press, 2009).

[74] See: Aaron E. Klein, *Trial by Fury: The Pollio Vaccine Controversy* (New York: Scribner's, 1972).

[75] I have taken a position on this process, see: Michael Boylan, "Medical Pharmaceuticals and Distributive Justice" *Cambridge Quarterly of Healthcare Ethics* 17.1 (Winter, 2008): 32–46.

cranial damage began to be treated by a developing field of neurosurgery. Following the founders such as Harvey Cushing (early 1900s who founded ways of treating brain tumors)[76] and Walter Dandy (who created the best practices for aneurysm clipping in 1937)[77] this sub-specialty of surgeons took on tasks that seemed miraculous as they saved lives and allowed the lost to regain life and activity.[78]

Likewise, general surgeons who specialized in oncology and cardiology also made strides in implementing surgeries for particular diseases—e.g., more targeted breast cancer surgery that began with double radical mastectomy with lots of tissue removed including muscle margins began limiting the scope of the surgery until finally it fixated upon lumpectomy that have significantly narrower surgical margins.[79]

One of the reasons that various sorts of cancer surgery became more widespread during the twentieth century is that more people were getting and dying from the disease than at the beginning of this era.[80] One reason for this is possibly because of the proliferation of new chemicals that entered the lives of those living in industrialized countries that used toxic chemicals in the fabrication process. These chemical trailings were dumped into pits that contaminated water supplies, or sent out waste into the air that contaminated the very air we breathe. Also, in this mixture, were fertilizers used for agriculture that also made their way into humans within the biome in which they were used to improve crop yields and profits.[81]

Heart surgeries in various forms in the first half of the twentieth century also made tremendous progress.[82] In 1910 Alexis Carrel began successful intrathoracic aortic and cardiac anastomoses in dogs.[83] In 1935 Claude Beck relieved angina pectoris by muscle pedicles, omentum, and pericardial fat inside the pericardium to

---

[76] See: Michael Bliss, *Harvey Cushing: A Life in Surgery* (Oxford: Oxford University Press, 2007).

[77] Mary Ellen Marmaduke, *Walter Dandy: The Personal Side of a Premier Neurosurgeon* (New York: Lippincott, 2003).

[78] For a range of these procedures see: Harutomo Hasegawa, Matthew Crocker, Pawanjit Singh Minhas, *Oxford Case Histories in Neurosurgery* (Oxford: Oxford University Press, 2013). I note also, that neurosurgeons also became involves with oncology—especially in spinal oncology—but this was in the latter part of the twentieth century.

[79] Two general accounts of note: Theo Wagener, *The History of Oncology* (Dordrecht: Springer, 2009): ch. 4; and D. de Moulin, *A Short History of Breast Cancer* (Cham, Switzerland: Springer, 2013).

[80] https://pubmed.ncbi.nlm.nih.gov/29961900/ (accessed August 10, 2022).

[81] Some accounts of air and water pollution are set out in Michael Boylan, ed. *Environmental Ethics*, 3rd ed. (Oxford: Wiley-Blackwell, 2022).

[82] Some general accounts include: Thomas Morris, *The Matter of the Heart: A History of the Heart in Eleven Operations* (New York: Thomas Dunne, 2018); [an account of Dr. C. Walton Lillehei] G. Wayne Miller, *King of Hearts: The True Story of the Maverick who Pioneered Open Heart Surgery* (New York: Crown, 2002); and Gabriel Brownstein, *The Open Heart Club: A Story about Birth and Death and Cardiac Surgery* (New York: Public Affairs, 2019).

[83] Alexis Carrel VIII, "On Experimental Surgery of the Aorta and Heart" *Annals of Surgery* 52.1 (1910): 83–95.

increase blood supply to the heart.[84] This was further improved by Arthur Vineberg in 1946 when he implanted the left internal thoracic artery directly into the front wall of the left ventricle. This reduced the symptoms of angina and was used for almost 30 years.[85]

Big Nature in Medicine

Public Health and Pandemics

The most obvious example of "big Nature" in medical care would be a public health crisis.[86] In the early twentieth century, the most consequential public health crisis was the post-World War I pandemic—sometimes called the Spanish Flu: 1918–1919: The so-called Spanish Flu actually showed its first symptoms in the United States at a military base in Kansas, March 11, 1918.[87] It then went to Europe—probably through troop movements during the last phases of World War I. Very soon, thereafter, the King of Spain, Alfonso XIII, contracted the disease (though some say he got scarlet fever instead). He recovered. After that, the disease was pinned upon Spain—who, in turn, pinned it upon France. However, the first case was in the United States! Why is it so important to *nationalize* the blame for a virus that carries no passport?[88]

According to David Killingray, though the British Empire had dealt with epidemics before through yellow fever, sleeping sickness, small pox, cholera, and the bubonic plague, they were unprepared for effective public health responses to this strain of flu.[89] The infection rate was much higher (up to 40% of the world's population) and there was more consequential human harm (21–50 million dead).[90] Most at risk were the young and healthy (so that World War I soldiers fit into this category).

---

[84] C.S. Beck, D.S. Leighninger, et al., "Some New Concepts of Coronary Heart Disease; Results after Surgical Operation" *Journal of the American Medical Association* 168 (1958): 2110–17.

[85] J.L. Thomas, "The Vineberg Legacy: Internal Mammary Artery Implantation from Inception to Obsolescence" *Texas Heart Institute Journal* 28 (1999): 107–113.

[86] For further information on the grounds of public health policy, see Michael Boylan, ed. *Public Health Policy and Ethics* (Dordrecht: Kluwer, 2005); _____. *International Public Health Policy and Ethics* (Dordrecht: Springer, 2008) and _____. *International Public Health Policy and Ethics* 2nd ed. (Cham, Switzerland: Springer, 2023).

[87] For a discussion of the historical spread of the disease—especially in the British Empire—see: David Killingray, "A New 'Imperial Disease': The Influenza Pandemic of 1918–1919 and its Impact on the British Empire" *Caribbean Quarterly* 49.4 (2003): 30–49.

[88] This behavior exists today as (at the initial writing of this essay) President Trump on multiple occasions blames Covid-19 on China: https://abcnews.go.com/Politics/chinas-ambassador-us-slams-trump-covid-19-blame/story?id=72187153 (accessed August 25, 2020). Cf. Terence Ranger and Paul Sacks, eds. *Epidemics and Ideas: Essays on the Historical Perceptions of Pestilence* (Cambridge: Cambridge University Press, 1992).

[89] Killingray, *loc. cit.*

[90] The standard figures of the Spanish Flu pandemic were 21–24 million, but work done by Niall P.A.S. Johnson and Jeurgen Mueller, "Updating the Accounts: Global Mortality of the 1918–1919

In the case of this pandemic, the first wave was not too far from normal in infection rates and human harm causing civil reactions to be rather restrained. It was the second wave of the viral infection that occurred in the autumn hitting Asia (especially in India and then moving east) that was the deadliest. Again, Killingray believes that part of the cause of this was the British shipping industry that carried the flu from Europe to India and to Hong Kong.

The communicability of the disease was very high—especially for particular sub-populations. *The British Medical Journal*[91] estimates that incidence rates among native populations around the world was as high as 80% in some locales. In general, white populations in the Commonwealth Nations and the United States was as high as 60%. Among the military, the British Navy was around 30% and the British Army around 20%. (Interestingly enough, those military who had been exposed to poisonous gas during the war had their incidence rate drop to 4.7%). This can be compared to similar "in season" flu populations today that have up to an 8.3% incidence rate.

From the same source, the probability of human harm (in this case fatality) was greater than 2.5% overall for those who became sick with this strain of the flu. This can be compared to .1% for flu outbreaks in general populations today.

Further, the human harm continues beyond mere fatalities. In one study that focused upon the U.S. population 22+ years after the outbreak, the aftermath of pregnant women who were affected brought forth offspring who exhibited: 1. Reduced educational attainment; 2. Increased physical disabilities; 3. Lower modal incomes throughout life; and 4. Decreased socio-economic status (due to #1-#3).[92]

Concerning civic, public health responses, though there were sporadic efforts that varied city-by-city, no effective, comprehensive public health measures were undertaken even though there had been prior influenza epidemics in 1889–1890. Some examples of what *was* done in the United States includes first New York City where the sick were (in many cases) quarantined from the healthy. Public health "runners" went about the city taking notes on which areas of the city seemed to be worse. The city's board of education engaged in health education about hygiene and what symptoms to note about sickness (and what course of action to take). New laws concerning public hygiene (such as anti-spitting laws) were enforced.[93] Then, as infections and deaths mounted, schools were closed.[94] African Americans were given systematic, inferior treatment due to attitudes concerning biological inferiority

---

'Spanish' Influenza Pandemic" *Bulletin of the History of Medicine* 76.1 (2002): 105–115, put the figure rather higher—to at least 50 million.

[91] Tom Jefferson and Eliana Ferroni, "The Spanish Flu Through BMJ's Eyes" *British Medical Journal* 399.7735 (2009): 1397–1399.

[92] Douglas Almond, "Is the 1918 Influenza Pandemic Over? Long-Term Effects of In Utero Influenza Exposure in the Post-1940 U.S. Population" *Journal of Political Economy* 114.4 (2006): 672–712.

[93] Francesco Aimone, "The 1918 Influenza Epidemic in New York City: A Review of Public Health Response" *Public Health Reports* 125, supplement 3 (2010): 71–79.

[94] Alexandra Minna Stern, et al. "School Closures" *ibid.*: 63–70.

and general racism. They were put into segregated, sub-standard hospitals without state-of-the-art equipment. Because of this, their mortality rates exceeded their percent of the population.[95]

In San Antonio, Texas (a major city in that era) the public health response was three-fold: ban children from schools, quarantine the sick, and prohibit large public gatherings. This was moderately effective.[96,97]

In Pittsburgh, Pennsylvania (a manufacturing city at the time one-third the size of Philadelphia), there were inconsistent standards. This was caused by a political tug of war between the governor and the mayor. Once people showed signs of clear sickness, they were home quarantined. But this wasn't adequate as Pittsburgh soon has the worst rate of infection and fatality. The general population was sicker than normal before the flu due to the manufacturing plants and the smoke they produced. Public gatherings were permitted including publicly sponsored concerts. Housing was crowded with low-paid mill workers. The public sanitation system was sub-par with open sewers that had excrement floating about. Once again, African American populations were hit the worst. This public health response was a terrible failure.[98]

One lesson from this pandemic is that certain medical threats to health are best served via a broad, public health response. Since the medical machinery works at both the one-to-one interaction between patient and physician (the clinical level) and at the general community level, it is important to have guidelines on which threats should be treated by which apparatus. For example, in mid-century it was decided in the Western World that cigarette smoking was *not* being well-handled via the clinical approach. Therefore, there was a successful transition to the public health approach.

For future health challenges, we must agree upon criteria that will target "little Nature" in a clinical approach or criteria that will demand "big Nature" public health step in. As we have seen recently in the COVID-19 pandemic, this can be very controversial.[99]

---

[95] Vanessa Northington Gamble, "'There wasn't a lot of Concern in those days' African Americans, Public Health, and the 1918 Influenza Epidemic" *ibid.*: 114–122.

[96] Ana Luisa Martinez-Catsam, "Desolate Streets: The Spanish Influenza in San Antonio" *The Southwestern Historical Quarterly* 116.3 (2013): 286–303.

[97] Michael Boylan, "Introduction" in Michael Boylan, ed. *Ethical Public Health Policy Within Pandemics: Theory and Practice in Ethical Pandemic Administration* (Cham, Switzerland: Springer, 2022): 3–12.

[98] James Higgins, "'With Every Accompaniment of Ravage and Agony:' Pittsburgh and the Influenza Epidemic of 1918–1919" in *The Pennsylvania Magazine of History and Biography* 134.3 (2010): 263–286.

[99] For an example of one way to think about this see: Rosemarie Tong, "Taking on Big Fat" in Michael Boylan, ed. *Public Health Policy and Ethics* (Dordrecht: Kluwer, 2005): 39–48.

Conclusion

As in Chaps. 6 and 7, this chapter highlights large advances in general science and in their application to medicine, in particular. There always is a lag between the advances in general science and how they might be applied in medicine. There is also a philosophical response to what the structure of the scientific/medical paradigms *mean*.

As per the evaluative structure of this book, this takes on two dimensions. In the first is the presentation of n/Nature in one of its three forms. Since much of both big and little Nature are built upon mathematically-grounded models, this suggests a $nature_2$ approach.

The second dimension contains the philosophical explanatory approach that began with the logicist thesis, which fell prey to self-reference and how this harmed the formal axiomatic structure of the underlying systems. The replacement, logical empiricism, sought to interpret both big Nature and little Nature in a systematic, reductionistic way that yields *completeness, coherence, simplicity* and *elegance*. However, this explanatory framework will be called into question once we reach Chap. 9.

Chapter 9
Technology and Nature: A Marriage or a Divorce?

Abstract This chapter examines the development of general science from mid-twentieth century to the beginning of the twenty-first century. Then it segues to the big Nature/little Nature in medicine during the same period with a particular focus on the growing role of technology in the practice of medicine. This new partner in the practice of medicine has some upsides and downsides. The evolving role of philosophy of science during this period is also set out. Finally, Public Health policy and practice is examined as an example of Big Nature in medicine.

Keywords Big Nature · Little Nature · Particle physics · Dark matter · Black holes · Stephen Hawking · Cosmology · Einstein · Newton · Maxwell · Quantum Gravity · Loop Quantum Gravity · String Theory · Philosophy of Science · Technology and Medicine · Crispr · International Public Health · Pandemics

Introduction

We left our last chapter with two distinct conceptions of Nature: big Nature and little Nature. Each sought to describe what n/Nature was (most essentially). In one way this seems like setting up a war between the two, concerning priorities. However, since the earliest times in the Western Tradition, n/Nature was described (chapters #1, #2, #3, #4, and #5), there was always an assumption that n/Nature was singular. This seemed necessary to satisfy *completeness, coherence, simplicity,* and *elegance.*

The *objects* of the rules were obviously plural, but the *rules themselves* were united in some way. In ancient Greek Philosophy this was a contentious question.[1]

---

[1] Some perspectives on this from the perspective of ancient Greek philosophy: Michael Stokes, *One and Many in Presocratic Philosophy* (Washington, D.C.: Center for Hellenic Studies, 1971);

© The Author(s), under exclusive license to Springer Nature
Switzerland AG 2025
M. Boylan, *A Philosophical History of Western Medicine*,
https://doi.org/10.1007/978-3-031-97806-7_9

In the context of the mid-twentieth century, there was also a desire to unify natural accounts. This has occurred on both the micro and macro levels, and among some practitioners, to a unification of the two realms—though this is often done from the perspective of one of the two realms: little Nature or big Nature.

Little Nature in General Science

The unification of the various directions in little Nature began with attacks upon the so-called *standard account* (which initially was quantum mechanics). In the standard account, as it stood in the late 1950s and early 1960s, it was set out that: (A) the particles of atoms consisted in a nucleus, and then electrons and protons (consistent with the Bohr model of the atom as existing in concentric *clouds* about the nucleus). Depending upon the ratio of these positive and negative charges, the atom will have a proportionate valence that corresponds to this combination; (B) the operating forces that operate between atoms and molecules are: 1. The strong force, 2. The weak force, 3. Gravity, and 4. Electromagnetism.

Beginning in the early 1960s both the (A) and (B) assumptions are called into question. First, the description of the particles changed. The idea of what is primary in particle physics goes back to the ancient Greeks in which there were the atomists (Democritus and Leucippus) and those who asserted that *properties* rather than very little particles were at the basic level (Thales, Empedocles, Anaximander, Anaximenes, et al.—see Chap. 1). This continued in the seventeenth century when Newtonian mechanics favored Atomism versus Cartesian Vortex Theory.[2]

When considering these two approaches from our evaluative criteria, the approaches that put *properties* as primary are *simpler* than those in which particles are the most primitive. This is because the way "properties" exist (in the latter instance), is generally described as emerging from the particles: thus, creating a two-step presentation, at least (1. Properties exist as a derived phenomenon + 2. Particles exist that combine to create the properties). This could be contrasted to a one-step presentation (1. Properties, as primitive, going forward). However, to many, putting the properties as primitive, seems to be anti-empirical (either nature$_2$ or nature$_3$ [as an anti-realist construct to "save the phenomena" only]).

From the (A) point of view, in the early 1960s quark theory emerged as a refinement of the Bohr model of the atom. In this description of the lowest level of particles/properties we have two large classes: Leptons and Hadrons. Leptons are subject to the weak nuclear force and are the lightest sub-atomic particle group. They consist in electrons, muons, and neutrinos. Hadrons are subject to the strong nuclear

---

Edward C. Halper, *One and Many in Aristotle's Metaphysics* (Las Vegas: Parmenides Press, 2005); and Necip Fikri Alican, *One over Many: The Unitary Pluralism of Plato's World* (Albany: SUNY Press, 2022).

[2] For an account of this dispute see: Daniel Garber, *Descartes' Metaphysical Physics* (Chicago: University of Chicago Press, 1992): chapts. 3–5.

force and are heavier and are divided into Baryons and Mesons. Examples of Baryons are protons and neutrons. There are also anti-Baryons which are unstable and can decay into protons. Mesons contain a quark and an anti-quark. Examples are pions (the lightest) and kaons (any group of four mesons with a quantum number called *strangeness*).[3] Quarks are defined through the *properties* they exhibit—such as "spin" that can be "up (electrons)" or "down (electron-neutrino)" then "bottom (tau)" or "top (tau-neutrino)" and "strange (muon)" or "charm (muon-neutrino)."

Quark theory essentially moves to a "measurable property" basis of particle physics. And the "measuring" is done by expensive colliders that have been constructed in the U.S.A. and Europe (CERN).[4] Obviously, this part of particle research exploration is dependent upon very expensive measuring devices that are built with certain conceptions in mind (their hypothesis) of what they are looking for and whether they have found it. This risks the philosophical fallacy of *begging the question* (assuming what you are seeking to prove).

When we bring in technology into the study of Nature in a big way, this is always a risk of assuming what you are trying to empirically prove with measuring devices that assume what they are trying to prove.[5] Part of the problem here is the uncritical use of technology in scientific experimentation. Machines (especially those with complicated physical "readouts") are thought to be the messengers of objective truth. However, scientists, in their longing to make significant *discoveries*[6] can create devices that will assist them to *confirm* their hypotheses. They can also read the data *sympathetically* in order to *confirm* their hypotheses.[7]

These problems may be mitigated if we change our emphasis from *confirmation* to *falsification*. From the falsification point of view, scientists testing a hypothesis

---

[3] For an overview of the quark group of entities see: Francisco J. Yndurain, *The Theory of Quark and Gluon Interactrions* 4th ed. (Cham, Switzerland, 2010).

[4] See: Paul Halpern, *Collider: The Search for the World's Smallest Particles* (Oxford: Wiley-Blackwell, 2010) and Pauline Gagnon, *Who Cares About Particle Physics: Making Sense of Higgs Boson, the Large Hadron Collider and CERN* (Oxford: Oxford University Press, 2018).

[5] One instance of this in the history of science was the "discovery" of N-rays in the early twentieth century. The measuring device (constructed on the assumption that N-rays existed) showed these N-rays to exist when they really did not. For an account of this episode see: M.J. Nye, "N-Rays: An Episode in the History and Psychology of Science" *Historical Studies in the Physical Sciences* 11.1 (1980): 125–156.

[6] This is the "psychology angle." Sometimes, a researcher is "so sure" their hypothesis is correct that they falsify data in a study so that they might be able to "prove" what they believe to be the case. In the biomedical realm this happened with the lumpectomy procedure for breast cancer—that turned out to be correct, but the data from the study did not support the hypothesis. See: https://www.baltimoresun.com/news/bs-xpm-1994-03-13-1994072002-story.html (accessed August 18, 2022).

[7] I discuss one "much discussed" thought experiment called the Raven's Paradox in Michael Boylan, *The Good, The True, and The Beautiful* (London: Continuum, 2008): 92–97. Another broader take on the "confirmation problem" can be found in Richard Boyd, "Confirmation, Semantics, and the Interpretation of Scientific Theories" in Richard Boyd, Philip Gasper, and J.D. Trout eds. (Cambridge, MA: Massachusetts Institute of Technology Press, 1993): 3–36.

seek to set up their study such that they can define the empirical data input that would show their hypothesis to be *wrong*. This was the approach of Karl Popper.[8] The notion behind this is that if a scientist is ready to admit that the hypothesis *could be* wrong, then they would want to create parameters that would show whether this was the case. This approach might be psychologically more efficacious for researchers because it structures "truth" front and center rather than "accolades for confirming a scientific discovery."[9] If a researcher structured their project such that there were clear evidence gathering which, if found to be the case, would *falsify* the hypothesis and that path was followed with as much rigor as the gathering of evidence to *confirm* the hypothesis, then the cause of *completeness* and *coherence* would be better served.

From the (B) vantagepoint, there is the issue of a possible *fifth force*.[10] The search for a fifth force has increased in recent decades due to two discoveries in cosmology which were not explained by the big Nature theories of the time. In the first case it has been calculated that there is considerably more mass in the universe than the standard model of the time (pre-1960s) had predicted. This had to be made up with another source of mass. So, in the second instance, cosmologists put forward a theory that most of the mass of the universe is accounted for by an unknown form of matter called dark matter and that exists in empty space, and that there is a consequential accompanying force that can be called dark energy.[11]

Some cosmologists account for this possible fifth force as arising from undiscovered particles and their interactions while others put the causal components upon the big bang theory and a constantly expanding universe (inflation) that includes the creation of black holes which (perhaps in concert with dark matter and the consequent energy) may set out another account of gravity (one of the initial 4 forces in the standard account).[12]

---

[8] Karl Popper, *The Logic of Scientific Discovery* (London: Urwin Hyman, 1985 [1959]): 133–161.

[9] Of course, if the data do not come in as expected, there is no guarantee that it is necessarily false. The data may have been incorrectly processed and even if properly processed, the causal relations may not be any clearer than they would be under the "confirmation" approach. This has been a contentious discussion. For more on this see: Hilary Putnam, "The 'Corroboration' of Theories" from *The Library of Living Philosophers*, Vol XIV, *The Philosophy of Karl Popper,* Paul A. Schlipp, ed. (LaSalle, IL: Open Court, 1974): 221–240.

[10] Allan Franklin and Ephraim Fishbach, *The Rise and Fall of the Fifth Force: Discovery, Pursuit and Justification in Modern Physics.* 2nd ed. (Cham, Switzerland: Springer, 2016).

[11] For a discussion of dark matter and dark energy see: P. James E. Peebles, "Dark Matter" *Proceedings of the National Academy of Sciences of the United States of America* 112.40 (2015): 12246–12248; and David N. Spergel, "The Dark Side of Cosmology: Dark Matter and Dark Energy" *Science* n.s. 347.6226 (2015): 1100–1102.

[12] For a discussion of dark matter, dark energy, black holes and their effect on a theory of gravity see: Alicia Suskin Ostriker, "Dark Matter and Dark Energy" in *Waiting for the Light* (Pittsburgh: University of Pittsburgh Press, 2017): 17–27; Anthony Zee, "A Friendly Contest Between the Four Interactions" from *On Gravity: A Brief Tour of a Weighty Subject* (Princeton: Princeton University Press, 2018): 9–14; and Dan Hooper, "Radically Rethinking Dark Matter" from *At the Edge of Time: Exploring the Mysteries of Our Universe's First-Seconds* (Princeton: Princeton University Press): 143–157.

Stephen Hawking was born on January 8, 1942 in Oxford. He came from a family of physicians. He received a first-class B.A. degree in physics from Oxford in 1962. In 1966 he was awarded his Ph.D. in applied mathematics and theoretical physics from Trinity Hall, Cambridge.

In 1963 he was diagnosed as having ALS (amyotrophic lateral sclerosis). Despite his disability, his work advanced the understanding of black hole dynamics by using the model of thermodynamics as template for black hole dynamics. In 1971 he put forth an account of the big bang that included the standard *life history of a star* which, if it is massive enough in its final phase, will explode and then contract. In this process it collapses inwardly propelled by its weight intensifying its gravity such that numerous quantities of mass might be condensed into the space occupied by a proton.[13] These repositories he called mini-black-holes which had to be ruled by both general relativity and quantum mechanics (which gave a nod at how big Nature and little Nature might be combined). But this could contain a problem: the general reversibility of physics and the seeming irreversibility of this black hole birthing process. "Reversibility" is important because it is embedded in the elemental symmetries of space, time, and causality. What is "done" can always be "undone." But if black holes imply (by their birthing story) an irreversibility, then this irreversibility might mean that black holes do not preserve information.[14] This would make the behavior of black holes to be contrary to the way general relativity otherwise operates. This seeming contradiction is called *the black hole paradox*.

The paradox is even more puzzling when one considers quantum entanglement in which two sub-atomic particles can be linked to each other despite vast separations in 3-d space.[15] This new wrinkle asks the question that if nothing (i.e., no information) can leave a black hole because of irreversibility, then what happens to two sub-atomic particles that are entangled when one is inside a black hole and the other is separated from it? It would seem that there must be an answer to the irreversibility problem, which is at the center of the black hole paradox. Hawking's solution was that the black holes are not irreducibility contained, but emit small amounts of electromagnetic radiation until they collapse and release what is inside. As to entanglement, when one particle is pulled inside the black hole, the other escapes at the event horizon, and when the black hole collapses, the previous entanglement relation is maintained.[16]

Hawking also gave support to John Archibald Wheeler's "no-hair theorem" which asserts a stationary black hole solution to the Einstein-Maxwell equation of

---

[13] Stephen Hawking, "Gravitational Radiation from Colliding Black Holes" *Physical Review Letters* 26.21 (1971): 1344–1346.

[14] David Finkelstein, "Past-Future Asymmetry of the Gravitational Field of a Point Particle" *Physical Review* 110.4 (1958): 965.

[15] John Bell, "On the Einstein Podolsky Rosen Paradox" *Physics* 1.3 (1964): 195–200; cf. A. Einstein, N. Rosen, and B. Podolsky, "Can Quantum-Mechanical Description of Physical Reality be Considered Complete?" *Physical Review* 47.10 (1935): 777.

[16] For a revision of his 1974 position see: Stephen Hawking, "Breakdown of Predictability in Gravitional Collapse" *Physical Review* 14.10 (1976): 2460–2473.

gravitation and electromagnetism through a process that uses mass, electric charge, and angular momentum—beyond these, a black hole has no other properties: this is an example of *simplicity* in scientific theory.[17]

Hawking believed that because the mass density of the black holes was so large, this created a strong gravitational force that would attract adjacent matter by negatively curving spacetime. This process re-defining the force of gravity (normally a member of the big Nature tribe) with the pinhole of the black hole which was the driving force. Thus, physical science is united via the driving force of a re-defined quantum arena.

Big Nature in General Science: Cosmological Choices

As mentioned in the previous sections, the big bang theory, with its ongoing inflationist consequences, is now part of the new standard account. As we saw in Chap. 1, there are three positions one can take in general cosmology: (a) there was a beginning to the universe and that beginning either keeps on going or it hits a stopping point after which things hang on for a spell and then they possibly disintegrate to an end (one reading of Empedocles and of modern entropy); (b) the universe is and was always the same way (with the implication that it forever will be the same—Aristotle; and (c) there is a cyclic nature to the universe as it is created, dies and then is re-created again[18] (this cosmological perspective is consistent with Empedocles (Chap. 1) and is consistent with several views of Hinduism[19]).

Many of these cosmological n/Nature$_2$ theories have as their aim either: (a) a new causal account of the entities in big Nature such that it could also accommodate little Nature or (b) a new causal account of little Nature so that it could accommodate big Nature or (c) modifications to both such that a grand, unified theory would result. Certainly, there is both a *simplicity* and an *elegance* in having big Nature and little Nature explained through a single account. In the 60-year period between the 1960s and 2020s, there were several attempts at doing this: 1. Followers of Einstein who wanted to make General Relativity the driving force of a unified approach (choosing the (a) approach). 2. Stephen Hawking and his followers who wanted to follow the strategy of (b)—as per above;[20] 3. Quantum Gravity (and its

---

[17] Charles W. Misner, Kip S. Thorne, John Archibald Wheeler, *Gravitation* (San Francisco: W.H. Freeman, 1973): 875–877.

[18] Sometimes this contraction after the inflation is called *the big crunch* and some find evidence for this in the cosmic microwave background (CMB)—see: Dylan L. Jow, Douglass Scott, "Re-evaluating Evidence for Hawkings' Points in the CMB" *Journal of Cosmology and Astrophysics* 3 (2020): 1–10.

[19] For the Hinduism slant see: Malcolm Stewart, *Sacred Geometry of the Starcut Diagram:Number, Proportion, and Cosmology* (Rochester, VT: Inner Traditions, 2022).

[20] Stephen Hawking, *The Theory of Everything: The Origin and Fate of the Universe* (Mumbai: Jaico Publishers, 2006).

sub-categories of Loop Quantum Gravity and String Theory)[21] which also adopt the strategy of (b). Here is a brief review of each approach (starting with #1 and moving to #3 as we've already discussed #2). Note also that a few theorists who were keen on both big and little Nature as equal driving forces (c).[22]

1. Followers of Einstein.

After 1915 and the publication of his general theory of relativity, Einstein's approach was to combine gravity and electromagnetism. He had two models that he wanted to contend with. The first was Newton's approach which created the structure of a universal law of everything: $F = G\, m_1\, m_2/r^2$ (The force of gravity, F, equals the universal Gravitational Constant, G, times the masses of two bodies (m_1 and m_2) divided by the square of the distance between them (r^2). This statement was structurally *complete, coherent, simple,* and *elegant*: the criteria for a model scientific law. Einstein wanted to create a similar exemplar. Second, was the model of James Clerk Maxwell who combined the two major forces of his time: electricity and magnetism (including light) into one account of electromagnetism.[23] (Of course, Maxwell's work was before the discovery of the strong and weak atomic forces.)

With these two models in mind, Einstein sought to make general relativity the driving force of the unification of his well-received account of big Nature with the developing quantum little Nature. Initially, Einstein was concerned with the uncertainty dimension of quantum mechanism. This led to his letter to Max Born in 1926: "Quantum mechanics is very impressive. But an inner voice tells me that it is not yet the real thing. The theory produces a good deal but hardly brings us closer to the secret of the Old One. I am at all events convinced that *He* does not play dice."[24] This move connected him to Newton's approach. The big Nature approach was what was *most real* to him (empirically) and the little Nature had to follow the same rules since all Nature is uniform, after all. Since the weak and strong atomic forces were not well-known, he could take this stance and try to make it all about *gravitation* as redefined by general relativity and the *field theory* which it engendered (which had the effect of marginalizing quantum theory as a separate Natural account in its own right).[25]

2. Stephen Hawking (see above).

---

[21] Steven Carlip, "Quantum Gravity: A Progress Report" *Reports on Progress and Physics* 64.8 (2001): 885–942; and Sundance O. Bilson-Thompson, Fotini Markopaulou, Lee Smolin, "Quantum Gravity and the Standard Model" *Classical and Quantum Gravity* 24.16 (2007): 3975–3984.

[22] Steven Weinberg, *Dreams of a Final Theory: The Scientist's Search for the Ultimate Laws of Nature* (New York: Knopf, 2011).

[23] For a general overview of Maxwell's life and work see: Basil Mahon, *The Man Who Changed Everything: The Life of James Clerk Maxwell* (Chichester: Wiley, 2004).

[24] Abraham Pais, *Subtle is the Lord: The Science and Life of Albert Einstein* (Oxford. Oxford University Press, 1982): 443.

[25] Albert Einstein, "On the Generalized Theory of Gravitation" *Scientific American* 182.4 (1950): 13–17.

3. Quantum Gravity. One logical response to unifying big Nature and little Nature is to incorporate the two in an account of Gravity. Now this *not* something *new*. Newton and Einstein thought that it was all about Gravity and that the best way to explore these dynamics was through the big Nature perspective. Hawking looked at big Nature from the micro-perspective of condensed black holes. It took quantum gravity to link the big and the small in a way that intended to give each their legitimacy. At the writing of this book there are two major contenders for the best explanation of quantum gravity: loop quantum gravity and string theory.

Loop Quantum Gravity

Since the quantum gravity approach begins with the background independence of quantum mechanics and general relativity (over and against using general relativity, alone, to describe everything). It contends that we need a theory that will employ perspectives of each to generate a comprehensive standpoint. The quantum standpoint calls for position, time, momentum, and energy to be expressive in *wave function* terms that also are characterized by *uncertainty*. The quantum perspective is highly background dependent while general relativity is background independent: it allows one to calculate the changing shape of the coordinate system itself.

One important step in this direction was the work of John Archibald Wheeler and Bryce S. DeWitt.[26] In their work, they make several key moves. First, the notion of "geons" posits gravitational-electrodynamic *entities*. This is a reification of the fields of gravity and electrodynamics. These "waves" have a very small mass (from 10^{-39} g. to 10^{-57} g.). They also posit that spacetime is foliated into surfaces that allow the observer to calculate the changing shape of the vector field's coordinate system. Whether this dynamic is quantifiable in a "non-question-begging way" is controversial.

When referencing Wheeler and DeWitt in the context of ADM Formalism[27](the impetus to create geometric depictions of motive field interactions [Hamiltonian Mechanics]), then one can set out an *abstract space of spaces*: 3-D spatial metrics are cut out of 4-D spacetime slices giving an account of motion through time. As you move through this coordinate system, the geometry of space changes—resulting in a quantum equation for the fabric of space. Thus, little Nature *describes* big Nature and is not a candidate for "The Theory of Everything," but rather is an account of "Unifying Existing Accounts." The standard accounts of general

---

[26] John Wheeler, "Geons" *Physical Review* 97 (1955): 511; _____. *Geometric Dynamics* (New York: Academic Press, 1962) and Bryce S. DeWitt, "Quantum Theory of Gravity" *Physical Review* 160 (1967): 1113. Their concurrent work is sometimes characterized as the Wheeler-DeWitt equation.

[27] R. Arnowitt, S. Deser, C. Misner, "Dynamical Structure and Definition of Energy in General Relativity" *Physical Review* 116.5 (1959): 1322–1330.

relativity and quantum mechanics are brought into one framework and vocabulary that is consistent.

If the geons-approach of Wheeler/DeWitt turns out *not* to be solvable, then a new tack needs to be explored—perhaps this is loop gravity. The logic here is to go *past* the space of metrics into a space of assumed connections. These "connections" provide us with the "loops" (aka *spin networks*) that represent the coordinate systems that are connections (mathematical functions that tell you *how* (something like a vector) changes as it moves between two points in space).

A contrast can be made to the standard account in parallel transport in which (as a vector moves in a path in negatively curved space) the vector, itself, rotates. The amount of rotation encodes information about the changes in geometry along its path. This breaks-up the process into *quanta*. By analogy, if the localized part of the field of connections contains all the information about spacetime, then maybe we can represent these connections (quanta) as *being* spacetime. Einstein tried to take this general path in the early 1950s, but was unsuccessful. What Loop Gravity does is to question whether the geometric depiction of spacetime is non-continuous (instead of Einstein's assumption that it was continuous). This non-continuity constitutes the quanta and the way these quanta are connected are the "loops."

This theory provides one account of how quantum gravity might provide a way to unify quantum mechanics and general relativity (little Nature and big Nature).

String Theory

Another way to characterize quantum gravity for the same goal is via string theory. String Theory posits fundamental, one-dimensional entities (with length, but no width or depth) called strings.[28] From the perspective of Big Nature, space, the scale of effect-observation extends to the macro: thus, a string can resemble a particle from the standard account having mass and charge that flow from the vibrational state of the string. For example, these vibrational states can (from the macro perspective) appear as a graviton (a particle from quantum mechanics that carries gravitational force). It is because of this possible macro perspective that string theory can be a contender-account of quantum gravity and thus be the perspective of the unifying entity of big and little Nature. An advantage to this is the plasticity of the account. Since the strings can be various in their arrangement as they vibrate and configure themselves with loose ends, and medial stretching that can encourage joining in various patterns, constrained only by the requirement of super-symmetry.

---

[28] Two opposite perspectives on the efficacy of string theory can be found in: Edward Witten, "String Theory Dynamics in Various Dimensions" *Nuclear Physics B* 443.1 (1995): 85–126 and Peter Woit, *Not Even Wrong: The Failure of String Theory and the Search for Unity in Physical Law* (New York: Basic Books, 2006).

The ultimate claim is to be able to be a single-level account of everything. If it is the correct account, then it will be primitive, in itself. But because it is hard to empirically verify in a non-question begging way, it is clearly a nature$_2$ account.

In comparing these two candidates for an *explanans* for quantum gravity, we can say that loop quantum gravity emphasizes the *quantum* (bits of spacetime) to geometrically describe a two-dimensional array in which spin networks work their way through coordinates in a specified fashion that gives a foundational account of that which it is trying to account for (general relativity and quantum mechanics).[29] This approach aspires to *completeness, coherence,* and *simplicity* (three of our four categories for judging a scientific theory).

String theory, on the other hand, redefines the mode of discourse in a significant way. If it is correct, then it will more completely account for everything, and because it depends upon super-symmetry it aspires to contain three of our four categories: *completeness, coherence,* and *elegance*. Thus, in a simple, formal fashion, one could contend that the difference is between the *simpler* loop quantum gravity theory versus the more *elegant* string theory account. These two criteria are essentially nature$_2$.

In the end, it will probably depend upon what empirical justification or falsification models turn up in the future to satisfy the nature$_1$ dimension.

Changes in the Philosophy of Science

In Chap. 8 we examined how philosophy had reacted to developments in science and in medicine by reverting to models of reductionism and mathematics to create an axiomatic-structured account of n/Nature. In its prime form it would consist of nature$_2$ (via logic & mathematics) → nature$_1$ and this follows the general pattern set out in Part Two of this volume—where nature$_2$ represented mathematics and formal logic instead of divinity and magic as set out in Part One.

However, the completeness and coherence criteria that are emphasized in the logical empiricist response described in Chap. 8, run into a problem with the explanandum of biology. It may be the case that in general, physical science one might aspire to the *certainty* that logical empiricism proposed.

From my vantage point as a philosopher/historian of science, the situation at the beginning of the twentieth century and the publication of *Principia Mathematica* and *Principia Ethica* by Cambridge professors Russell, Whitehead, and Moore there was a decided move to situate the philosophical profession into an inquiry of *ordered pairs*. By "ordered pairs" here I mean a single methodological approach that unites a small group of scholars in the humanities (only classics is smaller) around a set of questions to be discussed via conceptual and linguistic analysis

---

[29] The reader is encouraged to look back to Descartes in Chap. 7 who set out a geometrically-based vortex theory to account the foundation of everything.

(sometimes backed-up by formal logic or mathematics)—such as the *logicist thesis*, or the conceptual analysis of the word 'good'—as well as an accepted method for doing so. The advantage of a tight-knit community is that there is a lot of positive interaction among its members. It has been shown that positive interaction among like-minded thinkers can press the advancement of various problems beyond what single thinkers can do alone (cf. the Field Prize folk and modern research laboratories).[30] And so things progressed in UK, US, Canada, and Australia. Through a handful of journals and prestigious group meetings such as UCL's Aristotelian Society, the cadre of colleagues hotly debated the accepted questions of the day. There was a general sense that though the methodology of logical empiricism was very narrow, that people could engage arguments from the different areas of philosophy and that people read widely beyond the micro problem that engaged them at that time (because *everyone* would employ the same methodology).

There were outliers, to be sure. *Process and Reality,* by Alfred North Whitehead was thought to be a bit odd. Whitehead was always defending this work to his students, but because Whitehead had already earned his stripes with his collaborative work with Russell and was known to Albert Einstein, philosophers in the profession talked about Whitehead's work using process philosophy in terms of Leibnitz and his interest in mathematics. The "process thinking" that led to process theology (through followers such as Hartshorne) was not discussed in polite company.

What philosophy aspired to in this time period was "a committee of the whole." When the group decided that science was the hot topic, the logical empiricists arose and took ascendency. Talk of "black ravens" and counterfactual confirmation by reference to the Washington Monument gave everyone a chuckle. The goal at every philosophy conference was to think up an outrageous thought experiment that made people both laugh and say, "Well, there's something to that, you know." This showed imagination (a necessary condition for genius—the philosopher's stone that transfixed ordinary discourse into the divine discourse—among atheists, of course). It was an interesting community dynamic that was based partly on camaraderie as well as cut-throat competition. Everyone wanted to be the last one standing. When a question-and-answer session was engaged, it could never be ended under the conditions that philosopher A said x and philosopher B said ~x. No. Either x or ~x was correct and thus a rhetorical jousting contest was engaged that could last for hours—even after the official time-period of the conference or the invited public talk.

But then there were the outliers. Ian Mueller, Alan Donagan, Stephen Toulmin, William Wimsatt and Ted Cohen formed one renegade group while other groups were led by Quine, Kuhn, Feyerabend, Hull, and others (elsewhere)—more on that shortly. These were the parents of the *un*-ordered pairs.

Before we go there, let us consider the 1950s. In the United States this was not a period of socially supported speculation about the nature of first principles and their

---

[30] I'm thinking of the Polymath Project to find a new proof to the Hayes-Jewell Theory by Fields winner Timothy Gowers, cf. Timothy Gowers and M. Nielsen, "Massively Collaborative Mathematics" *Nature* 461.7266 (2009): 879–881.

causes.[31] Instead, as John McCumber has persuasively argued, American philosophers tried to escape social responsibility and possible attack by the McCarthy-inspired Communist-hunters, by pretending that everything that they presented was value-neutral.[32] It was just like the logical-empiricists' dream: matters of fact, abstract axioms and theorems, all tied together in Ramsey sentences. Rudolph Carnap was one of the captains of the fleet. The fantasy was this: If we ascend Aristotle's line of predicables[33] to the highest levels of "substance," "quality," "quantity," "position," "posture," "place," et al., then we have escaped. Who could question the discourse about primary substance? No one can accuse us of talking Commie-lingo, because we have been very careful that we are talking about nothing that has an application to the world we live in. No black-list for American philosophers! What they discussed didn't pertain at all to the lived-world of the *hoi polloi*.[34]

This was the world of the *ordered pairs*.[35] Like porcelain bibelots set in a row upon a mantel-piece for all to see and admire, the ordered-pair-shared-community-worldview sought an insularity that would protect it from the world of action and human events.

Well, obviously this wasn't entirely the case. There were some philosophers who stepped outside the academy to fight for civil rights and economic equality—but from the point of view of the academy the *ordered pairs* greatly outnumbered the *unordered pairs*. But this was about to change.

Enter the philosophy of biology. When Hilary Putnam and Paul Oppenheim authored their famous defense of reductionism in, "Unity of Science as a Working Hypothesis" in 1958[36] they were writing toward the zenith of the logical empiricist tradition in the philosophy of science. The tradition began with a vision of unity for all science. It began with the logicist conjecture that led to the unity of science hypothesis through its insistence on thorough reductionism: if human action could be fully described via psychology and sociology and these, in turn, could be fully

---

[31] Metaphysics, I,i.

[32] John McCumber, *Time in the Ditch: American Philosophy and the McCarthy Era* (Evanston, IL: Northwestern University Press, 2001).

[33] Aristotle, *Categories*, III, ff.

[34] I might note here that this was the attitude of Martin Heidegger that he expressed to his student Hannah Arendt (that philosophy in Germany could be "value-free" so that Heidegger could write "value-free" philosophy and collaborate with the Nazis)—see my fictive narrative depiction of their debate in "Eichmann and Heidegger in Jerusalem" in Michael Boylan and Charles Johnson, *Philosophy: An Innovative Introduction—Fictive Narrative, Primary Texts, and Responsive Writing* (Boulder, CO: Westview, 2010, rpt. Routledge, 2020): 234–244.

[35] I call these "ordered" pairs because they rigidly set out what constitutes *doing philosophy* along the line of the logical empiricists, *only*. I create an argument against this in Michael Boylan, "Reflections on Reshaping Philosophy and the Emergence of Un-Ordered Pairs" in Wanda Teays, ed. *Reshaping Philosophy: Michael Boylan's Narrative Fiction* (Cham, Switzerland: Springer, 2002): ch. 1.

[36] Hilary Putnam and Paul Oppenheim, "Unity of Science as a Working Hypothesis" from *Minnesota Studies in the Philosophy of Science*, vol II, ed. H. Feigl, M. Scriven, and G. Maxwell (Minneapolis: University of Minnesota Press, 1958): 3–36.

described by biology, and biology was nothing but chemistry, and (of course) chemistry was only applied physics and physics was only applied mathematics and mathematics was applied logic, then everything depended upon logic. Thus, Aristotle's fundamental question in the *Metaphysics* whether there is a master science of everything has been answered—signed, sealed, and delivered! It was a perfect set of ordered pairs. Well, we've done philosophy! Now what's next? A rousing game of tennis anyone?

Things looked pretty neat to Putnam and Oppenheim at the writing of their influential article, but then those pesky philosophers of biology came in and spoiled the picnic. First of all, evolutionary theory had the messiness constant diversity within species that yielded various fitness coefficients within particular environments. The fact that within evolutionary theory, upper-level constraints of the environment could have causal influence upon which genetic-expression through controlling what would be more or less successful at reproducing and surviving. This was "top-down" causation and thus was anti-reductionistic. Second, genetic epistasis, the environmentally influenced expression of low-level genetic expression was empirically verified and anti-reductionistic.

Philosophers of biology also brought in holism & feedback loops. One important dispute fed these questions: the unit of selection debate. If the reductionistic bent of the logical empiricists was correct, then the causal momentum would always be "bottom → up." But the unit of selection debate brought in a more complex causal account.[37] This hurt deductive reductionism in a big way. Evolutionary theory with its potential for "top → down" causation[38] ushered in an entirely new way of understanding causal efficacy—as per the pioneering work in the mathematics of population biology that codified this non-reductionistic approach.[39]

These questions were not confined to the philosophy of science, either. The move against the absolute reductionism of logical empiricism led to other approaches, such as post-modernism and philosophy of literature which both came forward to offer alternatives to doing philosophy after the fall of logical empiricism as the "default way to do philosophy in the Western Tradition."[40]

---

[37] Two key articles in this regard are: Eliott Sober and Richard Lewontin, "Artifact, Cause, and Genic Selection" *Philosophy of Science* 49 (1982): 157–180 and Kim Sterelny and Philip Kitcher, "The Return of the Gene" *Journal of Philosophy* 85.7 (1988): 339–361). David Hull's, *Philosophy of Biological Science* (Englewood Cliffs, N.J.: Prentice Hall, 1974) was also an important work during this period.

[38] The "top" part of this causal account was the "evolutionary pressure" that changes in particular environments made upon reproducing populations over time giving some an "advantage" over others within the species—thus favoring one genotype over another. cf. Richard Dawkins, *The Selfish Gene* (Oxford: Oxford University Press).

[39] A good example of this is the work of Edward O. Wilson and William H. Bossert, *A Primer of Population Biology* (Sunderland, M.A.: Sinauer Publishers, 1971).

[40] For several accounts along these lines see: Wanda Teays, ed. *Reshaping Philosophy: Michael Boylan's Narrative Fiction* (Cham, Switzerland: Springer, 2022).

Technology and Medicine

One of the consequences of the development of general science (both the big Nature and the little Nature) was the development of general science's work on little Nature in genetic engineering.[41] If the big changes described at the beginning of the twentieth century to treat infectious disease via vaccination (little Nature) and surgery (big Nature) that with a new appreciation for hygiene was able to confront cancer and heart disease.

There was one source of disease that had not yet been approached: genetically-caused diseases and organ disorders. Since the cause in this instance is in realm of little Nature (defective genes that will either kick-in at birth (Spina Bifida, Cystic Fibrosis, Fragile X Syndrome, Tay-Sachs), or express themselves on a time-fuse (such as Huntington's Disease, Parkinson's Disease, or Alzheimer's Disease). Genetically caused conditions seem to be ripe for genetic engineering. Because the process of epigenesis is so complicated, it is difficult (at this stage of scientific knowledge) to: (a) treat these at the pre-conception-level via IVF to either "fix" the defective gene or "knockout" factors that might inhibit normal genetic expression.; (b) treat the genetic disorder before it becomes expressed—either in utero or early in post-birth development. In principle, (b) should be easier to accomplish than (a) because it falls more into the *treatment* phase of medicine—albeit with a shorter timeline for diseases/disorders that are expressed shortly after birth than those that are expressed later in life.

The science behind genetic engineering came from several sources. First, in 1990 was the Human Genome Project (the mapping of the Human Genome) which was undertaken. It was completed in 2003 by NIH (the United States' National Institutes of Health).[42] To some this harkened after the allure of reductionism, as we saw among the logical empiricists.[43] The Human Genome Project's director, Francis Collins, predicted that within ten years, there would be personalized drugs to treat diseases.[44] The takeaway was not to clone lots of sheep or other mammals, nor even to clone complex organs like the heart or liver for transplant. Rather, the promise was to clone tissue that could be used in treatment of a diseased individual, e.g., cloning islet cells that function in insulin secretion. It was a heady time in which little Nature was on a role in emerging medical treatment. This proved to be a more daunting task than it was first imagined to be.[45]

---

[41] For an account for some of these issues see: Michael Boylan and Kevin Brown, *Genetic Engineering: Science and Ethics on the New Frontier* (Upper Saddle River, N.J.: Prentice Hall, 2002).

[42] https://www.nih.gov/news-events/nih-research-matters/first-complete-sequence-human-genome (accessed September 8, 2022).

[43] A.I. Tauber and S. Sarkar, "The Human Genome Project: Has Blind Reductionism Gone Too Far?" *Perspectives in Biology and Medicine* 35 (1992): 220–235.

[44] Francis Collins, "Medical and societal consequences of the human genome project." *New England Journal of Medicine* 341 (1999): 28–37.

[45] Boylan and Brown (2001): ch. 3.

However, simultaneously, another possibility came onto the scene—using cloning to alter germline cells in order to "fix" genetic disorders. This became a line of contention with genetic engineering. When one miscalculates with one patient in treatment, the worst that can happen is that the patient dies. However, germline miscalculations could affect many more—depending upon when the possible (unforeseen painful/mortal consequence) was expressed. If ethical judgments on human life take into account the *numbers* of individuals hurt/killed (as utilitarians do), then germline genetic engineering is more *risky* and potentially more unethical than treatment of a single individual.[46]

The first cloning of a mammal occurred on July 5, 1996 when Dolly, the sheep, was born. The event was announced to the world on February 22, 1997.[47] At least two things became immediately clear: (a) genetic engineering had developed to the point where it could positively be used in *treatment* for human subjects; and (b) genetic engineering had developed to the point where it might be used to "solve" genetic disorders. In the former case, one might query a patient who has a fatal condition in which there are no other viable options. The patients might make a decision for themselves that they would accept the risk—since death or some other very deleterious outcome was virtually certain: the dynamics of one patient and one risk is easier to examine than germline intervention in which many people might be hurt. And this is because of the complication involved in the *expression* of genes via epistasis.[48] This complication is beyond current capacity for exact calculation (in principle) and may be an example of the "three-body problem" in Newtonian Physics in which the calculus cannot handle the dynamics of three-masses and their own forces interacting in even a stable environment.[49]

Because of this, this philosopher has held, on the principle of precautionary reason, that genetic engineering is only justified for *treatment* of somatic symptoms and not for alteration of the germline.

Another powerful tool that has more recently (2012) come to the fore and has the potential to make significant genetic alterations in somatic treatment or germline intervention is Crispr Technology.[50] Crispr technology allows the editing of DNA strands that code for specific proteins that eventually have expression in phenotypes through epistasis. The result potentially allows medical researchers to alter phenotypic traits via genetic alteration. Sahotra Sarkar states in a forthcoming essay that:

---

[46] For a popular example of this see the Trolley Problem—one version I present in Boylan, *Basic Ethics,* 3rd ed (New York and London: Routledge, 2021): 180.

[47] For details on this see the University of Edinburgh's website on this: https://dolly.roslin.ed.ac.uk/events/index.html (accessed on September 8, 2022).

[48] One account of this that follows the general philosophy of this is David Hull, *Philosophy of Biological Science* (Englewood Cliffs, N.J.: Prentice Hall, 1974):23–42.

[49] For a brief description see: June Barrow-Green, "The Three Body Problem" in Timothy Gowers, June Barrow-Green, Imre Leader, eds. *The Princeton Companion to Mathematics* (Princeton, N.J.: Princeton University Press, 2008): 726–728.

[50] Much of the material I mention on Crispr comes from Sahotra Sarkar, *Cut and Paste Genetics: A Crispr Revolution* (New York and London: Rowman and Littlefield, 2021).

However, much of the hype about CRISPR arises from the potential it provides for editing human genes not only in targeted somatic cells but, more easily, in the germ-line by intervention at the zygotic stage of development. The latter ability raises the possibility of removing disease-implicated genes from the human population. But the popular press—and a questionably-informed section of the bioethics community—has also argued that CRISPR had made the prospect of genetic enhancement of future human beings possible.[51] The questions, in both cases, are, first, whether the anticipated outcomes scientifically plausible, and, second, whether attempts to use human germ-line manipulation for these ends ethically permissible.[52]

Thus, one manner in which technology enters into the realm of medical practice is in genetic alteration. Whether the design is for somatic treatment or genetic alteration, will be an issue of current and future discussion. One Chinese scientist went in the direction of genetic engineering respecting the implantation of gene-altered embryos into a woman (germline engineering). This broke Chinese law and offended many international scientists working in CRISPR research. The physician/scientist, He Jiankui, was sentenced to 3 years in prison for his activity.[53]

As mentioned above, this author (on the principle of precautionary reason) believes that at this time, we do not know enough about possible effects of this intervention to engage in germline intervention to prevent genetic disease—much less to engage in human enhancement.

AI, Technology, and Modern Surgery

An issue of big Nature in medicine is the use of AI (artificial intelligence) and computer-assisted surgery. In Chap. 7 we saw the great leap forward that germ theory and antiseptics had on making invasive surgery a viable part of medicine after more than two thousand years. At the beginning of the twentieth century infectious disease was the biggest cause of death in the West. But infectious diseases were greatly lessened by vaccines.

What became the new big killers were: coronary disease and cancer. These were successfully engaged via advances in surgery. Even very fine macroscopic forays into the brain, spine, heart, lungs, and coronary system became possible because of

---

[51] For a background on human enhancement from the side of proponents see: Julian Savulescu and Nick Bostrom, eds. *Human Enhancement* (Oxford: Oxford University Press, 2009). It should be noted that an opposition position is taken by this author in Michael Boylan, "The Ethical Limits of Science" in *Journal of Philosophical Ressearch* 30 (2005): 15–26.

[52] This question is engaged by Sahotra Sarkar, "Crisper and Cut and Paste Genetics" and critics Janella Baxter and Stuart Newman in Michael Boylan, ed. *International Public Health Policy and Ethics,* 2nd ed. (Cham, Switzerland: Springer, 2023): part three.

[53] https://www.science.org/content/article/chinese-scientist-who-produced-genetically-altered-babies-sentenced-3-years-jail (accessed September 9, 2022).

advances in surgery.[54] This has generally been due to the skill and intelligence of the surgeon.

At the writing of this book (2022), there has been a move toward using AI and the subsequent technologies to assist (and in some cases *replace*) the human guiding the medical intervention process.[55] There can be problems in this because of the presentation of unique problems that occur during the surgical process.[56]

Public Health—Basic Principles

It has been my position that the moral imperative behind public health policy ought to be a generalized form of the individual, clinical duty each physician takes on from the Hippocratic Oath (see Chap. 4). Sometimes, public health mandates are merely seen from the standpoint of the benefit of the wealthy and powerful in a particular nation or region. I have argued elsewhere that this is the wrong strategy. All agents must be considered as equal and their rights claims to well-being and health are of equal weight against any other agent or group of agents (within the nation or region).[57] As my generating proposition, I use my Shared Community Worldview Imperative.[58] This is a deontological justification for helping others in need—both near at hand and far away in distant lands.

Public Health—Pandemics

A Brief View from Select Pandemics

Using the assessment criteria above, let's take a brief look at some of the more recent pandemics that express themselves through one or more of the categories to have caused excessive human harm and the public health responses to the same.

---

[54] A brief example of some of these advances can be found in: Rifat Latfi, Peter Rhee, Rainer W.G. Gruessner, eds. *Technological Advances in Surgery, Trauma, and Critical Care* (Cham, Switzerland: Springer, 2016); Satish N. Nadig, Jason A. Wertheim, eds. *Technological Advances in Organ Transplantation* (Cham, Switzerland: Springer, 2018); and Ashok K. Hemal and Mani Menon, eds. *Robotics in Genitourinary Surgery* (Cham, Switzerland: Springer, 2018).

[55] See Rita Manning, "AI in Healthcare: Ethical Issues" in Michael Boylan and Wanda Teays, eds. *Ethics in the AI, Technology, and Information Age* (New York and London: Rowman and Littlefield, 2022): 169–180.

[56] See the discussion of self-driving cars in by Jens Kipper and Sven Nyholm: *ibid*: 181–206.

[57] I have argued this position in Michael Boylan, ed. *Public Health Policy and Ethics* (Dordrecht: Springer, 2004), Michael Boylan, ed. *International Public Health Policy and Ethics* (Dordrecht, Springer, 2008), Michael Boylan, ed. *Ethical Public Health Policy Within Pandemics* (Cham, Switzerland: Springer, 2022), and the second edition of *International Public Health Policy and Ethics*, forthcoming 2023.

[58] Michael Boylan, *Natural Human Rights: A Theory* (Cambridge: Cambridge University Press, 2015): ch. 6.

Spanish Flu: 1918–1919: The so-called Spanish Flu actually showed its first symptoms in the United States at a military base in Kansas, March 11, 1918.[59] It then went to Europe—probably through troop movements during the last phases of World War I. Very soon, thereafter, the King of Spain, Alfonso XIII, contracted the disease (though some say he got scarlet fever instead). He recovered. After that, the disease was pinned upon Spain—who, in turn, pinned it upon France. However, the first case was in the United States! Why is it so important to *nationalize* the blame— for a virus that carries no passport?[60]

According to David Killingray, though the British Empire had dealt with epidemics before—through yellow fever, sleeping sickness, small pox, cholera, and the bubonic plague, they were unprepared for effective public health responses to this strain of flu.[61] The infection rate was much higher (up to 40% of the world's population) and there was more consequential human harm (21–50 million dead).[62] Most at risk, were the young and healthy (so that World War I soldiers fit into this category).

In the case of this pandemic, the first wave was not too far from normal in infection rates and human harm (causing civil reactions to be rather restrained). It was the second wave of the viral infection that occurred in the autumn hitting Asia (especially in India and then moving east). Again, Killingray believes that part of the cause was the British shipping industry that carried the flu from Europe to India and to Hong Kong.

The communicability of the disease was very high—especially for particular sub-populations. *The British Medical Journal*[63] estimates that incidence rates among native populations around the world was as high as 80% in some locales. In general, white populations in the Commonwealth Nations and the United States was as high as 60%. Among the military, the British Navy was around 30% and the British Army around 20%. (Interestingly enough, those military who had been exposed to poisonous gas during the war had their incidence rate to the flu dropped to 4.7%). This can be compared to similar "in season" flu populations today that have up to an 8.3% incidence rate.

---

[59] For a discussion of the historical spread of the disease—especially in the British Empire—see: David Killingray, "A New 'Imperial Disease': The Influenza Pandemic of 1918–1919 and its Impact on the British Empire" *Caribbean Quarterly* 49.4 (2003): 30–49.

[60] This behavior exists today as (at the initial writing of this essay) President Trump on multiple occasions blames Covid-19 on China: https://abcnews.go.com/Politics/chinas-ambassador-us-slams-trump-covid-19-blame/story?id=72187153 (accessed August 25, 2020). Cf. Terence Ranger and Paul Sacks, eds. *Epidemics and Ideas: Essays on the Historical Perceptions of Pestilence* (Cambridge: Cambridge University Press, 1992).

[61] Killingray, *loc. cit.*

[62] The standard figures of the Spanish Flu pandemic were 21–24 million, but work done by Niall P.A.S. Johnson and Jeurgen Mueller, "Updating the Accounts: Global Mortality of the 1918–1919 'Spanish' Influenza Pandemic" *Bulletin of the History of Medicine* 76.1 (2002): 105–115, put the figure rather higher—to at least 50 million.

[63] Tom Jefferson and Eliana Ferroni, "The Spanish Flu Through BMJ's Eyes" *British Medical Journal* 399.7735 (2009): 1397–1399.

From the same source, the probability of human harm (in this case fatality) was greater than 2.5% overall for those who became sick with this strain of the flu. This can be compared to .1% for flu outbreaks in general populations today.

Further, the human harm continues beyond mere fatalities. In one study that focused upon the U.S. population 22+ years after the outbreak, the aftermath of pregnant women who were affected brought forth offspring who exhibited: 1. Reduced educational attainment; 2. Increased physical disabilities; 3. Lower modal incomes throughout life; and 4. Decreased socio-economic status (due to #1-#3).[64]

Concerning civic, public health responses, though there were sporadic efforts that varied city-by-city, no effective, comprehensive public health measures were undertaken even though there had been prior influenza epidemics in 1889–1890. Some examples of what *was* done in the United States includes first New York City where the sick were (in many cases) quarantined from the healthy. Public health "runners" went about the city taking notes on which areas of the city seemed to be worse. The city's board of education engaged in health education about hygiene and what symptoms to note about sickness (and what course of action to take). New laws concerning public hygiene (such as anti-spitting laws) were enforced.[65] Then, as infections and deaths mounted, schools were closed.[66] African Americans were given systematic, inferior treatment due to attitudes concerning biological inferiority and general racism. They were put into segregated, sub-standard hospitals without state-of-the-art equipment. Because of this, their mortality rates greatly exceeded their percent of the population.[67]

In San Antonio, Texas (a major city in that era) the public health response was three-fold: ban children from schools, quarantine the sick, and prohibit large public gatherings. This was moderately effective.[68]

In Pittsburgh, Pennsylvania (a manufacturing city at the time one-third the size of Philadelphia), there were inconsistent standards. This was caused by a political tug of war between the governor and the mayor. Once people showed signs of clear sickness, they were home quarantined. But this wasn't adequate as Pittsburgh soon has the worst rate of infection and fatality. The general population was sicker than normal before the flu due to the manufacturing plants and the smoke they produced. Public gatherings were permitted including publicly sponsored concerts. Housing was crowded with low-paid mill workers. The public sanitation system was sub-par

---

[64] Douglas Almond, "Is the 1918 Influenza Pandemic Over? Long-Term Effects of In Utero Influenza Exposure in the Post-1940 U.S. Population" *Journal of Political Economy* 114.4 (2006): 672–712.

[65] Francesco Aimone, "The 1918 Influenza Epidemic in New York City: A Review of Public Health Response" *Public Health Reports* 125, supplement 3 (2010): 71–79.

[66] Alexandra Minna Stern, et al. "School Closures" *ibid.*: 63–70.

[67] Vanessa Northington Gamble, "'There wasn't a lot of Concern in those days' African Americans, Public Health, and the 1918 Influenza Epidemic" *ibid.*: 114–122.

[68] Ana Luisa Martinez-Catsam, "Desolate Streets: The Spanish Influenza in San Antonio" *The Southwestern Historical Quarterly* 116.3 (2013): 286–303.

with open sewers that had excrement floating about. Once again, African American populations were hit the worst. This public health response was a terrible failure.[69]

HIV: 1981 onwards: The second pandemic that I wish to gloss is the HIV/AIDS pandemic. According to United States' Center for Disease Control, from 1981 to 2006, 22 million people died of AIDS (500,000 in the United States). In 2005 2.8 million died from AIDS worldwide and 4.1 million were newly infected.[70] As of 2019 the total number of deaths due to HIV/AIDS worldwide is 32.7 million (with 690,000 in 2019 alone).[71] This is one measure of the quantity of human harm that has to be seen in the context of the diminished life-style of those on HIV antiretrovirals.[72]

On the issue of communicability, data from Canada gives some important details.[73] Anal sex-receiving semen has an infection rate of 1.4% (or 1–71 episodes); anal sex-giving semen has an infection rate of .11% (or 1–750 episodes). Vaginal sex-receiving semen has an infection rate of .08% (or 1–1250 episodes), vaginal sex-giving semen has an infection rate of .04% (or 1–2500 episodes). There are no statistics in this report on oral sex. Therefore, on this side of the pandemic analysis, communicability is lower than the 1918 Influenza pandemic. There is also an easy prophylactic device, the condom, which can reduce risk even further.

On the side of human harm, there is the mortality rate for those who contract HIV and develop AIDS. At its height in the 1980s (pre-treatment), those who contracted HIV in the United States were studied by the University of California, San Francisco Medical Center. In this study those who retained HIV and then developed AIDS died at up to a 50% rate.[74] This is a very high mortality rate for a pandemic. Now that antiretroviral treatment is available in much of the world, this mortality rate has dropped precipitously. In 2019 there were an estimated 38 million with HIV/AIDS (36.2 million adults and 1.8 million children). 1.8% of these infected individuals died.[75]

Treatment today via pre-exposure prophylaxis (PREP) and via post-exposure prophylaxis (PEP) can greatly lessen the transmission of HIV among people who are HIV negative. For those who are HIV positive, there are medications

---

[69] James Higgins, "'With Every Accompaniment of Ravage and Agony:' Pittsburgh and the Influenza Epidemic of 1918–1919" in *The Pennsylvania Magazine of History and Biography* 134.3 (2010): 263–286.

[70] The United States Center for Disease Control report as of 2006: https://www.cdc.gov/mmwr/preview/mmwrhtml/mm5531a1.htm (Accessed August 29, 2020).

[71] Source is the United Nations: https://www.unaids.org/en/resources/fact-sheet (Accessed August 29, 2020).

[72] https://www.hiv.gov/hiv-basics/staying-in-hiv-care/other-related-health-issues/other-health-issues-of-special-concern-for-people-living-with-hiv (Accessed August 29, 2020).

[73] https://www.catie.ca/en/pif/summer-2012/putting-number-it-risk-exposure-hiv.

[74] http://hivinsite.ucsf.edu/InSite?page=kb-01-03#S1.4X (accessed August 31, 2020).

[75] https://www.hiv.gov/hiv-basics/overview/data-and-trends/global-statistics (accessed August 31, 2020).

(antiretroviral therapy, ART) that can lower the levels of HIV so that the risk of transmission to others via sex is greatly diminished.[76]

The story of civic public health responses to the HIV/AIDS pandemic varied by country. One of the major problems facing public health officials was that in most countries in the world there are strong social taboos concerning sex outside of marriage and (at this time especially) concerning gay sex. These taboos created problems for public health officials to effectively instigate a public health response.

Donald P. Francis did a retrospective of the first 30 years of the HIV/AIDS public health reactions in the United States (and some of these problems listed resonate with some other countries).[77] The United States lacked an early understanding of epidemiology of infectious disease at the political levels. Because of this ignorance, a rapid public health response (critically necessary for all epidemics and pandemics) was lacking. People got side-tracked into focusing on issues of sex, homosexuality, and intravenous drug users as a moral issue. Some religious zealots even proclaimed that this plague was a penalty from God for those "sins" of the victims (another version of *blame the victims*).[78] This re-invigorated the ancient quarrel between science and religion. The result was a public policy paralysis as politicians did not want to anger their base by letting a scientific medical question be addressed by public health scientists.

The U.S.A. CDC (Center for Disease Control) showed early on that there was a problem that needed immediate intervention, but the U. S. President, Ronald Reagan, did not understand the role of a President in a public health crisis. The refusal to listen to public health experts made the problem worse than it had to be and increased the number of lives lost. There was even a reticence for some time at recommending condoms for those engage in sex in high-risk groups. This was because of the "religious right," who did not want to seem like they were encouraging sex—except among heterosexual married people in their churches, who might expand those churches and their money and power.

---

[76] https://www.hiv.gov/hiv-basics/hiv-prevention/reducing-sexual-risk/preventing-sexual-transmission-of-hiv (accessed September 1, 2020).

[77] Donald P. Francis, "Commentary: Deadly AIDS Policy Failure by the Highest Levels of the U.S. Government: A Personal Look Back 30 Years Later for Lessons to Respond Better to Future Epidemics" *Journal of Public Health Policy* 33.3 (2012): 290–300. For a brief scan of international reaction see: Zaryab Iqbal and Christopher Zorn, "Violent Conflict and the Spread of HIV/AIDS in Africa" *The Journal of Politics* 72.1 (2010): 149–162; Patrica Rodney, Yassa Ndjakani, et al. "Addressing the Impact of HIV/AIDS on Women and Children in Sub-Saharan Africa" PEPFAR, The U.S. Strategy" *Africa Today* 57.1 (2010): 64–76; James Enoch and Peter Piot, "Homan Rights in the Fourth Decade of HIV/AIDS Response An Inspiring Legacy and Urgent Imperative" *Health and Human Rights* 19.2 (2017): 117–122; Patrick Michael Eba, "Towards Smarter HIV Laws, Considerations for Improving HIV-Specific Legislation in Sub-Saharan Africa" *Reproductive Health Matters* 24.47 (2016): 178–184; Md. Nazrul Islam Monday and Mahendran Shitan, "HIV/AIDS Epidemic in Malaysia: Trend Analysis from 1986–2011" *Southern African Journal of Demography* 16.1 (2015): 36–56.

[78] A.R. Jonsen and J. Stryker, eds. *National Research Council on Monitoring the Social Impact on the AIDS epidemic* (Washington, D.C.: National Academic Press, 1993): https://www.ncbi.nlm.nih.gov/books/NBK234566/ (accessed September 3, 2020).

The tale of the tape here is a very bad start, and a gradual change over time that has largely put the pandemic into a controlled—yet still present and virulent—mode. There is continued hope for the future (40 years later) going forward.

Ebola: 1976, 2014, 2019: The next pandemic that we will briefly examine in this chapter are the various Ebola outbreaks that began in in the Democratic Republic of the Congo in 1976 in a village near the Ebola River. Since then, they have spread to Uganda, Gabon, Guinea, Liberia, Sierra Leone and Nigeria. The original transmission was between animals (bats and non-human primates to humans). The original mode of transmission is thought to have been the eating of infected bats and non-human primates. There is no data on how many exposures in this mode will cause human infection, however; the disease between humans is *highly* contagious—so we might be able to extrapolate here, as well. The principal outbreak to date was 2014–2016, which is touched upon below. As of 2019 another outbreak has begun in the Democratic Republic of the Congo and is still in its early phases.

The human-to-human transmission occurs from contact with bodily fluids: sweat, blood, urine, feces and then having that contact enter the blood stream via open cuts, contact with eyes, ears, nose, and mouth. Contact with corpses of infected individuals is also significant. This can cause problems due to social funeral customs and "showing respect to the deceased" by touching or kissing the corpse. Overall, the infection rate per 100,000 in West Africa was 100 → 300 per 100,000.[79] Even healthcare workers (in Sierra Leone), who take significant caution in treating patients, were infected in the 2014 epidemic period at 5.4%. This shows how contagious the disease is.[80] The disease is not thought to be transmitted via airborne, aerosol contact.

Sexual contact with an infected person is contagious when giving or receiving semen (slightly higher risk). Those who are exposed in these ways have a high rate of infection (sometimes > 90%).[81] When this is paired with mortality rates that are generally 60%-70%, but can approach 90% in some circumstances; the Ebola Virus Disease (EVD) is very dangerous among infectious diseases.[82] Once infected, there are three stages of the disease. A. (after 2–21 days) fever, headache, myalgia; B. (just after stage one) gastro phase: diarrhea, vomiting, abdominal pains; C. (just after stage two) collapse of neurological functions and massive internal bleeding. These three phases generally take 1–3 weeks + after first symptoms. Death generally

---

[79] Incident Management System: Ebola Epidemiology Team on Morbidity, "Update: Ebola Virus Disease Epidemic—West Africa 2014" *Morbidity and Mortality Weekly Report* 63.50 (2014): 1199–1201.

[80] Peter Kilmarx, Kevin Clarke, et al. "Ebola Virus Disease in Healthcare workers in Sierra Leone, 2014" *Morbidity and Mortality Weekly Report* 63.49 (2014): 1168–1171.

[81] Ebenezer T. Durojaye and Gladys Mirugi-Mukundj, "The Ebola Virus and Human Rights Concerns in Africa" *African Journal of Reproductive Health* 312.3 (2015): 48–55.

[82] Nicholas J. Beeching, Manuel Fenech, and Catherine F. Houlihan, "Ebola Virus Disease—Clinical Review" *British Medical Journal* 349 (2014): 1–15. This is a comprehensive, clinically-oriented article.

occurs quickly—generally before a month is finished. After the second week, the recovery rate increases significantly (up to 75%).[83]

The only real treatment is fluid and electrolyte replacement—low effective rate.

The combination of high communicability with high rate of mortality that proceeds rapidly, makes Ebola a formidable disease. In the time-span of April to October in 2014 in the countries of Senegal, Mali, Nigeria, Guinea, Sierra Leone, and Libera there were 17, 145 cases and 6070 deaths (35% mortality rate).[84]

The civil, public health reaction to the disease has varied. In Canada (and by extension the United States) the African pandemic was presented sometimes as an issue of national security and at other times as one of public health.[85]

Within Africa, the center of the various outbreaks, some popular beliefs have gotten in the way of effective public health measures. These beliefs include: (a) God will protect me against EVD; (b) traditional healers will be more effective than modern doctors; (c) people can become infected via airborne transmission; (d) it is a Western bioterrorism experiment against African nations.[86]

Because there is little continuity between countries for a coordinated public health response, there is no standard position on: 1. The compassionate use of experimental drugs for treatment along with new, untried vaccines; 2. There needs to be more money allocated towards infection control; 3. There needs to be a long-term commitment toward creating an international public health infrastructure for the future.[87]

At the writing of this essay the Ebola pandemic has started again (where it first began) in the Democratic Republic of Congo, 2019. The public health measures just set out have limited the number of cases involved and have contained the death rate: As of 23 July 2020—250.000 exposure contact tracings; 200,000 tests administered; 3481 cases reported resulting in 2299 deaths (66%), a decrease from the 2014 outbreak.[88]

This latest flare-up shows that proper public health responses can curb the numbers of cases as well as the mortality. Sometimes there are those who will not trust modern science, but these outliers must be overcome for the health of the larger community. Since outbreaks are likely for the future, this should be a measure of

---

[83] *Ibid.*

[84] Musa Abubukar Kana, Olufunmilayo Y Elegba, et al. "Ebola Viewed Through a Lens of African Epidemiology" *Journal of Epidemiology and Community Health* 70.1 (2016): 6–8.

[85] Brittany Humphries, Martha Radke, Sophie Lavzier, "Comparing 'Insider' and 'Outsider' News Coverage of the 2014 Ebola Outbreak" *Canadian Journal of Public Health* 108.4 (2017):1–15.

[86] Lonzozou Kpanake, Komlantsè Gossou et al. "Misconceptions about Ebola Virus Disease Among Lay People in Guinea: Lessons for Community Education" *Journal of Public Health Policy* 37.2 (2016): 100–172.

[87] Christian A. Gericke, "Ebola and Ethics: Autopsy of a Failure" *British Medical Journal* 350 (2015): 1–2.

[88] https://www.who.int/emergencies/diseases/ebola/drc-2019 (World Health Organization, accessed September 21, 2020).

support for those who seek to be resolved to use tried and true techniques for reigning in a virulent pandemic.

SARS: 2003/ MERS: 2012: severe acute respiratory syndrome (SARS) caused by the SARS corona virus was first detected by the World Health Organization (WHO) in February 2003.[89] It was spread, as most corona viruses, by airborne saliva aerosol. The infection rate indoors is around 3%. The incubation period is 2–7 days but that can expand to 10 days.[90] On March 12, 2003 the WHO issued a health alert describing a new unrecognizable flulike disease that was infecting healthcare workers. By August there were 8422 cases and 916 deaths (11%) from 29 countries—far higher than the flu .1% in 2019 in the U.S.[91] This turned out to be a rehearsal for subsequent outbreaks of H5N1 and H1N1. This offered an opportunity for public health agencies around the world to prepare themselves. Some did (particularly in east Asia) and most did not.[92]

In the time where the pandemic was just beginning (February 2003), a 64-year-old physician was treating a patient with atypical pneumonia in his home in Guangzhou, China. The physician developed symptoms of a respiratory complaint but felt well enough to travel to Hong Kong to go sightseeing with his brother-in-law. Little did he know that he was in the early stages of "severe acute respiratory syndrome." He became a spreader. The resulting pandemic caused WHO to issue new public health guidelines that helped ameliorate the later H1N1 and H5N1 outbreaks.[93]

Various other developments of similar viruses originated in Saudi Arabia, MERS (also a corona virus). Initially, the infection began among camel workers. With many of the same symptoms as SARS, the initial infection rate was about the same (though males were slightly higher), but the mortality rate was higher, 65% versus 10% for SARS.[94]

The public health reaction to these two outbreaks has been varied. There have been no concerted international efforts because unless bodies pile up in very high amounts, it can be difficult to garner general response among nations.[95] One way to

---

[89] https://www.who.int/health-topics/severe-acute-respiratory-syndrome#tab=tab_1 (accessed September 24, 2020).

[90] *Ibid.*

[91] https://www.cdc.gov/flu/about/burden/2018-2019.html (accessed September 24, 2020).

[92] David Koh and J. Sing, "Lessons from the Past: Perspectives on Severe Acute Respiratory Sydrome" *Asia Pacific Journal of Public Health* 23.3 (2010): 132–136.

[93] Isabelle Nuttall and Christopher Dye, "The SARS Wake-UP Call" *Science* 339.6125 (2013): 1287–1288.

[94] Zie Zhou, Hin Chu, Cum Li, et al. "Active Replication of Middle East Respiratory Syndrome Coronavirus and Aberrant Induction of Inflamatory Cytokines and Chemokines in Human Macrophages: Implications for Pathogenesis" *Journal of Infectious Diseases* 209.9 (2014): 1331–1342 and Owen Dyer, "MERS Will Spread while Natural Host is Unknown" *British Medical Journal* 346.7914 (2013): 6.

[95] This point is made by: Vincent Rollet, "Framing SARS and H5N1 as an Issue of National Security in Taiwa: Process, Motivations, and Consequences" *Extrême-Orient, Extrê-Occident* 37

measure how well the nations of the world took these pandemics as a wake-up call to create policies and institutional capabilities is to see which nations in the world have done comparatively better in yet one more novel corona virus outbreak.

COVID-19: At the initial writing of this chapter in October, 2020 we were at another link in the chain of novel corona viruses that we have just touched on. SARS-CoV-2 causes COVID-19. Like the other corona viruses, SARS-CoV-2 attacks the respiratory system, but may also affect other organs as well. The communicability so-far with the virus is 2.4% in the U.S.A. with a mortality rate among the infected group of 2.8%. World statistics for comparison are not as accurate for a variety of reasons.[96] As noted above, the infection rates and mortality rates of those infected is more than double the flu in the United States. Because of variants in the outbreak from the delta to the omicron variants (and their sub-variants), it is not useful to note total mortality because this is subject to change. But within the first 9 months total deaths in the U.S.A. are around 208,000 and worldwide around 995,000.[97] Since then the numbers have expanded to over a million in the U.S.A. and 6.5 million worldwide.[98]

One critical factor in trying to create a strategy for treatment and for a vaccine is to discover just how SARS-CoV-2 reacts within human hosts. William A. Haseltine, who worked extensively on the HIV treatments, set out viruses into two group: (a) those that are "hit and run" like polio—if you can survive it, there is long-term acquired immunity so that you won't get it again; and (b) that are "catch and keep it" like HIV or the common herpes zoster that begins as chicken pox and lives in the spine and may later cause shingles.[99] This is an important facet of understanding SARS-CoV-2 (and its variants), and is, at the writing of this essay, unknown.

What can be commented on is the initial public health reaction to SARS-CoV-2.

The countries that are doing best at the initial stages of the COVID-19 crisis are those that took the SARS and MER pandemics seriously. Countries such as Japan,

---

(2014)" 141–170; and Sophie Arie, "Would Today's International Agreements Prevent Another Outbreak like SARS?" *British Medical Journal* 348 (2014): 1–3.

[96] https://www.cdc.gov/coronavirus/2019-ncov/hcp/faq.html (accessed October 2, 2020). The world rates which are far less in the infection rate but around the same for the death rates of those infected need to be tempered by the fact that infection numbers are not readily available in all parts of the world because of lack of infrastructure and because sometimes there are political pressures to under-report infection numbers. Though the gross fatalities in the U.S.A. and the world have risen significantly, the CDC is now emphasizing vaccination rates in order to cut the rate of transmission. 99.5% of the deaths recently in the U.S. (summer, 2021) have been among the unvaccinated: https://www.bing.com/search?FORM=AFSCVH&PC=AFSC&q=cdc+coronavirus+update (accessed July 17, 2021).

[97] *Ibid.* Of course, the U.S. deaths are now (July, 2021) over 600,000 and world deaths over 4,000,000.

[98] https://covid19.who.int/ (accessed September 26, 2022).

[99] William A. Haseltine, "What We Learned from AIDS: Lessons from Another Pandemic for Fighting COVID-19" *Scientific American* 323.4 (October, 2020): 37–41, cf. William A. Haseltine and Flossie Wong-Staal, "The Molecular Biology of the AIDS VIRUS" *Scientific American* 259.4 (October, 1988): 52–63.

South Korea, and Germany come to the fore.[100] Also, some countries in Africa have also learned and are better prepared.[101]

Concerning general public health responses in this early phase, it is remarkable that there has been a lack of international cooperation to a global pandemic.[102] Part of this may be attributed to then-President of the United States, Donald Trump, who was intent on pursuing an isolationist foreign policy. As in other pandemics, there is also an unproductive "name-calling" blaming the virus upon China: "the Chinese virus." This has exacerbated prejudice on those of Asian descent within the United States and harmed these individuals unfairly.[103]

One last problem (that was also mentioned above in other public health infectious disease pandemics) is the dissemination of accurate information to the populations of the world's countries. One set of misinformation concerns the nature of the disease and how it should be confronted. Another is wild facts about some international nefarious agent who may be instigating this outbreak for national defense purposes.[104]

Conclusion to Pandemics

In this brief, select history of the pandemics over the past century, there are certain facts that stand out. First, it is important that governments in the countries affected recognize that this is a medical emergency that should be handled by health care professionals: doctors, nurses, and epidemiologists. Second, bringing in local traditions and religious leaders who connect it to supernatural dynamics are not helpful (as was said by the Hippocratic writer of *The Sacred Disease* many centuries ago). Still, this continues to happen. Third, using the crisis to demonize other countries (as an "indirect attack") or social groups within a country (such as LBGTQ individuals during the HIV/AIDS pandemic—that is still ongoing) is counterproductive to the task of keeping everyone safe and getting on the other side of the pandemic.

Unfortunately, many political leaders around the world think only of how impotent they are and have no vision of getting around denial and moving forward to

---

[100] For a comparative list see: https://time.com/5851633/best-global-responses-covid-19/ (Accessed 4 October, 2020). For some background on the Asian nations see: Cyrus Staniec, "Cruising into a Global Pandemic (Early Lessions from the Indian Ocean)" *Phalanx* 53.2 (2020): 46–49.

[101] Piere Somse and Patrick M. Eba, "Lessons from HIV to Guide COVID-19 Responses in the Central African Republic" *Health and Human Rights* 22.1 (2020): 371–374.

[102] Jeffrey D. Sachs, "Perspective COVID-19 and Multilateralism" *Consilience* 22 (2020): 1–5; and Jeffrey D. Sachs, "COVID-19 and Multilateralism" *Horizons* (2020): 30–34.

[103] https://www.washingtonpost.com/nation/2020/03/20/coronavirus-trump-chinese-virus/ (accessed October 1, 2020).

[104] For some examples see: Gabriel A. Fuentes, "Federal Detention and 'Wild Facts' during the COVID-19 Pandemic" *The Journal of Criminal Law and Criminology* 110.3 (2020): 441–476 and Michael Jackson and Paul Lieber, "Countering Disinformation; Are We Our Own Worst Enemy?" *The Cyber Defense Review* 5.2 (2020): 45–56.

actively confront the public health crisis through ethically and scientifically based policies that are laser-focused on the real problem of ridding ourselves of the public menace at hand.

An Assessment of Technology and the Mission of Medicine

The ethical mission of medicine can best be determined by the Hippocratic Oath, whether applied individually at the clinical level or at the group level in public health. When technology (whether it be directed at big Nature or little Nature) can bring about the cure of disease or the treatment from severe accidents, then this advances health, that lies at the practical mission of medicine (see also the Appendix). Though there are some possible issues on how *fast* these developments should come forth, it seems to this author that (barring any *per se* ethical breeches seen from the perspective of the principle of precautionary reason) we should move forward in a deliberate way in the direction of curing disease and extending health. I am not enamored with the project of "enhancing the human species that is already healthy." But this is a topic for debate.

Thus, ends the logical order section of Part Two.

Conclusion

In this chapter (in the context of Part Two of the book), I have tried show how the n/Nature$_2$ has evolved over time—especially since the seventeenth century from an account of the origins and causes of *the material* from being *divinely centered* in the ancient world, to being *mathematically centered* in the modern context. Both are *supernatural, nature$_2$*. Therefore, both require empirical verification by nature$_1$ to be fully epistemologically justifiable.

In the contemporary realm, detractors (whether "in principle" or "in practice") take on a necessary role of critics, nature$_3$. From David Hume's *mitigated skepticism* of the eighteenth century to Karl Popper of the twentieth century (thinking generally about possible falsification) to the critics of genetic engineering (including Crispr) in the twenty-first century, n/Nature$_3$ adherents fulfill an important role in keeping the status quo and alterations to the status quo within intellectual and ethical guardrails.

When I was working on my earlier book on genetic engineering in the early 2000s, I interviewed a number of scientists who were doing bench research on cloning and I did not find a single one who would admit that there should be *any* limits in genetic engineering or even in scientific inquiry, in general (save for Kevin Brown, the co author of the book on genetic engineering and a bench researcher engaged in cloning). The role of n/Nature$_3$ raises searching doubts to new protocols both from a practical and from an ethical point of view.

Thus, in the contemporary era (just as in previous eras discussed in the book) the three senses of n/Nature have proved to be essential in putting forth theories in general science and their application to medicine, which is the focus of this book (within the Western Tradition).

Also, the categories of *completeness, coherence, simplicity,* and *elegance* are useful in a meta-examination of the structure of the theories presented. In cases that, ceteris paribus, provide the same pragmatic outcomes, these criteria should come into play for theory evaluation (compare to Chap. 6). But empirically-based pragmatic outcomes ultimately need to prevail—especially in medicine when are talking about patients' lives and their medical well-being.

It is my hope that the distinctions and applications raised in this book will advance the discussion of how biomedicine should move forward and the vocabulary and categories by which it should be evaluated. It is a topic of great importance for us all.

Appendix

Health as Self-Fulfillment

Nothing is better than a diligent life.
　Ancient Roman adage
　Michael Boylan

Let me begin with a little story.[1] There was once a king named Agamemnon who was a general in a foreign war (on behalf of his brother). The war lasted a long time. When he finally returned (with a princess from the losing side who was now his concubine) he was killed by his wife (who had a consort of her own). The principal reason that Clytemnestra gave for killing Agamemnon was that he killed their daughter Iphigenia out of deer necessity dating back to a dispute with Artemis. In the middle of this tragedy Zeus comments:

> It is true that man's high health (*hygeia*) is not content with limitation. Sickness (*nosos*) chambered beats against a common dividing wall. It is human destiny to set a true course in life, yet this course may be dashed against the sudden reefs of disaster. (Aeschylus, "Agamemnon" ll. 1001–1007, my tr.)

So, what are we to make of our little tale? From the beginnings of the Western tradition in ancient Greece *health,* represented by the goddess *Hygeia,* stood within a context.[2] She was the daughter of Asclepius (god of medicine—who himself was the offspring of Apollo). Her siblings were *Eros* (god of love and directed desire), *Peitho* (goddess of eloquent persuasion), *Panakeia,* (goddess for all curing), *Iaso* (goddess of remedy and recuperation), Akeso (goddess of recovery), and Aglaea (goddess of natural beauty). Hygeia attended her father Asclepius (god of medicine) and palled

---

[1] This is, of course, my story, but it is loosely based upon Greek Mythology.
[2] These relations are often parsed differently. This is because there are discrepancies among the primary sources. Since this is not an essay on philology, I will present these characters in the context of my initial story.

around with Aphrodite (goddess of love, beauty, and sex). One day they had a feast to honor Hygeia's birthday. What began as panegyric for Hygeia quickly devolved into a dispute. Each sibling wanted *their* natures to be honored the most. This escalated into a fight concerning who father (Asclepius) loved best and who was grandfather's Apollo's favorite. Each sibling made his or her case (based upon their natures), but there was no agreement and in the end the party degenerated into a disaster as everyone exited—everyone except poor Hygeia, whose feast it was!

What a sad story. But dry your eyes, the tale has a message: Hygeia (health) is not best understood by *any* single sibling. Instead, we must understand health via a multi-layered presentation. Certainly, medicine is about assisting us all toward good health. This means that the *aim* of medicine is promoting health. Thus, 'health' is the foundational concern in medicine and medical ethics. But 'health' means different things in different contexts. There have been several popular paradigms that have been advanced in recent years about health. These can be roughly grouped into three categories: (a) a functional approach that clinicians might take to be based upon some understanding of physiology, (b) a public health approach based upon some group allocation of goods that are primary to human agency, and (c) a more subjective approach based upon some understanding of well-being. Let's take a quick look at all three and then move toward one particular understanding of subjectivism.

Functional Approaches to Health

The first approach to be examined is the functional approach (also called 'objectivism,' biostatistical theory (BST), functionalism/disfunctionalism, et al). The general approach of all these theories is to adhere to the methodological dream of the logical empiricists to create *value-free* science. BST has been advocated by Christopher Boorse (Boorse, 1975, 1976, 1977, 1997). BST takes a "body-state" to be diseased if it is operating below what is statistically normal for body-states of the same species, same sex, and same age (called the reference class). Health is thus defined as having all body-states operating at or above the normal efficiency level for the reference class (meaning there is an absence of disease). The intent is to make this a scientific measurement that is value free. One can measure body-states among humans (or among individuals in any other species) and create a normal range for functioning. The measurement device will thus produce data that are indicative of whether the individual in question is "normal" vis-à-vis her reference class. For example, there is the EQ5D system that uses five basic dimensions: mobility, self-care, usual activity, pain/discomfort, and anxiety/depression. There are three levels in each dimension (Kind). With all the permutations the EQ5D allows for 343 (3^5) health states—including death. However, the health-states approach has come under criticism because it assumes that each disease presents with identifiable symptoms. But this is not always the case. The measuring system is also environmentally insensitive. For example, mobility might mean different things if one lives in a relatively flat, handicapped accessible city rather than a rural mountainous region.

However, it is true that a part of medicine is always in search of a reference class based either upon health states or upon some sort of measurable data on the functioning of biological organs and systems against a background in which the patient seeks to score at the median level. When you get a blood test your platelet count, lipid count, etc. are all gauged in this way—likewise with other standard tests such as heart stress tests and the vaunted colonoscopy. So, in this way Boorse and others of this ilk seem to have identified one sibling of Hygeia (health) that was properly invited to the party.

But there are some problems, too. These follow from the interpretation of what has been shown. It is not the case that we have a value-fee general account. This is because the median is not equally valid as a reference for health in every reference class, which are all considered to be equal (Kingma). For example, if we were to identify the reference class of heavy drinkers, then the normal range of liver function would be different from the range of liver function among the tea-total population. It would seem odd to say that the median test results for liver function among the heavy drinkers would constitute health. In reply, Boorse might reply that the reference class of heavy drinkers is not the right kind of reference class. But to do this he would have to set forth a separate account of what makes a proper reference class. This would be grounded upon separate principles (another guest at the party). Also, other guests at the party might have their views of the measuring of health that is not physiologically grounded, e.g., evolutionary biologists or subjectivists (DeVito).[3] Therefore, if Boorse wants to modify his claim to the vantage point of one single guest at the party, then there is no problem. However, if the claim is grander (being about health, as such), then his argument fails in his design (Allmark, Hamilton).

A second approach in a similar vein is objectivism. According to this account life is the value and health is associated with an "uncompromised" life-span and disease with a compromise or shortened life-span (Lennox). Everything is based upon the notion that the summum bonum is achieving and surpassing the normal life-span. But this mistakes the wide-spread phenomena among many species of dying for the sake of the cohort in message warnings, and so forth—biological altruism (Sober and Wilson). Among humans, suicide and euthanasia are certainly sought by some suffering chronic, incurable pain. Also, there are those who willingly give up a kidney for another to save their lives (usually a family member—DeVito). These organ donors statistically lessen the donor's life-span. Thus, under the objectivist account such organ donations are instances of disease. Finally, there are those situations in which someone is in chronic pain to a high degree but the pain does not limit life expectancy. Can this person be said to be totally healthy? Thus, though there is an important insight about what mortality can tell us about general health and well-being, it is only one member of the party (Sen, ch. 7).

Finally, there is the functionalism/dysfunctionalism debate. This is generally carried on in the psychiatric community (Wakefield, 1992a, 1992b, 1993, 1997a, 1997b).

---

[3] Devito sets out additionally that if BST is correct, then any state that does not decrease life-span or fecundity is not a disease, e.g., flu, chronic pain, or bronchitis. Most would say that this is a major flaw with BST.

Health in this context requires an understanding of healthy and unhealthy mental states. This is interpreted as having a functional mental state (controlled by a functional mental mechanism). What does it mean to have a dysfunctional mental mechanism? There seems to be no way around this being a socially constructed concept. For example, homosexuality is on Wakefield's account (and Boorse's) judged to be a disorder or disease. This is totally based upon using *reproduction and mate selection* as the categories that define the reference group. Because homosexuals are a statistical minority of the population, they are "hurt" in these two areas because (a) they cannot naturally reproduce, and (b) they have fewer mates to choose from. Wakefield asserts that society would view these two conditions as a harm, therefore; homosexuals are "diseased" because of these two "dysfunctions." However, this is very problematic because it imports social values in the assessment of normal function. Where do these come from? Some theories of evolutionary biology claim that homosexuality is adaptive because it creates more nurturers and thus increases group-level fitness (Sober and Wilson). Also, what about the possibility that the social group is prejudiced? For example, in the United States (for much of its history) there were fewer opportunities for Americans of African descent than for Americans of European descent? Does this mean that to be descended from Africa and living in America constitutes a disease? This author says *no*.

Definitions of mental health as seen within a functional arena can be useful. Autistic individuals are clearly not able to socially function in mainstream society in a manner that is to their own best interests. But what about other examples—such as deafness? Some deaf people value the deaf community as more viable than the mainstream community (Savulescu). This is certainly an issue that deserves further attention.

The addition of mental health to the picture of health is more complicated. There are certainly cases in which some references to normal functionality as supplemented by social adaptability are useful for determining mental health, but there are other cases in which we must pull aside another guest at the party in order to get an accurate depiction of health.

The above attempts all fail in their quest to become the universal definition of health because they base their approaches upon a value free science. This was the dream of the logical empiricists (e.g., Bridgman, Hempel, 1950 & 1951, Carnap, 1956). They saw raw objective observation sentences that were explained by theory statements that were, themselves, derived solely from induction that was probability-based (thus purifying it from subjective value prejudices) as the holy grail of natural philosophy.

But the value-free dream of the logical empiricists has been largely rejected. This began with Quine in his "Two Dogmas of Empiricism" and continued with Kuhn's *The Structure of Scientific Revolutions* and ran its course via reformed-logical-empiricist Hilary Putnam in "The Collapse of the Fact/Value Distinction." This trend in the philosophy of science was partially driven by the philosophy of biology (because of biology's heavy clinical & lab observations) that exhibits its assumed values more clearly than the previous paradigm of the philosophy of physics (largely mathematically driven). Because science is not value-free, those theories of health that depend upon a grand attachment to the outmoded logical empiricist worldview will not deliver on their claim.

The universal designs of this objective functionalism approach also fail because of the nature of biology. Whereas physics is largely driven by a universalist model that can be falsified by one instance (such as showing that E is unequal to mc^2 or that F is unequal to ma), biology is not. Aristotle, the inventor of biology, saw this in his depiction of the aims of biology, *epi to polu* (for the most part, see Boylan, 1983). What this means is that there are many exceptions in biology that do not exhibit themselves in physics. For example, if a baby homo sapiens was born with only one hand, we would not say that the empirical generalization in biology that homo sapiens have two hands is false. This is because of the *for the most part* caveat. This variation between individuals within a species does not count against biological laws, but reinforces them via the theory of evolution that requires variation within a species so that fitness is higher, over a range of possible environments. This means that *diversity* and not *stasis* is the watchword. This variation extends to *health* as well (Hamilton). In the ancient Greek world of medicine acceptance of individual variation gave rise to various schools: *Empirics, Methodists, and Dogmatists* (Boylan, 2007, see also Chap. 5). Each school asserted a different emphasis upon how to acknowledge individual variation from the Empirics (the most) to the Dogmatists (the least). But even the Dogmatists (descending intellectually from the Hippocratic Writers and Aristotle) asserted that *for the most part* was the way to understand the aspirational goal of biological laws.

Thus, this section has tried to show what is right and what is not right in understanding human health through the lens of the value-free functionalism.

Public Health Approach

A very popular guest at Hygeia's party in modern times is the group perspective. Some authors such as Amartya Sen have conjectured that public statistics on group longevity says something about how happy and capable people are within a society (Sen, 2009). Figures about infant mortality, morbidity due to certain types of disease, epidemiological data on who the sick are and any common forms of causation yield important information on community health. Individuals in the community can be protected by evidence-based medical responses, but the focus is upon the group.

There are at least two ways to understand the public health perspective on human health. Both perspectives are not clinical with its focus upon the individual and the physician. Instead, the focus is upon groups of people and maintaining environmental conditions that will minimize the spread of infectious disease via clean air, water, sanitation, vaccination, and access to basic medical care. This can be called the thin theory of public health. It is largely based upon prudential self-interest understood collectively. There is another vision of public health that extends this vision to basic human rights—such as those enunciated in the United Nations Universal Declaration of Human Rights. This can be called the thick theory of public health. I have been an advocate for the latter vision and believe that its broader mandate can only be supported by an appeal to normative ethics (Boylan, 2004b).

The difference between these two approaches is that the thin theory of public health views a person as healthy if she isn't ill (defined as having recognizable bacteria or viruses attacking the body causing a loss of function leading to diminished productivity in the workforce). This is often extended in the thin theory to include workplace injury, accident, and response to war and natural disaster. This viewpoint concentrates upon negative physical influences of various sorts upon the body and its physical systems—viewed collectively via an identifiable social/community group.

In contrast, the thick theory sets out that there is more about being healthy than merely being not-diminished by one's physical systems—viewed via an identifiable social/community group. More is needed to demonstrate public health: namely various educational opportunities, human rights, and the ability to participate in one's community as an equal partner and to be able to strive towards one's vision of a life fulfilled (Boylan, 2004a).

What the public health (in either of its two forms) has going for it is that it identifies groups of individuals within a context. There are natural and social environments in which we all live. While the thin theory focuses upon the natural environment, the thick theory combines the social and the natural contexts that permit individuals to act purposively according to their vision of the good. By focusing upon target groups, social changes can positively affect health within that target group. Darrick Tovar-Murray completed a small demonstration project to explore the truth of this conjecture. The results of his project supported the hypothesis. As a result, now it is correct to contend that the context is not everything. One can live in an area that has a cholera outbreak and never get cholera. One can live under a repressive dictator and never get jailed for being an agent provocateur. Just because one lives in a bad natural or social environment does not ensure that they will be a victim. What these deleterious environmental conditions do is to increase the probability that something bad will occur, and that will affect one's ability to execute purposive action. The public health perspective is thus important because it *can* affect the context of our action.

However, we must be clear that it does not *guarantee* it. One may have a relatively good natural and social environment and still fall prey to a fatal disease or be the victim of an unjust action. As was argued above, we are working in the arena of statistical probabilities.

Given this structure of probabilities, one approach to public health is to view public preferences as a key determinant in creating the right sort of public health policy. Those who take this approach view public health much like economists view commercial choices in the marketplace. In the case of public health, the marketplace is perceived well-being. Thus, certain health states are to be preferred to others on the basis of their contribution to well-being (Broome). At the individual level this leads to indifference curve choices in which the general public chooses to go after one health state, S, over another S' because of its perceived link to health as understood through well-being (Hausman). This leads to a version of the social choice model for group evaluation of public health, that is not unlike Sen's famous model (Sen, 1970). Social choice takes individual evaluations of preference rankings as its primary data.

The bases of the preferences are subjective states and how we assess them. But unless we benchmark the bases of the assessment (say upon some generally agreed upon group of core capabilities—so-called "basic preferences"), then we risk people falsely judging how to rank various health states. This error can occur via *argumentum ad populum* (given our tendency as social creatures to jump on band wagons—especially in the era of social media).

Unfortunately, the average citizen is not an expert on health states or even on health-linked well-being, as seen on the long range. Socrates used to ask whether the horse trainer would be a better person to ask about horses than the average person (ditto the shoe maker, flute player, et al.). Just as general consumer preferences have often chosen the inferior over the superior, e.g., the first IBM mainframe over the UNIVAC, VHS video tape over BETA, the PC computer platform over Apple, etc., so also have policy advisors made very bad collective preference choices, e.g., the repeal of Glass-Steagall Act in the United States or the non-regulation of derivatives and credit default swaps in the financial services industry.

There is something very appealing about seeing health via the public health guest to the party (thin, thick, or governed by social choice), but this perspective is general: often good for policy, but possibly inaccurate for individuals. It is one important perspective, but as we have seen before, there are many guests at this party.

Subjectivist Approaches

The last group of party goers would be those who represent subjectivist approaches. These include the advocates of well-being broadly understood and well-being understood via the lens of self-fulfillment.

Well-being is a term that is used variously in different contexts. Derek Parfit suggests that there are three sorts of theories in this category: (a) *hedonistic theories*; (b) *desire-fulfillment theories*; and (c) *objective list theories*. The first path to well-being is merely to seek what one perceives will make her happier via some calculus that is made through a preference-hedonism model (cf. the public health preference model above). If X is thought to bring about more happiness than Y, X is preferred over Y. The very fact that the agent chooses X over Y indicates that the agent thinks that X will deliver the most happiness/pleasure. The criteria for this preference are calibrated via the personal worldview of the individual. Thus, Freud near the end of his life might prefer to forego pain-killing drugs in order to maximize mental lucidity. This is a hedonistic calculation based upon Freud's personal theory of value (Griffin). Such an account is relative to the chosen personal theory of value. Unless one has created meta-ethical value criteria to steer the process, it is subject to very wide relativistic swings—some of which are in direct contradiction.[4]

---

[4] I have created such criteria in Boylan (2004a) (ch. 2).

In *desire-fulfillment theories* the model works this way: we should seek a course of life that will fulfill as many desires as possible. Parfit calls this approach the *success* orientation. The agent decides for himself what approach will yield success and thus fulfillment of as many desires as possible (thus ensuring well-being). In the context of health, one might choose to be an exercise enthusiast because undergoing that strategy can satisfy more physical desires than any other alternative. However, this could be turned on its head if I turn out to die from an inherited disease (despite my careful exercise routine). The very structure of the *desire-fulfillment theory* is such that it operates on a conditional "p → q" structure (if p then q). However, if this model is faulty (as in the exercise example), then the conditional becomes contingent. This means that achieving the state "p" does not guarantee "q" (invalidating modus ponens) which also implies that one can fail to achieve q (~q), without assuming ~p (thus invalidating the logical rule of modus tollens). There may also be multiple ways of achieving q without invoking p. If "p = 2 hours daily of vigorous exercise" and "q = not being sick from a bacterial/viral or subject to an organ failure," then it is easy to see that one might exercise and not stay well in the sense of q. Also, one might be well in the sense of q and not exercise. The reason for this harkens back to the objective functionalist theories of health discussed above.

Objective list theories seek a paternalist path toward well-being. Under this approach one acts according to a set of criteria that are generally agreed to lead to the fulfillment of as many desires as possible (thus ensuring well-being). This is very much like the *success model* of the previous paragraph except that the origin of the strategy is commonly accepted maxims. However, the problem raised with the exercise example would still hold here. The only real difference is the origin of the strategic approach. However, when thinking about health, we can think about the difference between these approaches as one in which first the agent chooses their path that they think will yield as much happiness as possible. The source of the strategy is a list of value priorities and factual understandings as found in the personal worldview.

In the second case, the source of the strategies lies outside the agent—as well as the values and facts concerning the world (for example from the family physician). The agent then chooses to follow a regimen that is generally thought to improve one's chances of achieving "q."

There are many advocates of a third sense that uses well-being as a way of understanding and achieving health. However, there are some detractors, too. One important attack on well-being as a master value comes from Thomas (Tim) Scanlon.[5]

---

[5] Scanlon's argument on well-being works this way:

1. There are three uses of 'well-being': (a) the basis of individual decision-making (1st person), (b) the basis of a concerned benefactor's action (3rd person), (c) the answer to the 'why should I be moral' question (1st person)—Assertion [A] (P. 108)
2. WB 1-a is experientally important to us all—Fact [F] (p. 108)
3. WB 1-a is sometimes understood as fulfilling desire—F (p. 113)
4. Desire is not sufficient for rational choice—F (from Ch 1/ p. 114)
5. Desire and its fulfillment cannot give an account of WB sufficient for morality—2–4

Scanlon distinguished three uses of 'well-being': (a) the basis of individual decision-making (1st person), (b) the basis of a concerned benefactor's action (3rd person), (c) the answer to the 'why should I be moral' question (1st person). The first sense amounts to fulfilling desire. But rational choice (which should undergird theories of morality) cannot be based solely upon fulfilling desire—even rational desire expressed as a preference (see above). This is because of the connection with well-being that lacks the requisite boundaries.

These boundaries amount to the criteria of being choice worthy which might lead some to theories such as Rawls or Sen (the second sense enunciated). But these also fail because the boundary line between agent and recipient is not clear. This is especially true when dealing in cases involving mental health. It is often the case that a person with diminished mental health might think that her well-being is just fine

---

6. Rational desire understood as preference is often put forward as a ground for WB—F (p. 116)
7. The good is not dependent upon preference (informed desire) but the reasons that make it worthwhile—A (p. 119)
8. Rational desires understood as preferences cannot give an account of WB sufficient for morality—6–7
9. Some say that rational aims tied to WB create a motivation superior to desire—A (p. 121)
10. [Motivation is important to morality]—F
11. Fulfillment of rational desire (broadly and specifically) must be tied to WB or the desire wasn't rational—A (pp. 121–123)
12. Many rational desires have intrinsic aims (e.g., friendship and science) that are not connected to WB—A (p. 124)
13. Though WB has some connections with rational aims, it is not the sole source of determination—9–12 (p. 124)
14. WB 1-a does not have clear boundaries because it cannot account for why it is good—A (p. 127)
15. There is no limit to WB 1-a—A (p. 129)
16. [What has no limits has no boundaries]—F
17. WB 1-a has a boundary problem—14–16 (p. 129)
18. In choosing the best life, the 'most choice worthy' trumps 'well-being'—A (p. 131)
19. WB is not primary and sufficient—17–18
20. When one concentrates upon his own WB, he becomes selfish—A (p. 137)
21. [Being selfish is bad]—A
22. WB 1-a can be counterproductive—20–21 (p. 133)
23. WB 1-a is not a master value for morality—5, 8, 13, 17, 19, 22
24. WB 1-b is generally connected to morality via justice and benevolence, cf. Rawls and Sen—A (p. 139)
25. A benefactor may act to promote a choice worthy life over one based upon WB (e.g., artist or labor organizer)—A (p. 135)
26. WB 1-b implies a standard account5 of WB based upon promoting pleasure—A (p. 136)
27. The boundaries between the benefactor's and recipient's WB are unclear—A (p. 136)
28. The recipient does not have reason for merely promoting his pleasure—A (p. 136)
29. WB 1-b is not a master value for morality—24–28
30. WB 1-c would require one to justify moral principles on grounds that presuppose what people are entitled to—A (pp. 137–138)
31. Premise #30 involves a circular claim—F (p. 138)
32. WB 1-c is not a master value for morality—30–31
33. Well-being is not a master value for morality—1, 23, 29, 32

when most people (particularly her loved ones) might come to a different conclusion (e.g., those who cut themselves, or suffering from anorexia, bulimia, bi-polar disorder, et al.). The subjective basis of well-being can be skewed when the agent's assessment capacity is impaired. Therefore, because of these reasons, the third sense of well-being is circular. Thus, all three can fail and we must look further for a master underlying argument. Enter self-fulfillment.

Self-fulfillment

Self-fulfillment in the guise of functionalism has been raised before in the health debate (Allmark). This approach is generally tied to an understanding of Aristotle's *eudaimonia* as functionally "good souled" (where *soul* indicates a natural capacity of the human person, e.g., rationality). The idea of being "fulfilled" presupposes a standard that one works towards. The closer one gets to the terminus, the more fulfilled she is. The million-dollar question is: "What is the standard for homo sapiens?" Unless this question is answered, the self-fulfillment question devolves into the well-being question with all its various mazes of interpretation.

What makes this essay different is its connection of health to self-fulfillment as understood through an analysis of the personal worldview and my assumption that in life we all strive to achieve our vision of the good (Boylan, 2004a). I have written much on the personal worldview. The personal worldview is a compilation of all one's understandings of the world factually and normatively. I suggest a way to self-diagnose one's worldview via the personal worldview imperative, "All people must develop a single comprehensive and internally coherent worldview that is good and that we strive to act out in our daily lives (Boylan, 2004a)." There are four parts to this imperative: (a) a worldview must be comprehensive (leading to the development of the rational and affective good will); (b) a worldview must be internally coherent (deductively and inductively), (c) a worldview must connect to a normative theory of ethics; and (d) a worldview must be at least aspirational[6] and acted upon. I see this as a first order metaethical theory that would give direction to how one would develop their life. As I show elsewhere, I believe that this position dictates that everyone should adopt cooperative theories of justice in holistic ways of looking at the world (Boylan, 2004a, 2011). Let's examine these claims in relation to a common objector position.

The position I'm thinking about centers around a personal worldview concept of life being like a daily hunt in the jungle for food (Boylan, 2008). There are many snares in the jungle: wild animals might take *you* for food. You might be unsuccessful and eventually die in pursuing your quarry. Certainly, those who have the hunting-in-nature metaphor view, life as a continual struggle for survival. At any moment you

---

[6] I interpret 'aspirational' as being something that is attainable—though it may be very difficult to achieve. It is to be contrasted to 'utopian' which is a goal that (though laudable) is practically impossible to achieve. For example, eliminating the common childhood diseases (that have known vaccines) in the world is aspirational while totally eliminating poverty is utopian.

may lose your hunting skills so that you will be forever lost. There is none to help you. You are on your own to make it or fail. Success comes to those who deploy themselves most effectively.

Those who hold this worldview metaphor will be suspicious of those who are possible competitors. This is because they may take from them their daily kill. As this metaphor is translated into its modern referent, the animal carcasses become bank account balances. Everyone is after your money. The only way to protect yourself is to accumulate and hoard large amounts of reserve cash to protect you against a shortfall or an emergency (kind of like modern hunters putting meat into the deep freezer). The vision of being penniless on the street is constantly before the holder of this worldview as the worst case—*yet possible* scenario.

Sometimes the hunting metaphor is combined with a war metaphor as in Thomas Hobbes' depiction of the state of nature.

> ... there is no way for any man to secure himself, so reasonable, as Anticipation; that is, by force, or wiles, to master the persons of all men he can, so long, till he sees no other power great enough to endanger him (Hobbes).

For Hobbes the metaphor is of a state of nature (forest, hill, and dale) in which all are equal—though not identical (e.g., you may be able run faster than I, but I'm stronger than you: in the end the sums are equal). Because of this summative equality, and the fact of scarcity of resources, the result is that there is fierce competition that will inevitably lead to continual strife (war). This is the human condition according to Hobbes and is depicted via his state of nature metaphor.

Another fellow traveler is Friedrich Nietzsche who seeks to describe the basic psychological nature of human kind in order to give a causal account from the agent's point of view.

> Suppose, finally, we succeeded in explaining our entire instinctive life as the development and ramification of *one* basic form of the will--namely, of the will to power, as *my* proposition has it... then one would have gained the right to determine *all* efficient force univocally as--*will to power*. The world viewed from inside... it would be "will to power" and nothing else (Nietzsche).

For Nietzsche, the will to power is a psychological fact that finds metaphorical expression in *Beyond Good and Evil* and *On the Genealogy of Morals*. In some respects, it is a deeper account than Hobbes' because it gives specification of *why* we are acquisitive. It is because, at base, we are psychological egoists whose quest in life is to exert whatever influence we can upon the world. There is a trust that those who can assert the most influence will also be driven by a love of nobility (beauty) that will keep them in check from being utter tyrants. Of course, skeptics of the regulative power of nobility (beauty) will see this depiction as one that devolves to mere *kraterism* ("to each according to his ability to snatch it"). Under this sparser interpretation, Nietzsche falls into the tradition of the hunting/war metaphor. (The more generous interpretation would put Nietzsche on the edges of the metaphor, given the tempering force of nobility [beauty]).

In either case 'the will to power,' as metaphorical expression, is seen in the context of other writers who assert the same thing. Like Hobbes, Nietzsche can be connected to a vision of life on earth as a competitive contest. We are all engaged in seeking to extend ourselves over our environment and over others.

One practical consequence of the hunting metaphor of life is *laissez faire* capitalism. We all strive to gain the goods, and the science of economics is created to describe (and not prescribe) the process. Since everything is all wrapped up tight in a theory of human nature, what could be more correct? *This metaphorical expression measures our goodness in terms of competitive acquisition of goods—money, status, and power. Thus, our primitive drive to be good is satisfied by the garnering these goods in the highest amounts. The individual with the biggest heap at the end of the day is the winner!*

If self-fulfillment is a legitimate way to understand health (one of the guests at Hygeia's party), then the competitive model sets the measure by which we can assess whether we are healthy: how much power have we achieved via competition within our frame of reference. The more power, the healthier we are. If one accepts that self-fulfillment is a legitimate way to understand health and if the best candidate for the scorecard is power, then Q.E.D.

In contrast to the highly competitive personal worldview paradigm, I would put forth an alternative model:

> Imagine that each of us is on a quest to be good. We are seeking a means to be good that will make our world (and us) better through intellectual excellence (theoretical and practical reason) and emotional excellence (love), i.e., establishing the good will within ourselves. The quest may last a long time. The quest may end in failure. It is up to us to do our best to seek and obtain the object of the quest. In the process of our quest we may be required to undergo various ordeals and tests of our resolve and worthiness. It is the nature of human existence to sally forth on this quest and do our individual best at achieving the reward (though we may be humbled, scorned, and ridiculed in the process). This process thus represents a prescriptive view of a good human life. To be healthy is to be closer to the endpoint in the quest. The closer we are, the healthier we are.

The Cooperative Goodwill Thought Experiment

In the thought experiment an alternative is set forth to the competitive worldview standpoint: the cooperative worldview standpoint as exhibited by the focus upon creating a good will. Though expressed differently both Aristotle and Kant, it holds that we ought to try to acquire various excellences in character that would create habits in us such that our decision-making apparatus (our will) would become increasingly good (as measured by our internal assessment via the personal worldview imperative).

There are several reasons why each of us should prefer the cooperative personal worldview over the competitive worldview—especially concerning health. First, there is some medical evidence that being hypercompetitive can lead to several serious deleterious medical conditions (Freidman, Haukkala, et al., Al-Asadi). Second, the nature of the competitive worldview is a zero-sum game while the cooperative worldview is not. The very nature of the competitive worldview is that there are limited goods sought after by many. It is much like the child's game of musical chairs. There are four chairs and six children. They walk politely around the chairs while the music is playing, but when the music stops, it is a mad dash for the chairs. Children are pushed away as the hyper-aggressive win the day. Is this the community worldview we wish to promote? Will this lead to general public health? I think not. It will lead to a society of a few super-winners at the expense of the many. Such a worldview violates the personal worldview imperative because it violates the affective goodwill and the inductive understanding of consistency—as such it violates this first order metaethical principle. In addition, I have argued elsewhere that such an outcome is inherently unjust (Boylan, 2004a).

So, let's suppose that we proceed with a cooperative personal worldview that is in accord with the personal worldview imperative. What else is necessary to proceed along a path of health as self-fulfillment? To answer this, I would again foray to the ancient Western world and the biomedical writers—particularly the Hippocratic writers and Galen. What these writers found to be the case was that *balance* was the most critical factor to health. By balance they meant of course the balance between the four humors of the body: blood, phlegm, yellow bile (sometimes serum), and black bile (Gill, Hankinson). Of course, Aristotle advocated balance, too, in his doctrine of the mean (and so did Confucius with his concept of *li*—balance presented via the metaphor of dance). What these ancient writers understood was that an *essential key* to health is balance because it encourages the development of *sophrosune*, self-control (a master virtue when considering balance). Self-control is also crucial in achieving self-fulfillment. This is because deciding what one wants to do in life (constrained by the personal worldview imperative) is a process of reflection, self-control, and the creation of habits of excellence. These three menu items work together so that one can act autonomously toward a worthy goal in a balanced manner. The process looks like this:

1. One seeks balance to achieve self-control—basic fact of human nature
2. Self-control allows one to more successfully carry out the personal worldview imperative (reflection that leads to the rational and affective good will)—A[ssertion]
3. A developed rational and affective good will allows one to develop habits of excellence that are directed toward one's chosen life plan—A
4. A life plan chosen as per above will be the most choice-worthy path toward an agent's life goals (self-fulfillment)—1-3
5. Because the process begins with balance, B, this property continues through all the steps as a property of the agent—F[act]
6. Balance supports personal health—5

---

7. The choice-worthy path toward self-fulfillment supports personal health—4-6

The Self-Fulfillment Approach to Health

What this approach to health offers is a subjectivist path that has an objective structure (the personal worldview imperative) that can alleviate the common objections to well-being (the most prominent subjectivist theory discussed). Because of this, I think that it offers the best subjectivist understanding of health.

Conclusion

This essay began with Hygeia's party on Mount Olympus. There were many guests—each of whom had a legitimate right to be there (no counterfeit invitations). I have tried to interpret this literary conceit to discuss the problem of health pluralistically—showing the proper roles of various perspectives, but also arguing that none of them gives a complete account.

From the point of view of many readers of this essay, the most personally relevant understanding of health is subjectivist. In order to avoid the common complaints against well-being not having an adequate external structure, this essay has set one in place (the personal worldview imperative) within the context of self-fulfillment as a personal measure of health as we lead our lives. In the end, we must judge ourselves after the lines of *The Eumenides* (the final work of the *Oresteia*):

> *Home, home ever high aspiring,*
> *Daughters of Night, aged children, cavalier processional*
> *Bless these with silence ...*
> *There shall be peace between Pallas Athena and the guests.*
> *Zeus, all knowing, met with Fate to confirm it*
> *Let us sing as we make our exit.* (ll. 1033–1047, my tr.).[7]

Bibliography for Appendix

Aeschylus. (1972). In D. Page (Ed.), *Aeschyli Septem Quae Supersunt Tragoedias*. Clarendon Press.

Al-Asadi, J. N. (2010). Type A behaviour pattern: Is it a risk factor for hypertnesion? *Eastern Mediterranean Health Journal, 16*(7), 740–745.

Allmark, P. (2005). Health, happiness and health promotion. *Journal of Applied Philosophy, 22*(1), 1–15.

Boorse, C. (1975). On the distinction between disease and illness. *Philosophy of Public Affairs, 5*, 49–68.

Boorse, C. (1976). Wright on functions. *The Philosophical Review, 85*, 70–86.

Boorse, C. (1977). Health as a theoretical concept. *Philosophy of Science, 44*, 70–86.

Boorse, C. (1997). A rebuttal on health. In J. M. Humber & R. F. Almeder (Eds.), *What is disease?* Humana Press.

---

[7] I presented a version of this paper to the philosophy department at the University of Warsaw on September 21, 2023.

Boylan, M. (1983). *Method and practice in Aristotle's biology.* Rowman and Littlefield/UPA.
Boylan, M. (2004a). *A just society.* Rowman and Littlefield.
Boylan, M. (2004b). The moral imperative to maintain public health. In M. Boylan (Ed.), *Public health ethics.* Springer/Kluwer.
Boylan, M. (2007). Galen on the blood, pulse, and arteries. *Journal of the History of Biology, 40*(2), 207–230.
Boylan, M. (2008). *The good, the true, and the beautiful.* Continuum.
Boylan, M. (2011). *Morality and global justice: justifications and applications.* Westview.
Bridgman, P. (1927). *The logic of modern physics* (pp. 1–32). Macmillan.
Broome, J. (2002). Measuring the burden of disease by aggregating well-being. In C. Murray, J. Salomon, C. Mathers, & A. Lopez (Eds.), *Summary measures of population health* (pp. 91–113). World Health Organization.
Canguihem, G. (2008). "Health: Crude Concept and Philosophical Question" Translated by Todd Meyers and Stefanos Geroulanos. *Public Culture: Bulletin of the Project for Transnational Cultural Studies, 20*(8), 467–477.
Carnap, R. (1950). *Logical foundations of probability.* University of Chicago Press.
Carnap, R. (1956). *Meaning and necessity* (pp. 205–221). University of Chicago Press.
De Vito, S. (2000). On the value-neutrality of the concepts of health and disease: Unto the breach again. *Journal of Medicine and Philosophy, 25*(5), 539–567.
Friedman, M., & Rosenman, R. H. (1960). Overt behavior pattern in coronary disease. Detection of overt behavior pattern A in patients with coronary disease by a new psychophysiological procedure. *JAMA, 173*, 1320–1325.
Gill, C. (2010). *Naturalistic psychology in Galen and Stoicism.* Oxford University Press.
Griffin, J. (1977). Are there incommensurable values? *Philosophy and Public Affairs, 7*(1).
Hamilton, R. P. (2010). The concept of health: Beyond normativism and naturalism. *Journal of Evaluation in Clinical Practice, 16*, 323–329.
Hankinson, R. J. (Ed.). (2008). *The Cambridge Companion to Galen.* Cambridge University Press.
Haukkala, A., Konttinen, H., Laatikainen, T., et al. (2010). Hostility, Anger control, and anger expression as predictors of cardiovascular disease. *Psychosomatic Medicine, 72*, 556–562.
Hausman, D. M. (2006). Valuing health. *Philosophy and Public Affairs, 34*(3), 246–274.
Hempel, C. (1950). Problems and changes in empiricist criterion of meaning. *Revue Internationale de Philosophie, 11*, 41–63.
Hempel, C. (1951). The concept of cognitive significance: A reconsideration. *Proceedings of the American Academy of Arts and Sciences, 80*(1), 61–77.
Hobbes, T. (1997/1651). In R. E. Flathman & D. Johnson (Eds.), *Leviathan* (p. 60). W. W. Norton: chapter thirteen.

Kind, P. (1998). The EuroQuoL Instrument: An index of health-related quality of life. In B. Spiker (Ed.), *Quality of life and pharmacoeconomics in clinical trials* (2nd ed., pp. 191–201). Lippincott-Raven.

Kingma, E. (2007). What is it to be healthy? *Analysis, 67*(2), 128–133.

Kuhn, T. S. (1996/1962). *The structure of scientific revolution* (3rd ed.). University of Chicago Press.

Lennox, J. (1995). Health as an objective value. *The Journal of Medicine and Philosophy, 20*, 499–511.

Nietzsche, F. (1968/1886). Beyond good and evil. In *Basic Writings of Nietzsche*, tr. Walter Kaufmann (Vol. 36, p. 238). Modern Library: II.

Savulescu, J. (2002). Deaf Lesbians, "Designer Disability" and the future of medicine. *Journal of Medical Ethics, 325*, 771–773.

Scanlon, T. M. (2000). *What we owe each other* (pp. 108–138). Harvard.

Sen, A. (1970). *Collective choice and social welfare.* Holden-Day.

Sen, A. (1992). *Inequality re-examined.* Harvard University Press.

Sen, A. (2009). *The idea of justice* (pp. 164–167). Harvard University Press.

Sober, E. (1980). Evolution, population thinking and essentialism. *Philosophy of Science, 47*, 350–383.

Sober, E., & Wilson, D. S. (1998). *Unto others: The evolution and psychology of unselfish behavior.* Harvard University Press.

Tovar-Murray, D. (2010) Social health and environmental quality of life: Their relationship to positive physical health and subjective well-being in a population of Urban African Americans. *The Western Journal of Black Studies, 34*(3), 358–366.

Parfit, D. (1984). *Reasons and persons* (pp. 493–503). Oxford University Press.

Putnam, H. (2002). *The collapse of the fact/value dichotomy and other essays.* Harvard.

Van Orman Quine, W. (1953). *From a logical point of view.* Harvard University Press.

Rawls, J. (1971) *A theory of justice.* Harvard.

Wakefield, J. C. (1978). Four basic concepts of medical science. In P. D. Asquith & I. Hacking (Eds.), *Proceedings of the Philosophy of Science Association, 1*, 210–222.

Wakefield, J. C. (1992a). The concept of mental disorder: On the boundary between biological facts and social values. *American Psychologist, 47*, 373–388.

Wakefield, J. C. (1992b). Disorder as harmful dysfunction: A conceptual critique of DSM III-R;s definition of mental disorder. *Psychological Review, 99*, 232–247.

Wakefield, J. C. (1993). The limits of operationalism: A critique of Spitzer and Endicott's proposed operational criteria for mental disorder. *Journal of Abnormal Psychology, 102*, 160–172.

Wakefield, J. C. (1997a). Diagnosing DSM IV—Part I: DSM-IV and the concept of disorder. *Behavior Research & Therapy, 35*, 633–649.

Wakefield, J. C. (1997b). Diagnosing DSM IV—Part II: Eysenick and the Essentialist Fallacy. *Behavior Research & Therapy, 35*, 651–665.

Bibliography

Adkins, A. W. H. (1970). *From the many to the one: A study of personality and views of human nature in the context of Ancient Greek Society, values, and beliefs*. Cornell University Press.
Adler, C. L. (2014). Interstellar travel and relativity. In *Wizards, Aliens, and Starships: Physics and math in fantasy and scientific fiction* (pp. 176–187). Princeton University Press.
Aeschylus. (1972). In D. Page (Ed.), *Aeschyli Septem Quae Supersunt Tragoedias*. Clarendon Press.
Agostino, B. (1836). *Del mal del segno, calcinaccio o moscardino*. Societa Italiana di Biochemica.
Aimone, F. (2022). The 1918 Influenza Epidemic in New York City: A review of public health response. *Public Health Reports, 125*(supplement 3 (2010)), 71–79.
al Nafis, I. (1477). *A commentary on anatomy in Avicenna's 'Cannon of Medicine'* 3 volumes. (Padua: Pierre Manufer, 1477)—Library of Congress, USA, control number: 2021667077.
Al-Asadi, J. N. (2010). Type A behaviour pattern: Is it a risk factor for hypertension? *Eastern Mediterranean Health Journal, 16*(7), 740–745.
Alican, N. F. (2022). *One over many: The Unitary Pluralism of Plato's World*. SUNY Press.
Allais, L. (2010). Kant's argument for transcendental idealism in the transcendental aesthetic. *Proceedings of the Aristotelian Society, n.s., 110*, 47–75.
Allmark, P. (2005). Health, happiness and health promotion. *Journal of Applied Philosophy, 22*(1), 1–15.
Almond, D. (2006). Is the 1918 Influenza pandemic over? Long-term effects of in utero influenza exposure in the post-1940 U.S. population. *Journal of Political Economy, 114*(4), 672–712.
Alston, W. P. (2002). What metaphysical realism is not. In W. Alston (Ed.), *Realism and antirealism* (pp. 97–116). Cornell University Press.
Ambrose, C. T. (2014). Andreas Vesalius (1514-1564): An unfinished life. *Acta Medico-Historica Adriatica, 12*(2), 217–230.
Archibald, J. D. (2017). Private Musings then shared sketches. In *Origins of Darwin's evolution: Solving the species puzzle through time and place* (pp. 127–146). Columbia University Press.
Aristotelis. (1831). In I. Bekker (Ed.), *Arisottelis Opera*. Reimer.
Arnott, R. (2014). Healers and medicines in the Mycenaean Greek texts. In *Medicine and healing in the Ancient Mediterranean* (pp. 44–53). Oxbow Books.
Arnowitt, R., Deser, S., & Misner, C. (1959). Dynamical structure and definition of energy in general relativity. *Physical Review, 116*(5), 1322–1330.
Asmis, E. (1984). *Epicurus' Scientific Method*. Cornell University Press.
Attneave, F., & Frost, R. (1969). The determination of perceived tridimensional orientation by minimum criteria. *Perception and Psychophysics, 6*(6B), 391–396.

Aubrey, J. (1972). In O. L. Dick (Ed.), *Aubrey's brief lives*. Harmondsworth.
Avery, O. (1944). Studies on the chemical nature of the substance indicating transformation of pneumococcal types. *Journal of Experimental Medicines, 79*(2), 137–158.
Babb, J., & Littledog, P. (1999). *Border healing woman: The story of Jewel Babb as told to Pat Littledog* (2nd ed.). University of Texas Press.
Bacon, F. (1902 [1620]). In J. Devey (Ed.), *Novum Oeganum Scientiarum*. Collier.
Bacon, F. (1962 [1605]). *The advancement of learning*. O. Smeaton.
Bainbridge, D. (2018). *Stripped bare: The art of animal anatomy*. Princeton University Press.
Bainton, R. H. (2005). *Hunted heretic: The life and death of Michael Servetus, 1511-1553*. Blackstone.
Barbo, J., tr. (1836). *Biologia Molecolare*. rpt. Bonbée.
Barker, S. F. (1957). *Induction and hypothesis*. Cornel University Press.
Barlett, E. E. (2001). Did medical research routinely exclude women? An examination of the evidence. *Epidemiology, 12*(5), 584–586.
Barrow-Green, J. (2008). The three body problem. In T. Gowers, J. Barrow-Green, & I. Leader (Eds.), *The Princeton companion to mathematics* (pp. 726–728). Princeton University Press.
Barthes, R. (1995 [1967]). The death of the author. In S. Burke (Ed.), *Authorship: From Plato to Post-Modern: A reader*. Edinburgh University Press: Ch. 15.
Bartiomiej, A. L. (2019). Why we shouldn't pity Schrödinger's Kitty: Revisiting David Lewis' worry about quantum immortality in a branching multiverse. *Metaphysics, 20*(1), 117–136.
Basalla, G. (1962). William Harvey and the heart as a pump. *Bulletin of the History of Medicine, 36*, 467–470.
Bates, D. G. (1992). Harvey's account of his 'discovery'. *Journal of Medical History, 36*, 361–378.
Baxter, J., & Newman, S. (2023). Sahotra Sarkar, 'Crisper and Cut and Paste Genetics'. In M. Boylan (Ed.), *International public health policy and ethics* (2nd ed.). Springer: part three.
Bayer, G. (1997). Coming to know principles in 'Posterior Analytics II 19'. *Apeiron, 30*(2), 109–142.
Beck, C. S., Leighninger, D. S., et al. (1958). Some new concepts of coronary heart disease; results after surgical operation. *Journal of the American Medical Association, 168*, 2110–2117.
Bell, J. (1964). On the Einstein Podolsky Rosen Paradox. *Physics, 1*(3), 195–200.
Ben-Menahem, A. (2009). *Historical encyclopedia of natural and mathematical sciences*. Springer.
Bennett, J. (2017). *Life in the universe*. Pearson.
Bennett, J. (1966). *Kant's analytic*. Cambridge University Press.
Bernoulli, R. (1994). Paracelsus--Physician, reformer, philosophy, scientist. *Experientia, 50*(4), 334–338.
Bielik, L. (2018). Explanation, H-D confirmation, and simplicity. *Erkenntnis, 83*(5), 1085–1104.
Bilson-Thompson, S. O., Markopaulou, F., & Smolin, L. (2007). Quantum gravity and the standard model. *Classical and Quantum Gravity, 24*(16), 3975–3984.
Bitbol-Hespériès, A. (2000). Cartesian physiology. In S. Gaukroger, J. Schuster, & J. Sutton (Eds.), *Descartes natural philosophy* (pp. 349–382). Routledge.
Blackburn, S. W. (1969). Goodman's Paradox. In N. Rescher (Ed.), *Studies in the philosophy of science: American Philosophical Quarterly* (Vol 3, pp. 128–142).
Blickman, D. R. (1987). The role of plague in the 'Iliad'. *Classical Antiquity, 6*(1), 1–10.
Bliss, M. (2007). *Harvey Cushing: A life in surgery*. Oxford University Press.
Bloom, A. (1968). *The Republic of Plato*. Basic Books.
Bohr, N. (1913). On the constitution of atoms and molecules. *The London, Edinburgh, and Dublin Philosophical Magazine and Journal of Science, 26*(151), 1–25.
Bohr, N. (1921). Atomic structure. *Nature, 107*(2682), 104–107.
Boolos, G. (1998). *Logic, Logic, Logic*. Harvard University Press.
Boorse, C. (1975). On the distinction between disease and illness. *Philosophy of Public Affairs, 5*, 49–68.
Boorse, C. (1976). Wright on functions. *The Philosophical Review, 85*, 70–86.
Boorse, C. (1977). Health as a theoretical concept. *Philosophy of Science, 44*, 70–86.

Boorse, C. (1997). A rebuttal on health. In J. M. Humber & R. F. Almeder (Eds.), *What is disease?* Humana Press.
Born, M. (1926). Zur Quantenmechanik der Stossvorgänge. *Zietschrift für Physik, 37*, 863–867.
Botting, J. H. (2015). Smallpox and after: An early history of the treatment and prevention of infections. In *Animals and medicine: The contribution of animal experiments to the control of disease* (pp. 3–16). Open Book Publishers.
Boughn, S. (2013). Fritz Hasenöhrl and E = mc². *The European Physical Journal H, 38*, 261–278.
Bouras-Vallianatos, P., & Zipser, B. (Eds.). (2019). *Brill's Companion to the Reception of Galen*. Brill.
Boyd, R. (1993). In R. Boyd, P. Gasper, & J. D. Trout (Eds.), *Confirmation, semantics, and the interpretation of scientific theories* (pp. 3–36). Massachusetts Institute of Technology Press.
Boyd, R., Gasper, P., & Trout, J. D. (Eds.). (1993). *The philosophy of science*. M.I.T University Press: section III.
Boylan, M. (1980). Henry More and the spirit of nature. *Journal of the History of Philosophy, 18*(4), 395–405.
Boylan, M. (1981). Mechanism and teleology in Aristotle's Biology. *Apeiron, 15*(2), 96–102.
Boylan, M. (1982). The digestive and 'circulatory' systems in Aristotle's Biology. *Journal of the History of Biology, 15*(1), 89–118.
Boylan, M. (1983). *Method and practice in Aristotle's Biology*. Rowman and Littlefield/UPA.
Boylan, M. (1984). The galenic and hippocratic challenges to Aristotle's conception theory. *Journal of the History of Biology, 17*(1), 83–112.
Boylan, M. (1985). The place of nature in Aristotle's Biology. *Apeiron, 19*(1), 126–139.
Boylan, M. (1986). Galen's conception theory. *Journal of the History of Biology, 19*(1), 44–77.
Boylan, M. (2004a). *A just society*. Rowman and Littlefield.
Boylan, M. (2004b). The moral imperative to maintain public health. In M. Boylan (Ed.), *Public health ethics*. Springer/Kluwer.
Boylan, M. (Ed.). (2004c). *Public health policy and ethics*. Springer/Kluwer.
Boylan, M. (2005). The ethical limits of science. *Journal of Philosophical Research, 30*, 15–26.
Boylan, M. (2007). Galen on the blood, pulse, and arteries. *Journal of the History of Biology, 40*(2), 207–230.
Boylan, M. (2008). *The good, the true, and the beautiful*. Continuum.
Boylan, M. (2010). Aristotle the outsider. In M. Boylan & C. Johnson (Eds.), *Philosophy: An innovative introduction—Fictive narrative, primary texts, and responsive writing* (pp. 61–72). Westview.
Boylan, M. (2011). *Morality and global justice: Justifications and applications*. Westview.
Boylan, M. (2014). *Natural human rights: A theory*. Cambridge University Press.
Boylan, M. (2015). *The origins of Ancient Greek Science: Blood—a philosophical study*. Routledge.
Boylan, M. (2019). *Fictive narrative philosophy: How fiction can act as philosophy*. Routledge.
Boylan, M. (2020a). *The process of argument: An introduction* (3rd ed.). Routledge.
Boylan, M. (2020b). *Basic ethics* (3rd ed.). Routledge.
Boylan, M. (Ed.). (2022). *Living the good life: 'Virtue' and 'goodness' in the social and political philosophy of Ancient Greece: The philosophy of A.W.H. Adkins*. Cambridge Scholars Press.
Boylan, M. (2023). *International public health policy and ethics* (2nd ed.). Springer.
Boylan, M., & Brown, K. (2002). *Genetic engineering: Science and ethics on the new frontier*. Prentice Hall.
Boylan, M., & Johnson, C. (2020). *Philosophy: An innovative introduction—Fictive narrative, primary texts, and responsive writing*. Westview, rpt. Routledge.
Boylan, M., & Teays, W. (Eds.). (2022). *Ethics in the AI, Technology, and Information Age*. Rowman and Littlefield.
Bremmer, J. N. (2002). How old is the ideal of holiness (of mind) in the Epidaurian temple inscription and the Hippocratic oath? *Zietschrift für Papyrologie und Epigraphik, 141*, 106–108.
Brenner, A. (2017). Simplicity as a criterion of choice in metaphysics. *Philosophical Studies, 174*(11), 2687–2707.

Bridgman, P. (1927). *The logic of modern physics* (pp. 1–32). Macmillan.
Brill, S. (2015). Animality and sexual difference in the Timaeus. In *Plato's animals: Gadflies, horses, swans, and other philosophical beasts* (pp. 161–176). Indiana University Press.
Bronstein, D. (2012). The origin and aim of 'Posterior Analytics' II.19. *Phronesis, 57*(1), 29–62.
Broome, J. (2002). Measuring the burden of disease by aggregating well-being. In C. Murray, J. Salomon, C. Mathers, & A. Lopez (Eds.), *Summary measures of population health* (pp. 91–113). World Health Organization.
Brown, D. (2011). *A new introduction to Islam*. John Wiley.
Brown, H. (1975). Paradigmatic propositions. *American Philosophical Quarterly, 12*, 85–90.
Brownstein, G. (2019). *The Open- Heart Club: A story about birth and death and cardiac surgery*. Public Affairs.
Bud, R. (2009). *Penicillin: Triumph and tragedy*. Oxford University Press.
Burkert, W. (1962). *Weisheit und Wissenschaft*. H. Carl.
Burkert, W. (1987). *Greek religion*. Harvard University Press.
Burnet, J. (1902). *Platonis Opera*. Clarendon Press.
Burnet, J. (1930). *Early Greek Philosophy* (4th ed., p. 170). A&C Black.
Bury, R. G. (1989). In R. G. Bury (Ed.), *Outlines of Pyrrhonism*. Harvard University Press.
Busch, J. (2011). Scientific realism and the indispensability argument for mathematical realism: A marriage made in Hell. *International Studies in the Philosophy of Science, 25*(4), 307–325.
Cain, P. J., & Hopkins, A. G. (2016). *British Imperialism 1688-2015* (3rd ed.). Routledge.
Campbell, D. (2013). *Arabian medicine and its influence in the Middle Ages* (p. 220). Routledge.
Campbell, G. (2020). Empedocles. In *The Internet Encyclopedia of Philosophy* (accessed November 2, 2020).
Canguihem, G. (2008). Health: Crude concept and philosophical question. Translated by Todd Meyers and Stefanos Geroulanos. *Public Culture: Bulletin of the Project for Transnational Cultural Studies, 20*(8), 467–477.
Carder, K. C., & Carter, B. R. (2005). *Childbed fever: A scientific biography of Ignaz Semmelweis*. Routledge.
Carlip, S. (2001). Quantum gravity: A progress report. *Reports on Progress and Physics, 64*(8), 885–942.
Carnap, R. (1938–1955). Logical foundations of the unity of science. In O. Neurath, R. Carnap, & C. Morris (Eds.), *International encyclopedia of unified science: Volume 1* (pp. 42–62). University of Chicago Press.
Carnap, R. (1950). *Logical foundations of probability*. University of Chicago Press.
Carnap, R. (1956). *Meaning and necessity*. University of Chicago Press.
Carnap, R. (1966). In M. Garner (Ed.), *An introduction to the philosophy of science* (pp. 137–139). Basic Books.
Caron, L. (2015). Thomas Wilis, the restoration and the first works of neurology. *Medical History, 59*(4), 525–553.
Carpenter, A. D. (2010). Embodied intelligent (?) Souls: Plants in Plato's *Timaeus. Phronesis, 55*(4), 281–303.
Carrel, A. V. I. I. I. (1910). On experimental surgery of the aorta and heart. *Annals of Surgery, 52*(1), 83–95.
Carroll, S. M. (2019). *Spacetime and geometry: An introduction to general relativity*. Cambridge University Press.
Casullo, A. (2010). *A priori knowledge*. Wiley-Blackwell.
Chaldun, I. (2011). *Die Muqaddima Betrachtungen zur Weltgeschichte* (pp. 391–395). C.H. Beck.
Chalmers, D. (2018). Structuralism as a response to skepticism. *Journal of Philosophy, 115*(12), 625–660.
Chandler, A. D. (2004). *Shaping the industrial century: Pharmaceutical industries (Harvard Studies in Business History)*. Harvard University Press.
Chilcott, C. M. (1923). The platonic theory of evil. *Classical Quarterly, 17*(1), 27–31.
Child, J. M. (2005). In J. M. Child (Ed.), *The early mathematical manuscripts of Leibniz*. Dover.

Chisholm, R. (1977). *Theory of knowledge* (2nd ed.). Prentice Hall.
Chisholm, R. (1982). *The foundations of knowing*. University of Minnesota Press.
Christensen, T., Thompson, D. J., & Vandorpe trs, K. (2017). *Land and taxes in Ptolemaic Egypt: An edition, translation, and commentary for the Edfu Land Survey (P. Haun. IV 70)*. Cambridge University Press.
Christian, J., Vercler, L., & Tricot, N. (2020). Schrödinger's cat and the ethically untenable act of not looking. *American Journal of Bioethics, 20*(6), 40–42.
Church, A. (1936). A Note on the *Entscheidungsproblem*. *Journal of Symbolic Logic*, 40–41.
Church, A. (1986 [1944]). *An introduction to mathematical logic*. Princeton University Press.
Churchland, P. (1981). Eliminative materialism and the propositional attitudes. *Journal of Philosophy, 78*(2), 67–90.
Cimon, Ü. (2019). On saving the astronomical phenomena: Physical realism in struggle with mathematical realism in Francis Bacon, Al-Bitruji and Averroes. *Hopos, 9*(1), 135–151.
Clarke, S. (2010). Transcendental realisms in the philosophy of science: On Bheskar and Cartwright. *Synthese, 173*(3), 299–315.
Cleaery, J. (1998). 'Proclus' philosophy of mathematics. In G. Bechte & D. J. O'Meara (Eds.), *La Philosophie des mathematiques de l'Antiquité tardive* (pp. 85–101). Editions Universitaires Fribourg Suisse.
Cobb, M. (2004). *The Railways of Great Britain: A historical atlas*. Ian Allen.
Cockayne, E. (2002). Theophrastus Phillipus Aureolus Bombastus Von Hohenheim (Paracelsus)--a short biography. *British Journal of General Practice, 52*, 876.
Cohen, I. B. (1974). Newton's Theory v. Kepler's Theory and Galileo's Theory: An example of difference between a philosophical and a historical analysis of science. In Y. Elkana (Ed.), *The interaction between science and philosophy* (pp. 299–388). Humanities Press.
Cohen, I. B. (1978). *Introduction to Newton's 'Principia'*. Harvard University Press.
Cohen, I. B., & Smith, G. E. (2002). *The Cambridge Companion to Newton*. Cambridge University Press.
Cohen, I. B., & Westfall, R. S. (1995). *Newton: Texts, backgrounds, and commentaries, A Norton critical edition*. Norton.
Cohen, I. B., & Whitman, A. (Eds.). (1999 [1687]). University of California Press.
Collins, F. (1999). Medical and societal consequences of the human genome project. *New England Journal of Medicine, 341*, 28–37.
Coogan, T. P. (2013). *The Famine Plot: England's Role in Ireland's Greatest Tragedy*. St. Martins.
Cooper, G. M. (2017). *Galen, De diebus decretoriis, From Greek into Arabic: A Critical Edition, with Translation and Commentary, of Hunayn ibn Ishaq, Kitab ayyam al-buhran*. Routledge.
Cornell, D. M. (2016). Taking monism seriously. *Philosophical Studies, 173*(9), 2397–2415.
Cornford, F. M. (1923). Mysticism and science in the Pythagorean tradition. *Classical Quarterly, 17*(1, 1), –12.
Cornford, F. M. (1935). *Plato's cosmology*. Routledge.
Corning, P. (2012). The re-emergence of emergence, and the causal role of synergy in emergent evolution. *Synthese, 185*(2), 295–317.
Couprie, D. L. (2004). How Thales was able to 'Predict' a Solar Eclipse without the Help of Mesopotamian Wisdom. *Early Science and Medicine, 9*(4), 321–337.
Cunningham, A. (1997). *The anatomical renaissance: The resurrection of the anatomical projects of the ancients*. Scolar Press.
Curd, P. (2004). *The legacy of Parmenides: Eleatic motion and later presocratic thought* (2nd ed.). Parmenides Publishing: Part II.
Curd, P. (2013). Where are love and strife? Incorporeality in Empedocles. In *Early Greek Philosophy* (pp. 113–138). Catholic University Press of America.
Curie, E. (2013). *Madame Curie: A Biography, tr. Vincent Sheean*. Doubleday.
Dadhich, N. (2015). Einstein as Newton with space curved. *Current Science, 109*(2), 260–264.
Dahm, R. (2008). Discovering DNA: Friedrich Miescher and the early years of nucleic acid research. *Human Genetics, 122*(6), 565–581.

Darwin, C. (1958a [1859]). *The origin of species.* New American Library.
Darwin, C. (1958b [1887]). In N. Barlow (Ed.), *The autobiography of Charles Darwin.* Collins.
Darwin, C.. (1962 [1859]). *The Voyage of the Beagle.* The Natural History Library, Anchor Books.
Darwin, C. (1859). *On the origin of the species by means of natural selection, or the preservation of favored races in the struggle for life.* John Murray.
Darwin, C., & Wallace, A. R. (1858). On the tendency of species to form varieties; and on the perpetuation of varieties and species by natural means of selection. *Zoological Journal of the Linnean Society, 3*(9), 45–62.
Das, A. R. (2014). Re-evaluating the authenticity of the fragments of Galen's 'On the Medical Statements in Plato's Timaeus (Scorialensis grac. ϕ-III l. 11ff. 123r-126v). *Zeitschrift für Papyrolgie und Epigrapik, 192*, 93–103.
Davey, K. (2007). Aristotle, Zeno, and The Stadium Paradox. *History of Philosophy Quarterly, 24*(2), 127–146.
Davidoff, F., Hayne, B., Sackett, D., & Smith, R. (1995). Evidence-based medicine. *BMJ, 310*, 1085–1086.
Davidson, D. (2001). Davidson responds to adverbs of action. In *Essays on Davidson: Action and events* (2nd ed., pp. 163–180). Oxford University Press.
Dawkins, R. (1976). *The selfish gene.* Oxford University Press.
De Bianchi, S., & Wells, J. D. (2015). Explanation and the dimensionality of space, Kant's argument revisited. *Synthese, 192*(1), 287–303.
De Cruz, H. (2016). Numerical cognition and mathematical realism. *Philosophers' Imprint, 16*(16), e1–e13.
de Cusance Morant Saunders, J. B., & O'Malley, C. (Eds.). (1973). *The illustrations from the works of Andreas Vesalius of Brussels.* Dover.
de Laplace, P. (1812). *Théorie Analytique des Probabilités.* Courcier.
De Leemans, P. (2010). Aristotle transmitted: reflections on the transmission of Aristotelian scientific thought in the Middle Ages. *International Journal of the Classical Tradition, 17*(3), 325–333.
de Prado Salas, J. G. (Ed.). (2018). Whose purposes? Biological teleology and intentionality. *Synthese, 195*(10), 4507–4524.
De Vito, S. (2000). On the value-neutrality of the concepts of health and disease: Unto the breach again. *Journal of Medicine and Philosophy, 25*(5), 539–567.
Deichgräber, K. (1965). *Die griechische Empirikerschule* (2nd ed.). Weidmann.
Deichgräber, K. (1971). Aretaeus von Kappadokein als medizinischer Schriftsteller. *Abhandlungen der sächsichen Akademie der Wissenschaften, 63*, 3.
Demont, P. (2016). Remarques sur le Fableau de la médecine et d' Hippocrate chez Platon. *Jones and Rosen,* 61–82.
Descartes, R. (1897–1913 [1596–1650]). *Oeuvres,* translated by Charles Adam and Paul Tannery *VII.* Léopold Cerf.
Descartes, R. (2003 [1630]). *Treatise of Man,* Thomas Steele Hall, ed. and tr. Prometheus.
Devlin, K. (2010). *The unfinished game: Pascal, Fermat, and the seventeenth-century letter that made the world modern.* Basic Books.
DeWitt, B. S. (1967). Quantum theory of gravity. *Physical Review, 160*, 1113.
Dicks, D. R. (1959). Thales. *Classical Quarterly, 9*(2), 294–309.
Diels, H., & Kranz, W. (1951). *Die Fragmente der Vorsokratiker* (6th ed.). Weidmann.
Dijksterhuis, E. J. (1986 [1950]). *The mechanization of the world picture: Pythagoras to Newton,* tr. C. Dikshoorn. Princeton University Press.
Dioscorides, P. (2000). *De materia medica: Being an Herbal with many other Medicinal Materials,* tr. Tess Anne Osbaldeston, based upon the 1655 translation of John Goodyer. Ibis Press.
Distelzweig, P. (2014). Meam de motu & usu cordis & circuitu sanguinis sentiam: Teleology in William Harvey's *De Motu Cordis. Gesnerus, 71*(2), 258–270.
Distelzweig, P. (2015). The use of *Usus* and the function of *Functio*: Teleology and its limits in descartes physiology. *Journal of the History of Philosophy, 55*(3), 377–399.

Distelzweig, P. (2016). 'Mechanics' and mechanism in William Harvey's Anatomy: Varieties and limits. In P. Distelzweig, B. Goldbert, & E. Ragland (Eds.), *Early modern medicine and natural philosophy* (pp. 117–140). Springer.
Distelzweig, P., Goldberg, B., & Ragland, E. (2016). *Know Thyself: Early modern medicine and natural philosophy*. Springer.
Dobbs, B. J. T. (2022 [1991]). *The Janus Faces of Genius: The role of Alchemy in Newton's thought*. Cambridge University Press.
Duhem, P. M. (2015). *The electric theory of J. Clerk Maxwell*. Springer.
Dummett, M. (1981). *The interpretation of Frege's philosophy*. Harvard University Press.
Dyer, O. (2013). MERS will spread while natural host is unknown. *British Medical Journal, 346*(7914), 6.
Eba, P. M. (2016). Towards smarter HIV laws, considerations for improving HIV-specific legislation in Sub-Saharan Africa. *Reproductive Health Matters, 24*(47), 178–184.
Echeverria, F., & Chandia, M. (2002). Miguel Servet o Villanueva documentalmente riavarro de Tudela. *Grupos sociales en la historia de Navarra, 1*, 425–438.
Edelson, E. (1999). *Gregor Mendel and the root of genetics*. Oxford University Press.
Edelstein, L. (1967). *Ancient medicine* (pp. 65–85). Johns Hopkins University Press.
Editorial. (1993). Remembering Paracelsus (1493-1541). *Indian Journal of Physiology and Pharmacology, 37*, 169–170.
Editorial. (2021). What do we know about early management of sepsis and septic shock in Polish hospitals? *Healthcare Basel, 9*(2), 140.
Einstein, A. (1905a). *Eine neue Bestimmung der Moleküldimensionen*, Inaugrual Dissertation der University of Zurich, Bern, 1905.
Einstein, A. (1905b). Über Einendie Erzeugung und Verwandlung des Lichtes betreffenden heuristischen Gesichtspunkt. *Annalen der Physik, 322*(6), 132–148.
Einstein, A. (2006a [1905]). Zur Elektrodynamik bewegler Körpers. *Annalen der Physik, 322*(10), 891–921.
Einstein, A. (2006b [1905]). Ist die Trägheit eines Körpers von seinem Energieinhalt abängig. *Annalen der Physik, 323*(13), 639–641.
Einstein, A., & Gültekin, F. (2017 [1905]). Über die von der molekularkinetischen Theorie der Wärme geforderte Bewegung von in ruhenden Flüssigkeiten suspendierten Teilchen. *Annalen der Physik, 322*(8), 549–560.
Einstein, A. (1915). Die Feldgleichungen der Gravitation. *Sitzungsberichte der Preussischen Akademie der Wissenschaften zu Berlin*, 844–847.
Einstein, A. (1950). On the generalized theory of gravitation. *Scientific American, 182*(4), 13–17.
Einstein, A., Rosen, N., & Podolsky, B. (1935). Can quantum-mechanical description of physical reality be considered complete? *Physical Review, 47*(10), 777.
Elkins, C. (2022). *Legacy of violence: A history of the British Empire*. Knopf.
Enoch, J., & Piot, P. (2017). Human rights in the fourth decade of HIV/AIDS response: An inspiring legacy and urgent imperative. *Health and Human Rights, 19*(2), 117–122.
Erbse, H. (1974). *Scholia Graeca in Homeri Iliadem* (Vol. 3, pp. 222–223). De Gruyter.
Euclid. (1956 [4th century B.C.E.]). *The thirteen books of the elements*, 2nd revised edition, ed. and tr. Thomas L. Heath. Dover.
Ewing, A. C. (1938). *A short commentary on Kant's critique of pure reason*. University of Chicago Press.
Falkenstein, L. (1995). *Kant's intuitionism: A commentary on the transcendental aesthetic*. University of Toronto Press.
Faraone, C. A. (2011). Curses, crime detection and conflict resolution at the Festival of Demeter Thesmophoros. *Journal of Hellenic Studies, 131*, 25–44.
Feder, G. (1993). Paradigm lost: A celebration of Paracelsus on his quincentenary. *Lancet, 341*, 1396–1397.
Feinberg, G. (1966). Physics and the Thales problem. *The Journal of Philosophy, 63*(1), 5–17.

Felka, K. (2015). *Talking about numbers: Easy arguments for mathematical realism*. Verlag Vittorio Klostermann.
Fine, G. (2010). Aristotle and the 'Aporema' of the 'Meno'. *Bulletin of the Institute of Classical Studies*, supplement no. 107, *Aristotle and the Stoics 'Reading' Plato*, 45–71
Finger, S. (1994). *The origins of neuroscience* (pp. 67–69). Oxford University Press.
Finkelstein, D. (1958). Past-future asymmetry of the gravitational field of a point particle. *Physical Review, 110*(4), 965.
Firzharris, L. (2017). *The Butchering Art: Joseph Lister's Quest to transform the Grisley World of Victorian Medicine*. Farrar, Straus, and Giroux.
Flemming, R. (2003). Empires of knowledge: Medicine and health in the Hellenistic World. In A. Erskine (Ed.), *A Companion to the Hellenistic World*. Blackwell.
Fletcher, A. H. (1958). The life and medicine of Paracelsus (1493-1541). *Central African Journal of Medicine, 4*, 252–256.
Fodor, J. A. (1981). The mind-body problem. *Scientific American, 244*, 114–123.
Fontenrose, J. (1974). Work, justice, and Hesiod's five ages. *Classical Philology, 69*(1), 1–16.
Forbes, E., & Sylvanus Charles, T. H. (1853). *A history of British Mollusca and their Shells*. John Van Voorst.
Forshaw, R. (2014). Before Hippocrates: Healing practices in Ancient Egypt. In *Medicine, healing, and performance* (pp. 25–41). Oxbow Books.
Fowler, R. L. (2011). Mythos and Logos. *Journal of Hellenic Studies, 131*, 45–66.
Francia, S., & Stobart, A. (Eds.). (2014). *Critical approaches to the history of western herbal medicine*. Bloomsbury.
Francis, D. P. (2012). Commentary: Deadly AIDS policy failure by the highest levels of the U.S. government: A personal look back 30 years later for lessons to respond better to future epidemics. *Journal of Public Health Policy, 33*(3), 290–300.
Franklin, A., & Fishbach, E. (2016). *The rise and fall of the fifth force: Discovery, pursuit and justification in modern physics* (2nd ed.). Springer.
Fraser, P. M. (1972). *Ptolemaic Alexandria*, 3 vol. Clarendon Press, I (pp. 335–356).
Frede, D. (2001). *Not in the Book: How does recollection work? Proceedings of the Second Symposium Platonicum Pragense*. Oikúmené.
Frede, M. (1987). The ancient empiricists. In *Essays in ancient philosophy* (pp. 243–260). Clarendon Press.
Frege, G. (1884). *Die Grundlagen der Arithmetik: eine logisch-mathematische untersuchung über der Begriff der Zahl*. W. Koebner.
Frege, G. (1892). Über Sinn und Bedeutung. *Zietschrift für Philosophie und Philosophische Kritik, 100*, 25–50.
French, R. K. (1994). *William Harvey's natural philosophy*. Cambridge University Press.
Friebe, C. (2012). Twins' paradox and closed time line curves: The role of proper time and the presentist view of spacetime. *Zeitschrift für allgemeine Wissenschaftstheorie, 43*(2), 313–326.
Friedman, M., & Rosenman, R. H. (1960). Overt behavior pattern in coronary disease. Detection of overt behavior pattern A in patients with coronary disease by a new psychophysiological procedure. *JAMA, 173*, 1320–1325.
Friedman, M. (2012). Kantian geometry and spatial intuition. *Synthese, 186*(1), 231–255.
Frölich, H. (1879). *Die Militärmedicin Homers*. F. Enke.
Funkenstein, A. (2018). *Theology and the scientific imagination: From the Middle Ages to the seventeenth century* (2nd ed.). Princeton University Press.
Furth, M. (1970). Elements of Eleatic ontology. *Journal of the History of Philosophy, 6*, 111–132.
Gagnon, P. (2018). *Who cares about particle physics: Making sense of Higgs Boson, the Large Hadron Collider and CERN*. Oxford University Press.
Gahringer, R. E. (1963). Analytic propositions and philosophical truths. *Journal of Philosophy, 60*(17), 481–502.
Galen. (1821-1833). *Opera Omnia*. Karl Gottlob Kühn, ed. Cnoblochii.

Galen. (1985). An outline of empiricism. In *Three treatises on the nature of science*, tr. R. Walzer and M. Frede. Hackett.
Galera, A. (2017). The impact of Lamarck's theory of evolution before Darwin's Theory. *Journal of the History of Biology, 50*(1), 53–70.
Gamble, V. N. (2022a). 'There wasn't a lot of concern in those days' African Americans, Public Health, and the 1918 Influenza Epidemic. *Public Health Reports, 125*(supplement 3), 114–122.
Gamble, V. N. (2022b). 'There wasn't a lot of concern in those days' African Americans, Public Health, and the 1918 Influenza Epidemic. *Public Health Reports. Supplement, 3*, 114–122.
Gammal, E. S. (1995). Rhazes contribution to the development and progress of medical science. *Bulletin of Indian Institute of the History of Medicine, 25*(1/2), 135–149.
Gansom, T. (2013). Are colors representational? *Philosophical Studies, 116*(1), 1–20.
Garber, D. (2001). *Descartes embodied: Reading Cartesian philosophy through Cartesian science*. Cambridge
Garber, D. (1992). *Descartes' metaphysical physics*. University of Chicago Press.
Garrison, F. H. (1929). *An introduction to the history of medicine*. Saunders.
Gaskill, N. (2018). Epilogue after the color sense. In *Chromographia: American literature and the modernization of color* (pp. 239–250). University of Minnesota Press.
Gergely, S. (2010). A geometrical characterization of the twin paradox and its variants. *Studia Logica, 95*(1/2), 161–182.
Gericke, C. A. (2015). Ebola and ethics: Autopsy of a failure. *British Medical Journal, 350*, 1–2.
Gewirth, A. (1941). The Cartesian Circle. *Philosophical Review, 1*(4), 368–395.
Gewirth, A. (1970). The Cartesian Circle Reconsidered. *The Journal of Philosophy, 67*(19), 668–685.
Gibbon, E. (2010 [1776–1789]). *The Decline and Fall of the Roman Empire*. Everyman: volume 6.
Gill, C. (2010). *Naturalistic psychology in Galen and Stoicism*. Oxford University Press.
Gill, M. L. (1987). Matter and flux in Plato's 'Timaeus'. *Phronesis, 32*(1), 34–53.
Glazebrook, T. (2001). Zeno against mathematical physics. *Journal of the History of Ideas, 62*(2), 193–210.
Glenn, S. (2011). Proportion and mathematics in Plato's Timaeus. *Hermathena* 190 *Philosophy and Mathematics*, 11–27.
Golden, C. J. (2010). *Posting it: The Victorian Revolution in letter writing*. U. Press of Florida.
Goodman, N. (1949). The logical simplicity of predicates. *Journal of Symbolic Logic, 14*, 32–41.
Goodman, N. (1952). New notes on simplicity. *Journal of Symbolic Logic, 17*, 188–191.
Goodman, N. (1955). Axiomatic measurement of simplicity. *Journal of Philosophy, 52*, 109–122.
Goodman, N. (1979). *Fact, fiction, and forecast* (4th ed.). Harvard University Press.
Gould, J. (1873). *The Birds of Great Britain*. Taylor and Francis.
Gowers, T., & Nielsen, M. (2009). Massively collaborative mathematics. *Nature, 461*(7266), 879–881.
Graham, D. W. (2013). The geometry of the heavens. In *Science before Socrates: Parmenides, Anaxagoras, and the New Astronomy*. Oxford University Press.
Graham, D. W., & Barney, J. (2016). On the date of Chaerephon's visit to Delphi. *Phoenix, 70*(3/4), 274–289.
Graham, D. (2006). *Explaining the Cosmos: The Ionian tradition of scientific philosophy*. Princeton University Press.
Graham, G. S. (1965). *The politics of Naval Supremacy: Studies in British Maritime Ascendancy*. Cambridge University Press.
Gram, M. A. (1968). *Kant, ontology and the a priori*. Northeastern University Press.
Granger, H. (2008). The Proem of Parmenides Poem. *Ancient Philosophy, 28*(1), 1–20.
Graz, B., et al. (2010). Argemone Mexicana Decoction versus Artesunate—Artesunate-Amodiaquine for the Management of Malaria I Mali: Policy and Public Health Implications. *Transactions of the Royal Society of Tropical Medicine and Hygiene., 104*(1), 33–41.
Greenberg, M. J. (1933). *Euclidean and Non-Euclidean Geometries: Development and histories* (3rd ed.). W.H. Freeman and Co..

Greenstone, G. (2010). The history of bloodletting. *BCMJ, 52*(1), 12–14.
Gregory, A. (2003). Eudoxus, Callippus and the Astronomy of the 'Timaeus'. *Bulletin of the Institute of Classical Studies,* Supplement no. 78: *Ancient Approaches to Plato's "Timaeus",* 5–28.
Gregory, A. (2021). Anaximander and the Ionian Contribution: Brilliant but No Miricle. In H. R. Herda (Ed.), *Ex Ionia Scientia: Knowledge in Archaic Greece* (pp. 315–322). Harvard University Press.
Gregory, A. (2000). *Harvey's Heart: The discovery of blood circulation.* Icon Books.
Gregory, A. (2009). The Ancient Greeks and the Supernatural. In G. Arabatzis (Ed.), *Studies in supernaturalism* (pp. 11–38). Logos Verlag.
Gregory, A. (2014). William Harvey, Aristotle, and Astrology. *British Journal for the History of Science, 47*(2), 199–215.
Grice, P. (1969). Utterer's meaning and intentions. *Philosophical Review, 78,* 147–177.
Griffin, J. (1977). Are there incommensurable values? *Philosophy and Public Affairs., 7*(1).
Griffiths, J. G. (1956). Archaeology and Hesiod's five ages. *Journal of the History of Ideas, 17*(1), 109–119.
Grube, G. M. A., rev. Reeve, C. D. C. (1997). *Plato: Complete Works,* ed. John M. Cooper, associate editor D.S. Hutchinson. Hackett.
Grünbaum, A. (1989). The pseudo-problem of creation in physical cosmology. *Philosophy of Science, 56,* 373–394.
Guirgis, F., Padro, T., et al. (2020). Response to Editor Letter 'Admission Characteristics Predictive of In-Hospital Death from Hospital-Acquired Sepsis'. *Journal of Critical Care, Philadelphia, 56,* 319–320.
Gumpert, M. (1948). Vesalius. *Scientific American, 178*(5), 24–31.
Gutfreund, H., & Renn, J. (2020). *Einstein on Einstein: Autobiographical and scientific reflections.* Princeton University Press.
Gutherie, W. K. C. (1965). *A history of Greek philosophy* (Vol. 1). Cambridge University Press.
Guyer, P., & Wood, A. (2021). *Introducing Kant's critique of pure reason.* Cambridge University Press.
Hacking, I. (2012). The Lure of Pythagoreans. *The Jerusalem Philosophical Quarterly, 61,* 103–128.
Hall, R. A. (1992). *Isaac Newton: Adventurer in thought.* Blackwell.
Halper, E. C. (2005). *One and many in Aristotle's metaphysics.* Parmenides Press.
Halper, E. C. (2009). *One and many in Aristotle's metaphysics: Books Alpha-Delta.* Parmenides.
Halpern, P. (2010). *Collider: The search for the world's smallest particles.* Wiley-Blackwell.
Halpern, P. (2016). *Einstein's Dice and Schrödinger's Cat: How two great minds battled quantum randomness to create a unified theory of physics.* Basic Books.
Hamilton, E. (rpt. 2011 [1964]). *Timeless tales of gods and heroes.* Grand Central Publishing.
Hamilton, R. P. (2010). The concept of health: Beyond normativism and naturalism. *Journal of Evaluation in Clinical Practice, 16,* 323–329.
Hamlyn, D. W. (1956). Analytic truths. *Mind, 65,* 359–367.
Hankinson, R. J. (1987). Causes and empiricism: A problem in the interpretation of later Greek medical method. *Phronesis, 32*(2), 329–348.
Hankinson, R. J. (1990). *Galen on the therapeutic method: Books 1 and 2.* Oxford University Press.
Hankinson, R. J. (1995). The growth of medical empiricism. In D. Bates (Ed.), *Knowledge and the scholarly medical traditions* (pp. 41–59). Cambridge University Press.
Hankinson, R. J. (Ed.). (2008). *The Cambridge companion to Galen.* Cambridge University Press.
Hankinson, R. J. (2016). Galen on Hippocratic physics. In Jones and Rosen (pp. 421–444).
Hanson, A. (2001). Papyrology: Minding other people's business. *Transactions of the American Philological Association, 131,* 297–313.
Harding, R. (2010). *The emergence of Britain's Naval Supremacy: The war of 1739-1748.* Boydell Press.
Harman, P. M. (2001). *The natural philosophy of James Clerk Maxwell.* Cambridge University Press.

Harris, C. R. S. (1973). *The heart and vascular system in Ancient Greek medicine: From Alcmaeon to Galen*. Oxford University Press.
Harvey, E. D. (2002). Anatomies of rapture: Clitoral politics: Medical Blazons. *Signs, 27*(2), 315–346.
Harvey, W. (1639). *De motu cordis & sanguinis in animalibus*. Lugduni Batavorum.
Hasegawa, H., Crocker, M., & Minhas, P. S. (2013). *Oxford case histories in neurosurgery*. Oxford University Press.
Hatfield, G. (2006). The Cartesian Circle. In S. Gaukroger (Ed.), *The Blackwell Guide to Descartes' Meditations* (pp. 122–141). Blackwell.
Haugen, M. P., & Lämmerzahl, C. (2001). Principles of equivalence: Their role in gravitation physics and experiments that test them. *Gyros, 562*, 195–212.
Haukkala, A., Konttinen, H., Laatikainen, T., et al. (2010). Hostility, anger control, and anger expression as predictors of cardiovascular disease. *Psychosomatic Medicine, 72*, 556–562.
Hausman, D. M. (2006). Valuing health. *Philosophy and Public Affairs, 34*(3), 246–274.
Hawking, S. (1971). Gravitational radiation from colliding black holes. *Physical Review Letters, 26*(21), 1344–1346.
Hawking, S. (1976). Breakdown of predictability in gravitational collapse. *Physical Review, 14*(10), 2460–2473.
Hawking, S. (2006). *The theory of everything: The origin and fate of the universe*. Jaico Publishers.
Heath, T. L. (1922–1923). Greek geometry with special reference to infinitesimals. *The Mathematical Gazette, 11*, 252–253.
Heck, R. (2017). Self-reference and the language of arithmetic. *Philosophia Mathematica, 15*(1), 1–29.
Hegel, G. W. F. tr. J. Sibree (2004 [1837]). *The philosophy of history*. Dover.
Heisenberg, W. (1925). Über quantentheoretische Undeutung kinematischer und mechanischer Beziehungen. *Zeitschrift für Physik, 33*(1), 879–893.
Helmreich, G. (Ed.). (1893). *Galen, On the Natural Faculties, Kühn* (Vol. II). Teubner.
Hemal, A. K., & Menon, M. (Eds.). (2018). *Robotics in genitourinary surgery*. Springer.
Hempel, C. (1950). Problems and changes in empiricist criterion of meaning. *Revue Internationale de Philosophie., 11*, 41–63.
Hempel, C. (1951). The concept of cognitive significance: A reconsideration. *Proceedings of the American Academy of Arts and Sciences, 80*(1), 61–77.
Hershey, A., & Chase, M. (1952). Independent functions of viral protein and nucleic acid in growth of bacteriophage. *Journal of Genetic Physiology, 36*(1), 39–56.
Hesiod, *Works and* Day, Glenn W. Most, ed. and tr. (2006). Harvard University Press.
Heyck, T. W. (1980). From men of letters to intellectuals: The transformation of intellectual life in nineteenth-century England. *Journal of British Studies, 20*(1), 158–183.
Hieronymus, F. (1933 [1603]). *De venarum ostiolis*, facsimile edition, ed. K.J. Franklin. Charles C. Thomas.
Higgins, J. (2010). 'With Every Accompaniment of Ravage and Agony:' Pittsburgh and the Influenza Epidemic of 1918-1919. *The Pennsylvania Magazine of History and Biography, 134*(3), 263–286.
Hilbert, D. (1903). *Grundlagen der Geometrie*. Teubner.
Hippocrates. *Oeuvres Complètes* (1839). Émile Littré, ed. J.B. Baillière.
Hobbes, T. (1997/1651). In R. E. Flathman & D. Johnson (Eds.), *Leviathan*. W. W. Norton.
Holmes, B. (2007). The *Iliad's* economy of pain. *Transactions of the American Philological Association, 137*(1), 45–84.
Holmes, B. (2010). *The symptom and the subject: The emergence of the physical body in Ancient Greece*. Princeton University Press.
Holmes, B. (2014). Proto-sympathy in the *Hippocratic Corpus*. In J. Jouanna & M. Zink (Eds.), *Hippocratismes: Médecine, Religion, Société* (pp. 123–138). Belles-Lettres.
Hooper, D. (2019). Radically rethinking dark matter. In *At the edge of time: Exploring the mysteries of our universe's first-seconds* (pp. 143–157). Princeton University Press.

Howard, E. R. (2013). Joseph Lister: His Contribution to early experimental physiology. *Notes and Records of the Royal Society, 67*, 191–198.

Howell, R. (2013). Kant and Kantian themes in recent analytic philosophy. *Metaphilosophy, 44*(1/2), 42–47.

Hu, X.-S., et al. (2021). Multi-generational effects of different resistant wheat varieties on fitness of *Sitobion avenae* Hemiptera: Aphididae. *Journal of Insect Science, 21*(5). ch. 9.

Hughes, P. (2016/2017). The face of god: The Christology of Michael Servetus. *Journal of Unitarian Universalist History, 40*, 16–53.

Hull, D. (1974). *Philosophy of biological science*. Prentice Hall.

Hulskamp, M. (2016). On regimen and the question of medical dreams in the Hippocratic Corpus. In Jones and Rosen, (pp. 258–270).

Hume, D. (1977a [1748]). In E. Steinberg (Ed.), *An enquiry concerning human understanding*. Hackett.

Hume, D. (1977b [1752]). On the populousness of ancient nations rpt. *Population and Development Review* 3(3).

Hume, D. (2011 [1739]). In D. F. Norton (Ed.), *A treatise of human nature*. Oxford University Press.

Humphries, B., Radke, M., & Lavzier, S. (2017). Comparing 'Insider' and 'Outsider' news coverage of the 2014 Ebola Outbreak. *Canadian Journal of Public Health, 108*(4), 1–15.

Hutton, J. (1788). *Theory of the Earth*. Royal Society of Edinburgh.

Iglesias, U., Burgos, F., et al. (2020). Letter to the Editor: Admission characteristics predictive of in-hospital death from hospital-acquired sepsis from community-acquired sepsis: A comparison of community-acquired sepsis. *Journal of Critical Care, Philadelphia, 56*, 318.

Illiffe, R. (2007). *Newton: A very short introduction*. Oxford University Press.

Iqbal, Z., & Zorn, C. (2010). Violent conflict and the spread of HIV/AIDS in Africa. *The Journal of Politics, 72*(1), 149–162.

Isaacson, W. (2007). *Einstein: His Life and Universe*. Simon and Schuster.

Israelowich, I. (2016). The use and abuse of hippocratic medicine in the *Apology of Lucius Apuleius*. *Classical Quarterly, 67*, 635–644.

Jardine, L. (2002). *The curious life of Robert Hooke*. Harper.

Jefferson, T., & Ferroni, E. (2009). The Spanish Flu through BMJ's eyes. *British Medical Journal, 399*(735), 1397–1399.

Jenner, E. (1801). *An inquiry into the causes and effects of the Variolae Vaccinae*. D.N. Shury.

Jessrey, B. (2012). Joseph Lister (1827-1912): A pioneer of antiseptic surgery remembered, a century after his death. *Journal of Medical Biography, 20*(3), 107–110.

Jevons, F. R. (1962). Harvey's quantitative method. *Bulletin of the History of Medicine, 36*, 462–467.

Johnson, M. R. (2008). *Aristotle on teleology*. Oxford University Press.

Johnson, N. P. A. S., & Mueller, J. (2002). Updating the accounts: Global mortality of the 1918-1919 'Spanish' influenza pandemic. *Bulletin of the History of Medicine, 76*(1), 105–115.

Joly, R. (1961). Le question Hippocratic et le témoinage de Phèdre. *Revue des Études grecques, 74*, 194–223.

Jones, L. D., & Rosen, R. M. (Eds.). (2016). *Ancient conceptions of the Hippocratic*. Brill.

Jonsen, A. R., & Stryker, J. (Eds.). (1993). *National Research Council on Monitoring the Social Impact on the AIDS epidemic*. National Academic Press.

Joseph, H. W. P. (1906). *An introduction to logic*. Oxford University Press.

Jost, E., & Maor, E. (2014). *Beautiful geometry* (pp. 1–3). Princeton University Press.

Jouanna, J. (2012a). *Greek medicine from Hippocrates to Galen: Selected Papers*. Brill.

Jouanna, J. (2012b). The legacy of the Hippocratic treatise '*The Nature of Man:* The theory of the four humours'. In *Greek medicine from Hippocrates to Galen* (pp. 355–360). Brill.

Jouanna, J. (2012c). Egyptian medicine and Greek medicine. In J. Jouanna (Ed.), *Greek medicine from Hippocrates to Galen: Selected papers* (pp. 3–20). Brill.

Jouanna, J., & Zink, M. (Eds.). (2014). *Hippocratismes: Médecine, Religion, Société*. Belles-Lettres.

Jow, D. L., & Scott, D. (2020). Re-evaluating evidence for Hawkings' points in the CMB. *Journal of Cosmology and Astrophysics, 3*, 1–10.
Kaciiuba, H., & Grucza, R. (2011). Gender difference in thermoregulation. *Current Opinion in Clinical Nutrition and Metabolic Care, 4*(6), 533–536.
Kahn, C. H. (1960). *Anaximander and the origins of Greek cosmology* (pp. 29–32). Columbia University Press.
Kahor, J., Davis, B., et al. (2019). Association between state-mandated probilized sepsis and in-hospital mortality among adults with sepsis. *Journal of the American Medical Association, 322*(3), 240.
Kamtekar, R. (2009). Knowing by likeness in Empedocles. *Phronesis, 54*(3), 215–238.
Kana, M. A., Elegba, O. Y., et al. (2016). Ebola viewed through a lens of African epidemiology. *Journal of Epidemiology and Community Health, 70*(1), 6–8.
Kant, I. (1968 [1781, 1787]). *Kritik der reinen Vernunft*, 1. Auflage 1781; und 2. Auflage 1787. Akademie Textausgabe, rpt. Walter de Gruyter & Co.
Kasting, J. F., Kopparapu, R., Ramirez, R. M., & Harman, C. E. (2014). Remote life-detection criteria, habitable zone boundaries, and the frequency of earth-like planets around M and Late K Stars. *Proceedings of the National Academy of Sciences of the United States of America, 111*(35), 12641–12646.
Katz, V. J., & Parshall, K. H. (2014). From analytic geometry to the fundamental theorem of algebra. In *Taming the unknown: A history of algebra from antiquity to the early twentieth century* (pp. 247–288). Princeton University Press.
Keele, K. D. (1962). *William Harvey: The man, the physician, and the scientist*. Nelson.
Kelly, K. T., Genin, K., & Lin, H. (2016). Realism, rhetoric, and reliability. *Synthese, 193*(4)., Special Issue: Causation, Probability, and Truth—The Philosophy of Clark Glymour: 1191–1223.
Kettlewell, H. B. D. (1955). Recognition of appropriate backgrounds by the pale and black phases of Lepidoptera. *Nature, 175*(4465), 943–944.
Killingray, D. (2003). A new 'imperial disease': The influenza pandemic of 1918-1919 and its impact on the British Empire. *Caribbean Quarterly, 49*(4), 30–49.
Kind, P. (1998). The EuroQuoL Instrument: An index of health-related quality of life. In B. Spiker (Ed.), *Quality of life and pharmacoeconomics in clinical trials* (2nd ed., pp. 191–201). Lippincott-Raven.
Kingma, E. (2007). What is it to be healthy? *Analysis, 67*(2), 128–133.
Kiritani, O. (2013). Naming and necessity from a functional point of view. *Croatian Journal of Philosophy, 13*(37), 93–98.
Kirk, G. S., Raven, J. E., & Schofield, M. (1995). *The Presocratic Philosophers* (2nd ed.). Cambridge University Press.
Klein, A. E. (1972). *Trial by Fury: The Pollio vaccine controversy*. Scribner's.
Klein, J. (1992). *Greek mathematical thought and the origin of algebra*. Dover.
Kline, S. W. (1939). The first philosopher of the western world. *The Classical Journal, 35*(2), 81–85.
Koh, D., & Sing, J. (2010). Lessons from the past: Perspectives on severe acute respiratory syndrome. *Asia Pacific Journal of Public Health, 23*(3), 132–136.
Kotrc, R. F. (1981). The Dodecahedron in Plato's 'Timaeus'. *Rheinisches Museum für Philogie, neue folge, 124*(3/4), 212–222.
Kpanake, L., Gossou, K., et al. (2016). Misconceptions about Ebola Virus Disease among Lay People in Guinea: Lessons for community education. *Journal of Public Health Policy, 37*(2), 100–172.
Kragh, H. (2012). *Niels Bohr and the Quantum Atom: The Bohr Model of Atomic Structure, 1913-1925*. Oxford University Press.
Kripke, S. (1980). *Naming and necessity* (4th ed.). Harvard University Press.
Kudlien, F. (1969). Antike Anatomie und menschlicher Leichnam. *Hermes, 96*, 78–94.
Kuhn, T. S. (1957). *The Copernican Revolution*. Harvard University Press.
Kuhn, T. S. (1996/1962). *The structure of scientific revolution* (3rd ed.). University of Chicago Press.

Kullmann, W. (1974). *Wissenschaft und Methode: Interpretationem zur aristotelischen Theorie der Naturwissenschaft*. De Gruyter.

Kullmann, W. (1985). Gods and men in the *Iliad* and *Odyssey*. *Harvard Studies in Classical Philology, 89*(1985), 1–23.

Kupreeva, I. (2014). Galen's theory of element. *Bulletin of the Institute of Classical Studies*, No. 116 *Philosophical Theories in Galen*, 153–196.

Laks, A., & Most, G. W. (2016). *Early Greek Philosophy*. Loeb.

Laks, A., & Most, G. W. (2018). *The concept of presocratic philosophy: Its origins, development, and significance*. Princeton University Press.

Laquer, T. W. (2003). Sex in the Flesh. *Isis, 94*(2), 300–306.

Latfi, R., Rhee, P., & Gruessner, R. W. G. (Eds.). (2016). *Technological advances in surgery, trauma, and critical care*. Springer.

Lefkowitz, M. (1981). *The Lives of the Greek Poets*. Duckworth.

Lehoux, D. (2017). Observation claims and epistemic confidence in Aristotle's Biology. *Isis, 108*(2), 241–258.

Leith, D. (2008). The *diatritus* and therapy in Graeco-Roman Medicine. *Classical Quarterly, 58*, 581–600.

Lennox, J. (1995). Health as an objective value. *The Journal of Medicine and Philosophy, 20*, 499–511.

Lennox, J. (2014). Aristotle on the emergence of material complexity: *Meteorology* IV and Aristotle's Biology. *HOPOS, 4*(2), 272–305.

Lesher, J. H. (2010). 'Just as in Battle?' The Simile of the Rout in Aristotle's 'Posterior Analytics' II 19. *Ancient Philosophy, 30*(1), 95–105.

Leunissen, M. (2010). Nature as a good housekeeper. Secondary teleology and material necessity in Aristotle's Biology. *Apeiron, 43*(1), 117–142.

Leunissen, M., & Gotthelf, A. (2010). 'What's teleology got to do with it?' A reinterpretation of Aristotle's 'Generation of Animals V'. *Phronesis, 55*(4), 325–356.

Levene, T. (1919). The structure of yeast nucleic acid IV. Ammonia Hydrolysic. *Journal of Biological Chemistry, 40*, 415–434.

Levine, J. (2011). Russell and the transfinite. *Hermathena: Philosophy and Mathematics, 190*, 53–112.

Lewis, D. (2004). How many lives has Schrödinger's Cat: The Jack Smart Lecture Canberra, 27 June 2001. *Australasian Journal of Philosophy, 82*(1), 3–22.

Lewis, F. A. (2009). Parmenides' Modal Fallacy. *Phronesis, 54*, 1–8.

Lindberg, D. C. (1978). *Science in the Middle Ages*. University of Chicago Press.

Lindbert, D. C. (2010). *The beginnings of western science: The European scientific tradition in philosophical, religious and institutional context: Prehistory to AD 1450*. University of Chicago Press.

Linsky, B. (2002). The resolution of Russell's paradox in 'Principia Mathematica'. *Language and Mind, 16*, 395–417.

Lloyd, D. R. (2009). Triangular relationships and most beautiful bodies: 'On the Significance of (sic) at Timaeus 57d5 and the Number of Plato's Elementary Triangles'. *Mnemosyne, 62*, 11–29.

Lloyd, G. E. R. (1991). *Methods and problems in Greek Science* (pp. 194–223). Cambridge University Press.

Lloyd, G. E. R. (2004). *Ancient worlds: Modern reflections—Philosophical perspectives on Greek and Chinese Science and Culture*. Oxford University Press.

Lloyd, G. E. R. (2020). *Intelligence and intelligibility: Cross-cultural studies of human cognitive experience*. Oxford University Press.

Lloyd, G. E. R., Zhao, J. J., & Dong, Q. (Eds.). (2018). *Ancient Greece and China Compared*. Cambridge University Press.

Lobachevsky, N., & Papadopoulos, A, ed. and tr. (2010 [1855]). *Pangeometry*. European Mathematical Society.

Longrigg, J. (1981). Superlative achievement and comparative neglect: Alexandrian Medical Medical Science and Modern Historical Research. *History of Science, 19*, 155–200.
Lovci, R. (2008). *Michael Servetus, Heretic or Saint*. Prague House.
Lovejoy, A. O. (rpt. 1971 [1936]). *The great chain of being: A study of the history of an idea*. Harvard University Press.
Lyell, C. (1830). *Principles of geology*. John Murray.
Maalouf, A. (1989). *The Crusades through Arab Eyes*. Schocken.
Mackenzie, T. (2016). The contents of Empedocles' Poem: A new argument for the single poem hypothesis. *Zietschrift für Papyrobgie und Epigraphik, 200*, 25–32.
MacLachlan, B. (2012). *Women in Ancient Greece: A sourcebook*. Continuum.
Mahon, B. (2004). *The man who changed everything: The Life of James Clerk Maxwell*. Wiley.
Makin, S. (2014). Parmenides, Zeno, and Melissus. In J. Warren & F. Sheffield (Eds.), *The Routledge Companion to Ancient Philosophy* (pp. 34–48). Routledge.
Malthus, T. R. (1798). *An essay on the principle of population*. J. Johnson.
Manafu, A. (2014). How much philosophy is in chemistry? *Journal for General Philosophy of Science, 45*(1), 33–44.
Manning, R. (2022). AI in healthcare: Ethical issues. In M. Boylan & W. Teays (Eds.), *Ethics in the AI, technology, and information age* (pp. 169–180). Rowman and Littlefield.
Marchistto, E. A. (2007). The Theorem of Pappas: A bridge between algebra and geometry. *The American Mathematical Monthly, 109*(6), 497–516.
Markatos, K., Mavrogenis, A., Brilakis, E., et al. (2019). Abulcasis (936-1013): his work and contribution to Orthopedics. *International Orthopedics, 43*, 2199–2203.
Markel, H. (2021). *The secret of life: Rosalind Franklin, James Watson, Francis Crick and the discovery of DNA's Double Helix*. W. W. Norton.
Marmaduke, M. E. (2003). *Walter Dandy: The personal side of a premier neurosurgeon*. Lippincott.
Martin, S. B. (2016). *Parmenides' Vision: A study of Parmenides Poem*. University Press of America.
Martinez, A. A. (2011a). Thomson, Plum-Pudding and Electrons. In *Science secrets: The truth about Darwin's Finches, Einstein's Wife and other Myths* (pp. 147–163). University of Pittsburgh Press.
Martinez, A. A. (2011b). Galileo and the Leaning Tower of Pisa. In *Science Secrets* (pp. 1–12). University of Pittsburgh Press.
Martinez-Catsam, A. L. (2013). Desolate Streets: The Spanish Influenza in San Antonio. *The Southwestern Historical Quarterly, 116*(3), 286–303.
Matthaei, C. F. (1808). *Medicorum veterum et clarorum graecorum varia Opusucla*. Royal University.
Matthen, M. (1988). Empiricism and ontology in ancient medicine. *Apeiron, 21*, 99–121.
Maurizio, F. (2015). Transcendental realism. *The Monist, 98*(2), 215–232. The New Realism.
Maxwell, J. C. (1855). On Faraday's lines of force. *Transactions of the Cambridge Philosophical Society, 10*(1), 3.
Maxwell, J. C. (1865). A dynamical theory of electromagnetic field. *Philosophical Transactions of the Royal Society of London, 155*, 459–512.
Maxwell, N. (2017). Non-empirical requirements scientific theories must satisfy: Simplicity, unity, explanation, beauty. In *Karl Popper, Science, and Enlightenment* (pp. 125–142). UCL Press.
Maziarz, E. A., & Greenwood, T. (1968). *Greek mathematical philosophy*. Frederick Ungar.
McCarty, C. (2013). Paradox and potential infinity. *Journal of Philosophical Logic, 42*, 195–219.
McClive, C. (2009). Masculinity on trial: Penises, Hermaphrodites and uncertain bale body in early modern France. *History Workshops Journal, 68*, 45–68.
McCormmach, R. (1968). John Mitchell and Henry Cavendish: Weighing the stars. *British Journal for the Philosophy of Science, 4*(2), 126–155.
McCumber, J. (2001). *Time in the Ditch: American Philosophy and the McCarthy Era*. Northwestern University Press.

Meli, D. B. (2013). Early modern experimentation on live animals. *Journal of the History of Biology, 46*(2), 199–226.
Mendel, J. G. (1866 and 1870). *Versuche Über Pflanzenhybriden*. Wilhelm Engelmann.
Menniger, K. (1968). *The crime of punishment*. Viking.
Meyerhof, M. (1934). La découverte de la circulation pulmonaire par Ibn al-Nafis, Médecin arabe du Caire (XIIIe) siècle. *Bulletin l'Institut d' Egypte (Cairo), 16*, 33–46.
Micheli, G. (2014). Kant and Zeno of Elea: Historical precedents of the 'skeptical method'. *Revista de Filosofia, 37*(3), 57–64.
Mill, J. S. (1843). *A system of logic*. John W. Parker.
Miller, B. (2013). When is consensus knowledge based? Distinguishing shared knowledge from mere agreement. *Synthese, 190*(7), 1293–1316.
Miller, G. W. (2002). *King of hearts: The true story of the Maverick who pioneered open heart surgery*. Crown.
Miller, P. H. (2016). *Theories of developmental psychology* (6th ed.). Worth Publishers.
Misner, C. W., Thorne, K. S., & Wheeler, J. A. (1973). *Gravitation*. W.H. Freeman.
Mix, L. (2016). Nested explanation in Aristotle and Mayr. *Synthese, 193*(6), 1817–1832.
Moody, E. A. (1951). Galileo and Avempace: The dynamics of the leaning tower experiment. *Journal of the History of Ideas, 12*(2), 163–193.
Moquin-Tandon, A. (1866). *Éléments de botanique médicale*. J. B. Baillière et fils.
Morioka, M. (2001). Reconsidering brain death. *Hastings Center Report, 31*(4), 41–46.
Morris, T. (2018). *The matter of the heart: A history of the heart in eleven operations*. Thomas Dunne.
Mucz, M. (2011). *Baba's Kitchen Medicines*. University of Alberta Press.
Mueller, I. (1981). *Philosophy of mathematics and deductive structure in Euclid's elements*. M.I.T. Press.
Muggler, C. (1960). *La physique de Platon*. C. Klincksie.
Munoz, M., & Moritz, C. (2016). Adaptation to a changing world: Evolutionary resilience to climate change. In *How evolution shapes our lives: Essays on biology and society* (pp. 238–252). Princeton University Press.
Myerhof, M. (1935). Ibn an-Nafis and his theory of lesser circulation. *Isis, 23*(1), 100–120.
Nabri, I. A. (1983). El Zahrawi (936-1013 AD), The Father of Operative Surgery. *Annals of the Royal College of Surgeons of England, 65*, 132–134.
Nadig, S. N., & Wertheim, J. A. (Eds.). (2018). *Technological advances in organ transplantation*. Springer.
Nagel, E., & Newman, J. R. (1958). *Gödel's Proof*. New York University Press.
Naqvi, N. H. (2009). A medical classic: Al-Razi's treatise on smallpox and measles. *The Historical Medicine Equipment Society, 22*, 1–20.
Nehamas, A. (1990). Eristic, antilogic, dialectic: Plato's demarcation of philosophy from sophistry. *History of Philosophy Quarterly, 7*(1), 3–16.
Nehamas, A. (2002). Parmenidean Being/ Heraclitan Fire. In V. Caston & D. Graham (Eds.), *Presocratic philosophy: Essays in Honor of Alexander Mourelatos* (pp. 45–64). Aldershot.
Nelson, E. D. (2005). Coan promotions and the authorship of the Presbeutikos. In P. J. van der Eijk (Ed.), *Hippocrates in Context: Papers Read at the 11th International Hippocrates Colloquium, University of Newcastle upon Tyne 27-31 August 2002* (pp. 209–236). Brill.
Newman, W. R. (2019). *Newton the Alchemist: Science, Enigma, and the Quest for Nature's "Secret Fire"*. Princeton University Press.
Newton, I. (1999 [1687]). *The Principia*, tr. I. Bernard Cohen and Anne Whitman. University of California Press.
Nietzsche, F. (1968/1886). Beyond good and evil. In: *Basic writings of Nietzsche*, tr. Walter Kaufmann. Modern Library: II. 36, p. 238.
Noble, E., & de Castro, M. F. (2017). The ancient versus the modern continuum. *Formal Sciences and Philosophy: Logic and Mathematics*, 1343–1380.
Novaes, C. D. (2016). Reductio ad absurdum from a dialogical perspective. *Philosophical Studies, 173*(10), 2605–2628.

Nuttall, I., & Dye, C. (2013). The SARS Wake-UP Call. *Science, 339*(6125), 1287–1288.
Nutton, V. (2024). *Ancient medicine* (3rd ed.). Routledge.
Nutton, V., & Scarborough, J. tr. and eds. (1982). The preface of Dioscorides' *Materia Medica: Introduction, Translation and Commentary. Transactions and Studies of the College of Physicians of Philadelphia, 4*(3), 187–227.
Nutton, V., & Scarborough, J. tr. and eds. (1992). In A. Wear (Ed.), *Healers in the market place: Towards a social history of Graeco-Roman medicine* (pp. 15–58).
Nutton, V., & Scarborough, J., tr. and eds. (1993). Greek science in the sixteenth-century Renaissance. In: J. V. Field & F. James (Eds.), *Renaissance and revolution: Humanists, scholars, craftsmen, and natural philosophers in early modern Europe* (pp. 7–16). Cambridge University Press.
Nutton, V., & Scarborough, J., tr. and eds. (2008). Greek medical astrology and the boundaries of medicine. In: A. Akasoy, C. Burnett, & R. Yoeli-Tlalim (Eds.), *Astrology and medicine, east and west* (pp. 17–31). SISMEL-Edizioni del Galluzzo.
Nutton, V., & Scarborough, J., tr. and eds. (2022). *Renaissance medicine: A short history of medicine in Europe in the sixteenth century*. Routledge.
Nutton, V., & Scarborough, J. tr. and eds. (2024). *Ancient medicine* (3rd ed.). Routledge.
Nye, M. J. (1980). N-Rays: An episode in the history and psychology of science. *Historical Studies in the Physical Sciences, 11*(1), 125–156.
O'Brien, D. (2016). Empedocles on the identity of the elements. *Elenchos, 37*(1/2), 5–28.
O'Sullivan, B. (2006). The 'Euthyphro' Argument 9d-11 b. *Southern Journal of Philosophy, 44*(4), 657–675.
Obenchain, T. G. (2016). *Genius Belabored: Childbed fever and the tragic life of Ignaz Semmelweis*. University of Alabama Press.
Opsomer, J. (2015). Plutarch on the geometry of the elements. In M. Meeusen & L. van der Stock (Eds.), *Aspects of Plutarch's philosophy of nature* (pp. 29–56). Leuven University Press.
Osborne, C. (1987). Empedocles recycled. *Classical Quarterly, 37*(ns), 24–50.
Ostriker, A. S. (2017). Dark matter and dark energy. In *Waiting for the light* (pp. 17–27). University of Pittsburgh Press.
Otte, M. (2014). Mathematics, logics, and philosophy: The analytic/synthetic distinction in Kant: Bolzano and Peirce. *Logique et Analyse, 225*(57), 83–112.
Owen, G. E. L. (1957-58). Zeno and the mathematicians. *Proceedings of the Aristotelian Society, 58*, 199–222.
Pai-Dhungat, J. V., & Parikh, F. (2015). Paracelsus (1493-1541). *Journal of Association of Physicians of India, 63*, 28.
Pais, A. (1982). *Subtle is the Lord: The science and life of Albert Einstein*. Oxford University Press.
Paparazzo, E. (2015a). It's a world made of triangles: Plato's 'Timaeus' 53b-55c. *Archiv für Geschichte der Philosophie, 97*(2), 135–159.
Paparazzo, E. (2015b). A note on the construction of the equilateral triangle with scalene elementary triangles in Plato's 'Timaeus 54 A&B'. *The Classical Quarterly, 65*(2), 552–558. 54-n.s.
Papineau, D. (2004). David Lewis and Schrödinger's Cat. *Australasian Journal of Philosophy, 82*(1), 153–169.
Papineau, D. (2013). *Philosophical devices: Proofs, probabilities, possibilities and sets*. Oxford University Press.
Parfit, D. (1984). *Reasons and persons* (pp. 493–503). Oxford University Press.
Park, K. (1997). The rediscovery of the Clitoris: French Medicine and the Tribate 1570-1620. In D. Hillman & C. Mazzio (Eds.), *The body in parts: Fantasies of corporeality in Modern Europe* (pp. 171–193). Routledge.
Pascal, B. (1991 [1670]). *Pensées*, ed. P. Sellier. Bords.
Pasteur, L. (1939 [1886]). "Discours prononcé à Douai, le 7 décembre 1854 à l'occasion de l'installation solemnelle de la Faculté des Lettres de Douai et la Faculté des Sciences de Lille" rpt. in Vallery-Radot, ed., *Oeuvres de Pasteur*, Vol. 7. Masson & Co.

Pears, C. D. (2015). Congruency and Evil in Plato's 'Timaeus'. *Review of Metaphysics, 69*(1), 93–113.
Pearson, G. (Ed.). (1798). *An inquiry concerning the history of Cowpox, principally with a view to supersede and extinguish the Smallpox* (pp. 102–104). J. Johnson.
Pearson, J., & Sarangi, R. F. (1996). Evidence-based medicine. *BMJ, 312*, 380.
Peebles, P. J. E. (2015). Dark matter. *Proceedings of the National Academy of Sciences of the United States of America, 112*(40), 12246–12248.
Pelavski, A. (2014). Physiology in Plato's *Timaeus*: Irrigation, digestion, and respiration. *The Cambridge Classical Journal, 60*, 61–74.
Peres, A., & Terno, D. R. (2004). Quantum information and relativity theory. *Review of Modern Physics, 76*(1), 93–123.
Perkins, C. A. (1900). Experiments of J.J. Thomson on the structure of the atom. *Science, 12*(297), 368–370.
Pesta, D. (2014). Resurrection Vivesection: Michelangelo among the anatomists. *The Sixteenth Century Journal, 45*(4), 921–950.
Petersen, W. (1939). Divinities and divine intervention in the *Iliad*. *The Classical Journal, 35*(1), 2–16.
Peterson, M. A. (2011). Mathematics old and new. In *Galileo's Muse: Renaissance Mathematics and the Arts* (pp. 237–254). Harvard University Press.
Petit, C. (2014). What does Pseudo-Galen tell us that Galen does not? *Bulletin of the Institute for Classical Studies*, Supplement 114: *Philosophical Themes in Galen*, 269–290.
Phillips, A. C. (2003). *Introduction to quantum mechanics*. Wiley.
Piroddi, C., & Baruzzi, A. (2018). *The Montessori Method: Numbers*. N.Y. Union Square Kinds.
Pizzorno, J. E., & Murray, M. T. (2020). *Textbook on natural medicine*. Churchill Livingston.
Plato. (1900). In J. Burnet (Ed.), *Platonis Opera*. Oxford University Press.
Plett, P. C. (2006). Other Discoverers of Cowpox Vaccination before Edward Jenner. *Sudhoffs Archim, 90*(2), 219–232.
Popper, K. (1985 [1934]). *The logic of scientific discovery*. Urwin Hyman.
Pormann, P. E., & Savage-Smith, E. (2007). *Medieval Islamic Medicine*. Georgetown University Press.
Preus, A. (1975). *Science and philosophy in Aristotle's biological works*. G. Olms.
Preus, A. (Ed.). (2001). *Essays in Ancient Greek Philosophy IV: Before Plato*. SUNY Press.
Price, S. (2016). The Peripatetic Hippocrates and other Monists in Anonymus Londinesis. In Jones and Rosen (pp. 99–116).
Priou, A. (2018). Parmenides on reason and revelation. *Epoche: A Journal for the History of Philosophy, 22*(2), 177–202.
Puett, S. B., & David Puett, J. (2016). Astronomy & time reckoning. In *Renaissance Art and Science @ Florence* (pp. 147–181). Penn State University Press.
Putnam, H. (1974). The 'Corroboration' of theories. In P. A. Schlipp (Ed.), *The library of living philosophers, Vol XIV, The philosophy of Karl Popper* (pp. 221–240). Open Court.
Putnam, H. (1973). Reductionism and the nature of psychology. *Cognition, 2*(1), 131–146.
Putnam, H. (2002). *The collapse of the fact/value dichotomy and other essays*. Harvard.
Putnam, H., & Oppenheim, P. (1958). Unity of science as a working hypothesis. In H. Feigl, M. Scriven, & G. Maxwell (Eds.), *Minnesota studies in the philosophy of science* (Vol. II, pp. 3–36). University of Minnesota Press.
Rabel, R. J. (1990). Apollo as a model for Achilles in the *Iliad*. *American Journal of Philology, 111*(4), 429–440.
Rahman, S., Primero, G., & Marion, M. (Eds.). (2012). *The realism-antirealism debate in the age of alternate logics*. Springer.
Ranger, T., & Sacks, P. (Eds.). (1992). *Epidemics and ideas: Essays on the historical perceptions of pestilence*. Cambridge University Press.
Rangos, S. (2012). Empedocles on Divine Nature. *Revue de Métaphysique et de Morale 3 Les Dieux, Le Sacrifice et la Grâce*, 315–338.

Rawcliffe, C. (1996). *Sources for the history of medicine in Late Medieval England*. Medieval Institute Publications.
Rawlings, L. (2007). War and religion. In *The Ancient Greeks at War*. Manchester University Press. ch. 9.
Rawls, J. (1971). *A theory of justice*. Harvard.
Redfield, J. M. (1994). *Nature and culture in the Iliad: The tragedy of hector*. Duke University Press.
Reeder, P. (2015). Zeno's arrow and the infinitesimal calculus. *Synthese, 192*(5), 1315–1335.
Reichenbach, H. (1938). *Experience and prediction*. University of Chicago Press.
Remoff, H. (2016). Malthus, Darwin, and the descent of economics. *The American Journal of Economics and Sociology, 75*(4), 862–903.
Rescher, N. (2015). Pythagoras's number. In *A journey through philosophy through 101 Anectotes* (pp. 18–19). University of Pittsburgh Press.
Reymond, A. (1927). *History of the sciences in Greco-Roman antiquity*. Methuen.
Riemann, B. (1867). Über Die Hypothesen welche der Geometrie zuGrunde liegen. *Abhandlugen der Königlichen Gesellschaft der Wissenschaften zu Gôtingen, 13*, 1–15.
Riley-Smith, J., & Throop, S. A. (2022). *The Crusades: A history* (4th ed.). Bloomsbury Academic.
Ritchie, D. G. (1891). Darwin and Hegel. *Proceedings of the Aristotelian Society, old series 1, 4*(1), 55–74.
Roark, T. (2011). *Aristotle on time*. Cambridge University Press.
Roberts, P. (2017). An ecumenical response to color contrast cases. *Synthese, 194*(5), 1725–1742.
Rocca, J. (2002). The brain beyond Kühn: Reflections on 'Anatomical Procedures' Book IX. *Bulletin of the Institute of Classical Studies, Supplement* No. 77 *The Unknown Galen*, 87–100.
Rocca, J. (2008). Anatomy. In R. J. Hankinson (Ed.), *The Cambridge Companion to Galen* (pp. 242–262). Cambridge University Press.
Rodney, P., Ndjakani, Y., et al. (2010). Addressing the impact of HIV/AIDS on women and children in Sub-Saharan Africa. PEPFAR, the U.S. strategy. *Africa Today, 57*(1), 64–76.
Rodrigues, M. A., & Niaz, M. (2004). A reconstruction of the structure of the atom and its implications for general physics textbooks: A history and philosophy of science perspective. *Journal of Science Education and Technology, 13*(3), 409–424.
Roland, O. (1999). The Copenhagen interpretation. In *Understanding quantum mechanics* (pp. 41–54). Princeton University Press.
Roney, A. (2012). *The history of medicine* (p. 121). Rosen Publishing.
Röntgen, W., & Sarton, G. (1937). The discovery of X-rays. *Isis, 26*(2), 349–369.
Rose, L. E. (1965). The Cartesian Circle. *Philosophy and Phenomenological Research, 26*(1), 80–89.
Roselli, A. (2005). Areteo di Cappadocia lettore di di Ippocrate. In P. van der Eijk (Ed.), *Hippocrates in Context: Papers Read at the 11th International Hippocratic Colloquium, University of Newcastle on Tyne 27-31 August 2002* (pp. 413–432). Brill.
Ross, W. D. (Ed.). (1949). *Aristotle's prior and posterior analytics*. Oxford University Press.
Rudner, R. (1961). An introduction to simplicity. *Philosophy of Science, 28*, 109–119.
Ruf, H. (1969). Transcendental logic: An essay on critical metaphysics. *Man and World, 2*, 38–64.
Russell, B. (1908). Mathematical logic as based upon a theory of types. *American Journal of Mathematics, 30*(3), 222–262.
Rutherford, E., Aston, F. W., Chadwick, J., Ellis, C. D., Gamow, G., Fowler, R. H., Richardson, W. W., & Hartee, D. R. (1929). Discussion of the structure of the atomic nuclei. *Proceedings of the Royal Society of London, 123*(792), 373–390.
Sackett, D., Rosenbert, W. M. C., Muir Gray, J. A., Brian Haynes, R., & Scott Richardson, W. (1996). Evidence-based medicine: What it is and what it isn't. *BMJ, 312*, 71–72.
Salazer, C. F. (2000). *The treatment of war wounds in Graeco-Roman Antiquity*. Brill.
Salmon, W. (Ed.). (1970). *Zeno's Paradoxes*. Bobbs-Merril.
Salmon, W. (1971). Statistical explanation. In W. C. Salmon (Ed.), *Statistical explanation and statistical relevance* (pp. 29–88). University of Pittsburg Press.
Sangalli, A. (2009). *Pythatoras' revenge: A mathematical mystery*. Princeton University Press.

Santos, J. T. (2011). The role of thought in the argument of Parmenides' Poem. In *Parminides: Venerable and Awesome: Proceedings of the International Symposium* (pp. 251–270). Parmenides Publishing.

Sarkar, S. (2021). *Cut and paste genetics: A Crispr Revolution*. Rowman and Littlefield.

Sassi, M. M. (2018). Thales, father of philosophy? In *The beginnings of philosophy in Greece* (pp. 1–3). Princeton University Press.

Saunders, K. B. (1999). The Wounds in *Iliad* 13-16. *Classical Quarterly, 49*, 345–363.

Savage-Smith, E., & Portmann, P. E. (2007). *Medieval Islamic Medicine*. Georgetown University Press.

Savulescu, J. (2002). Deaf Lesbians, "Designer Disability" and the future of medicine. *Journal of Medical Ethics, 325*, 771–773.

Savulescu, J., & Bostrom, N. (Eds.). (2009). *Human enhancement*. Oxford University Press.

Scanlon, T. M. (2000). *What we owe each other* (pp. 108–138). Harvard.

Scarborough, J. (1971). Galen and the Gladiators. *Episteme, 2*, 98–111.

Sedley, D. (1982). On signs. In J. Barnes et al. (Eds.), *Science and speculation* (pp. 239–266). Cambridge University Press.

Segel, A. (2019). Pythagoreanism: A number of theories. *Philosopher's Imprint, 19*(26), e1–e19.

Semmelweis, I.. tr. K. Codell (1983). *Etiology, concept, and prophylaxis of childbed fever* (Vol. 2). University of Wisconsin Press.

Sen, A. (1970). *Collective choice and social welfare*. Holden-Day.

Sen, A. (1992). *Inequality re-examined*. Harvard University Press.

Sen, A. (2009). *The idea of justice* (pp. 164–167). Harvard University Press.

Sepper, D. L. (2015). Animal spirits. In L. Nolan (Ed.), *The Cambridge Descartes Lexicon* (pp. 26–28). Cambridge University Press.

Shankar, R. (2019). Special Relativity II: Some consequences. In C. T. New Haven (Ed.), *Fundamentals of Physics I: Mechanics, relativity and thermodynamics, expanded edition* (pp. 209–226). Yale University Press.

Shitan, M., & Monday, I. (2015). HIV/AIDS epidemic in Malaysia: Trend analysis from 1986-2011. *Southern African Journal of Demography, 16*(1), 36–56.

Sider, D. (1982). Empedocles' Persika. *Ancient Philosophy, 2*, 76–78.

Sigerist, H. E. (Ed.). (1996). *Paracelsus, four treatises of Theophrastus von Hohenheim called Paracelsus*. Johns Hopkins University Press.

Sina, I. (tr. 1930, rpt. 1973). *Al-Qanun fi al-tibb; The Cannon of Medicine*, Oskar Cameron Gruner, ed. tr. Ams Press, rpt. London.

Sina, I. (2009). *The physics of healing*, Jon McGinnis, ed. tr. Provo. Brigham Young University Press.

Singer, P. N., van der Eijk, P. J., & Tassinari, P. (2019). *Galen: Works on Human Nature, vol. 1 Mixtures (De temperamentis)*. Cambridge University Press.

Siraisi, N. G. (1990). *Medieval and Early Renaissance Medicine: An introduction to knowledge and practice*. University of Chicago Press.

Sisko, J. E., & Weiss, Y. (2015). A fourth alternative in interpreting Parmenides. *Phronesis, 60*, 40–59.

Smith, A. (1776). *An inquiry into the nature and causes of the wealth of nations*. W. Strahan and T. Cadell.

Smith, A. P. (2012). *Thinking from the One: Science and the ancient philosophical figure of one*. Edinburgh University Press.

Smith, J. F. (1985). The Russell-Meinong Debate. *Philosophy and Phenomenological Research, 46*(3), 305–350.

Smith, N. K. (1962 [1923]). *A commentary to Kant's "Critique of Pure Reason"* (2nd ed.). Humanities Press.

Smith, R. R. R. (1990). Late Roman Philosopher Portraits from Aphrodisias. *Journal of Roman Studies, 80*, 127–155.

Smith, W. D. (1979). *The Hippocratic Tradition*. Cornell University Press.

Snell, B. (1944). Die Nachrichten über die lehren des Thales und die Anfänge der griechischen Philosphie und Literaturgeschichte. *Philogus, 96*, 170–182.
Snow, J. (1855). *On the communication of Cholera*. John Churchill.
Snow, J. (1858). *On chloroform and other anesthetics*. John Churchill.
Snowden, F. M. (2019). The Germ Theory of Disease. In *Epidemics and society: From the black death to the present* (pp. 204–232). Yale University Press.
Soames, S. (2001). *The unfinished semantic agenda of naming and necessity*. Oxford University Press.
Sobel, D. (2011). *Galileo's Daughter: A historical memoir of science, faith, and love*. Bloomsbury USA.
Sober, E. (1975). *Simplicity*. The Clarendon Press.
Sober, E. (1980). Evolution, population thinking and essentialism. *Philosophy of Science, 47*, 350–383.
Sober, E. (2015). *Ockham's Razors: A user's manual*. Cambridge University Press.
Sober, E., & Lewontin, R. (1982). Artifact, cause, and genic selection. *Philosophy of Science, 49*, 157–180.
Sober, E., & Wilson, D. S. (1998). *Unto others: The evolution and psychology of unselfish behavior*. Harvard University Press.
Solbakk, J. H. (1993). *Forms and functions of medical knowledge in Plato*, Ph.D. Thesis. University of Oslo, pp. 170–203.
Solmsen, F. (1958). Aristotle and Pre-Socratic Cosmology. *Harvard Studies in Classical Philology, 6*(3), 265–283.
Solmsen, F. (1980). Hymn to Apollo. *Phronesis, 25*, 219–227.
Sorabji, R. (1980). *Necessity, cause, and blame*. Cornell University Press.
Sorenson, R. A. (2003). *A brief history of paradox*. Oxford University Press.
Spence, J. C. H. (2019). *Lightspeed*. Oxford University Press.
Spergel, D. N. (2015). The dark side of cosmology: Dark matter and dark energy. *Science, 347*(6226), 1100–1102.
Speyer, W. (1971). *Die literarische Fälschung im heidnischen und christlichen Altertum: ein Versuch ihrer Deutung*. Beck.
Splavski, B., Rotin, K., et al. (2019). Andreas Vesalius, the predecessor of neurosurgery: How his progressive scientific achievements affected his professional life and destiny. *World Neurosurgery, 129*, 129–202.
Sprenger, J. (2015). A novel solution to the problem of old evidence. *Philosophy of Science, 82*(3), 383–401.
Sterelny, K., & Kitcher, P. (1988). The return of the gene. *Journal of Philosophy, 85*(7), 339–361.
Stern, A. M., et al. (2022). School closures. *Public Health Reports, 125*(supplement 3), 63–70.
Stewart, M. (2022). *Sacred geometry of the Starcut Diagram: Number, proportion, and cosmology*. Inner Traditions.
Stokes, M. (1971). *One and many in presocratic philosophy*. Center for Hellenic Studies.
Stolberg, M. (2003). A woman down to her bones: The anatomy of sexual difference in the sixteenth and early seventeenth centuries. *Isis, 94*(2), 274–299.
Stow, J. (1956). *The survey of London*. Dent & Dent.
Strawson, P. (1966). *The bounds of sense*. Routledge.
Stringer, M. D., & Berker, I. (2010). Columbo and the Clitoris. *European Journal of Obstetrics and Gynecology and Reproductive Biology, 151*(2), 130–133.
Strobino, R. (2016). Per se, inseparability, containment, and implication. Bridging the gap between Avicenna's Theory of demonstration and logic of the predicables. *Oriens, 44*(3/4), 181–266.
Strollo, A. (2019). If I were Kripke . . . Attributable names and the necessary *a posteriori*. *Philosophical Forum, 50*(1), 116–134.
Sturz, F. W. (1805). *Empedocles Agrigentinus*. Van der Ben.
Sunderlal, P. (2019 [1929]). *British Rule in India* 4 vols. Sage Publishers.
Szabo, A. (1978). *The beginnings of Greek mathematics*. D. Reidel.

Tappenden, P. (2004). The Ins and Outs of Schrödinger's Cat Box: A response to Papineau. *Analysis, 64*(2), 157–164.

Tauber, A. I., & Sarkar, S. (1992). The Human Genome Project: Has blind reductionism gone too far? *Perspectives in Biology and Medicine, 35*, 220–235.

Teays, W. (Ed.). (2022). *Reshaping philosophy: Michael Boylan's Narrative Fiction*. Springer.

Tescusan, M. (2004). ed. and tr. *The fragments of the methodists. Methodism outside Soranus* (Vol. 1). Brill.

Thivel, A. (1981). *Cnide et Cos? Essai sur les Doctrines médicales dans la Collection hippocratique* (pp. 279–383).

Thomas, J. L. (1999). The Vineberg Legacy: Internal mammary artery implantation from inception to obsolescence. *Texas Heart Institute Journal, 28*, 107–113.

Thomson, J. F. (1954). Tasks and super-tasks. *Analysis, 15*, 1–13.

Thorndike, L. (1916). The True Roger Bacon I. *The American Historical Review, 21*(2), 237–257.

Thumiger, C. (2018). The professional audience of the Hippocratic *Epidemics*: Patient cases in hippocratic scientific communication. In P. Bouras-Valhanatos & S. Xenophontos (Eds.), *Greek medical literature and its readers* (pp. 48–64). Routledge.

Thumiger, C. (Ed.). (2020). *Holism in ancient medicine*. Brill.

Thurston, L., & Williams, G. (2015). An examination of John Fewster's role in the discovery of Smallpox. *Journal of the Royal College of Physicians Edinburgh, 45*(2), 173–179.

Tomilinson, H. (1989). *After truth: Post-modernism and the rhetoric of science in dismantling truth: reality in the post-modern world*. St. Martins.

Tong, R. (2005). Taking on big fat. In M. Boylan (Ed.), *Public health policy and ethics* (pp. 39–48). Kluwer.

Tovar-Murray, D. (2010). Social health and environmental quality of life: Their relationship to positive physical health and subjective well-being in a population of Urban African Americans. *The Western Journal of Black Studies., 34*(3), 358–366.

Tracy, T. (1969). *Physiological theory and the doctrine of the mean in Plato and Aristotle*. Mouton.

Trenn, T. J., Geiger, H., & Ernest Marsden, E. R. (1974). The Geiger-Marsden Scattering Results of Rutherford's Atom, July 1912-July 1913: The shifting significance of scientific evidence. *Isis, 65*(1), 74–82.

Trépanier, S. (2010). *Early Greek theology: God as nature and natural gods*. Edinburgh University Press.

Tröhler, U. (2015). Statistics and the British Controversy about the effects of Joseph Lister's System of Antisepsis for surgery: 1867-1890. *Journal of the Royal Society of Medicine., 108*(7), 280–287.

Turkeiltaub, D. (2007). Perceiving Iliadic Gods. *Harvard Studies in Classical Philology, 103*, 51–81.

Urquhart, A. (2016). Russell and Gödel. *Bulletin of Symbolic Logic, 22*(4), 504–520.

Van Arsdall, A., & Graham, T. (Eds.). (2017). *Herbs and healers from the Mediterranean through the Medieval West: Essays to honor John M. Riddle*. Routledge.

van der Eijk, P. J. (1966). Diocles and the Hippocratic writings on the method of dietetics and the limits of causal explanation. In R. Wittern & P. Pellegrin (Eds.), *Hippokratische Medzin und antike Philosophie*. Olms Weidmann.

van der Eijk, P. J. (1991). 'Airs, waters, places' and 'on the sacred disease': Two different religiosities? *Hermes, 119*(2), 168–176.

van der Eijk, P. J. (2005). *Hippocrates in context: Papers Read at the 11th International Hippocratic Colloquium, University of Newcastle on Tyne 27-31 August 2002*. Brill.

van der Eijk, P. J. (2014). Galen on the nature of human beings. *Bulletin of the Institute of Classical Studies, Supplement, 114*, 89–134.

van der Eijk, P. J. (2016). On 'Hippocratic' and 'Non-Hippocratic' medical writing. In L. Dean-Jones & R. M. Rosen (Eds.), *Ancient concepts of the Hippocratic* (pp. 17–47). Brill.

van Fraassen, B. C. (1968). Perception, implication and self-reference. *Journal of Philosophy, 65*(5), 136–152.

Van Orman Quine, W. (1953). *From a logical point of view*. Harvard University Press.
Van Orman Quine, W. (1960). *Word and object*. M.I.T. Press.
Van Orman Quine, W. (1963). On simple theories of a complex world. *Synthese, 14*, 103–106.
Van Orman Quine, W. (1970). *The web of belief* (2nd ed.). Random House.
Van Orman Quine, W. (1987). *Quiddities: An intermittently philosophical dictionary*. Harvard University Press.
Van Wyhe, J. (2007). Mind the gap: Did Darwin avoid publishing his theory for many years? *Notes and Records of the Royal Societies, 61*(2), 173–205.
Vassend, O. B. (2015). Confirmation measures and sensitivity. *Philosophy of Science, 82*(5), 892–904.
Vegetti, M. (1995). L'Épistémologie d'Érasistrate et la Technologie Hellénistique. *Ancient Medicine in its Socio-Cultural Context; Papers read at the Congress held at Leiden University, 13-15 April 1992* (Vol. 28, pp. 461–472).
Veller, C., Kleckner, N., & Novak, M. A. (2019). A rigorous measure of genome-wide shuffling that takes into account crossover positions and Mendel's Second Law. *Proceedings of the National academy of Science of the United States of America, 116*(5), 1659–1668.
Vesalius, A. (1555-1558). *On the fabric of the human body* (Vol. 7 vols). Joannes Oporinus.
Vieler, H. (1926). *Untersuchungen über Unabhängigkeit und Tragweite der Axiome der Mengenlehre in der Axiomatik Zermelos und Fraenkels*. W. Fr. Kaestner.
Vlastos, G. (1966a). Zeno's Race Course. *Journal of the History of Philosophy, 4*(2), 95–108.
Vlastos, G. (1966b). A note on Zeno's Arrow. *Phronesis, 11*, 3–18.
von Leibniz, G. W. F. (rpt. 1924 [1641–1686]). On geometrical method and the method of metaphysics. In A. R. Chandler (Ed.), *Discourse on metaphysics*, trans. George Montgomery. Open Court.
von Leibniz, G. W. F., & Gottfried Leibniz, G. (1948 [1765]). In R. Latta (Ed.), *New essays on the human understanding and other philosophical writings* (pp. 357–385). Oxford University Press.
von Staden, H. (1989). *Herophilus: The art of medicine in early Alexandria*. Cambridge University Press.
von Staden, H. (1992). The discovery of the body: human dissection and its cultural contexts in ancient Greece. *Yale Journal of Biology and Medicine, 65*, 223–224.
von Staden, H. (1996). In a pure and holy way: Personal and professional conduct in the Hippocratic Oath? *Journal of the History of Medicine and Allied Sciences, 51*(4), 404–437.
von Staden, H. (2002). Division, dissection, and specialization: Galen's 'On the Parts of the Medical Techne'. *Bulletin of the Institute of Classical Studies*, Supplement No. 77 *The Unknown Galen*, 19–45.
Wagener, T. (2013). *The history of oncology*. Springer, 2009: ch. 4; and D. de Moulin, *A short history of breast cancer*. Springer.
Wakefield, J. C. (1978). Four basic concepts of medical science. In P. D. Asquith & I. Hacking (Eds.), *Proceedings of the Philosophy of Science Association* (Vol. 1, pp. 210–222).
Wakefield, J. C. (1992a). The concept of mental disorder: On the boundary between biological facts and social values. *American Psychologist, 47*, 373–388.
Wakefield, J. C. (1992b). Disorder as harmful dysfunction: A conceptual critique of DSM III-R;s definition of mental disorder. *Psychological Review, 99*, 232–247.
Wakefield, J. C. (1993). The limits of operationalism: A critique of Spitzer and Endicott's proposed operational criteria for mental disorder. *Journal of Abnormal Psychology, 102*, 160–172.
Wakefield, J. C. (1997a). Diagnosing DSM IV—Part I: DSM-IV and the concept of disorder. *Behavior Research & Therapy, 35*, 633–649.
Wakefield, J. C. (1997b). Diagnosing DSM IV—Part II: Eysenick and the essentialist fallacy. *Behavior Research & Therapy, 35*, 651–665.
Wallace, Alfred Russel (1869) *The Malay Archipelago; The Land of Orang-Utan, and the Bird of Paradise.* : Macmillan.
Wallace, A. R. (1870). *Contributions to the theory of natural selection*. Macmillan.
Wallace, A. R. (1908). *My life: A record of events and opinions*. Chapman and Hull.

Watling, J. (1953). The sum of an infinite series. *Analysis, 13,* 39ff.
Wear, A. (Ed.). (1992). *Medicine in society.* Cambridge University Press.
Weatherall, J. O. (2014). What is a singularity in geometrized Newtonian gravitation? *Philosophy of Science, 81*(5), 1077–1089.
Weatherall, J. O. (2016). Are Newtonian gravitation and geometrized Newtonian gravitation theoretically equivalent? *Erkenntnis, 81*(5), 1075–1091.
Weber, B. H. (2011). Design and its discontents. *Synthese, 178*(2), 271–289.
Webster, C. (2015). Heuristic medicine: The Methodists and Metalepsis. *Isis, 106*(3), 657–668.
Wedgwood, R. A. (2013). A priori bootstrapping. In A. Casullo & J. Thurow (Eds.), *The a priori in philosophy* (pp. 226–246). Oxford University Press.
Wee, J. Z. (2017). Earthquake and epilepsy: The body geologic in the Hippocratic treatise *on the sacred disease.* In J. Z. Wee (Ed.), *The comparable body: Analogy and Metaphor in Ancient Mesopatamian, Egyptian, and Greco-Roman Medicine* (pp. 142–169). Brill.
Weightman, G. (2020). Sutton and Jenner: The legacy. In *The Great Inoculator: The untold story of Daniel Sutton and his medical revolution* (pp. 156–161). New Haven.
Weil, S., Bespaloff, R., & Broch, H. (rpt. 2005). *War and the Iliad,* tr. Mary McCarthy. New York Review of Books.
Weinberg, S. (2011). *Dreams of a final theory: The scientist's search for the ultimate laws of nature.* Knopf.
Weiss, R. A., & Esparza, J. (2015). The prevention and eradication of Smallpox: A commentary on Sloane (1755) 'An Account of Inoculation'. *Philosophical Transactions: Biological Sciences, 370*(1666), 1–11.
Welch, E. (2008). Self-predication in Plato's 'Euthyphro'? *Apeiron, 41*(4), 193–210.
Wellmann, M. (1895). *Die pneumatische Schule bis auf Archigens.* Weidmann.
West, J. B. (2008). Ibn al-Nafis, the pulmonary circulation, and the Islamic Golden Age. *Journal of Applied Physiology, 105*(6), 1877–1880.
Wheeler, J. (1955). Geons. *Physical Review, 97,* 511.
Wheeler, J. (1962). *Geometric dynamics.* Academic Press.
Whitehead, A. N., & Russell, B. (1910, 1912, 1913). *Principia Mathematica* 3 vols. Cambridge University Press.
Whitteridge, G. (1971). *William Harvey and the circulation of the blood.* Elsevier.
Wilberding, J. (2015). Plato's embryology. *Early Science and Medicine, 20*(2), 150–188.
Wilcox, M. (2012). Improved traditional medicines in Mali. *Journal of Alternative and Complementary Medicine., 18*(3), 212–220.
Wilczek, F. (2016). *A beautiful question: Finding nature's deep design.* Penguin.
Willard, M. B. (2014). Against simplicity. *Philosophical Studies,* 165–181.
Wilson, C. (1997). *The invisible world: Early modern philosophy and the invention of the microscope.* Princeton University Press.
Wilson, E. O., & Bossert, W. H. (1971). *A primer of population biology.* Sinauer Publishers.
Wilson, J. (2007). Newtonian forces. *British Journal for the Philosophy of Science, 58*(2), 173–205.
Witten, E. (1995). String theory dynamics in various dimensions. *Nuclear Physics B, 443*(1), 85–126.
Woit, P. (2006). *Not even wrong: The failure of string theory and the search for unity in physical law.* Basic Books.
Wolfe, H. E. (2012 [1945]). *Introduction to Non-Euclidean Geometry.* Dover.
Wollaston, T. V. (1856). *On the variation of species with especial reference to the Insecta.* John Van Voorst.
Woodford, P. (2016). Neo-Darwinists and Neo-Aristotelians: How to talk about natural purpose. *History and Philosophy of Life Sciences, 38*(4), 1–22.
Woodward, J. (2014). Simplicity in the best systems account of the laws of nature. *British Journal for the Philosophy of Science, 65*(1), 191–213.
Woolf, R. (2002). Consistency and *Akrasia* in Plato's *Protagoras. Phronesis, 47*(3), 224–252.
Wright, T. (2013). *Circulation: William Harvey's Revolutionary Idea.* Vintage Books.
Yndurain, F. J. (2010). *The Theory of Quark and Gluon Interactrions* (4th ed.). Cham.

Zanker, A. T. (2013). Decline and Parainesis in Hesiod's Race of Iron. *Rheinisches Museum Für Philologie, neue folge, 156*(1), 1–19.

Zee, A. (2003). *Quantum field theory in a nutshell*. Princeton University Press.

Zee, A. (2018). A friendly contest between the four interactions. In *On gravity: A brief tour of a weighty subject* (pp. 9–14). Princeton University Press.

Zhou, Z., Chu, H., Li, C., et al. (2014). Active replication of Middle East Respiratory Syndrome Coronavirus and Aberrant induction of inflammatory cytokines and chemokines in human macrophages: Implications for pathogenesis. *Journal of Infectious Diseases, 209*(9), 1331–1342.

Index

A
AAA-First Figure Syllogism, 83
Abortion, 65
Abu Bakr Mohammad bin Zakariya Al-Razi (aka, Rhazes), 123
Abulcasis wrote his famous work: "Al Tasreef Liman 'Ajaz Aan Al-Taleef'", 124
ADM Formalism, 224
Aëtius, 14, 15, 17, 30
African American populations, 236
African Americans, 213
Against the Cnidian Sentences, 61
Agreement, difference, or a joint method of agreement and difference, 177
AI and the subsequent technologies, 233
AIDS, 6
Air, 16, 20, 77, 176
Air, fire, and water, 77
Air, water, and places, 49, 51, 185
AI, Technology, and Modern Surgery, 232–233
A lack of international cooperation to a global pandemic, 242
Alasdair MacIntyre, 65
Alchemy, 124, 146
al-Dakhwar, 126
Alexander, 81
Al-Hawi Fi al-Tibb (known in Latin as Liber Contiens), 124
Almagest, 143
Al-Razi, 122, 123
Al-Zarawi (aka, Abulcasis), 123, 124
Alzheimer's Disease, 230
Amyntas II, 81
Analogy, 14
Analytic propositions, 147
Analytic v. synthetic distinction, 147
Anatomical Procedures, 100
Anaxagoras, 47, 165
Anaximander, 164, 218
Anaximenes, 21, 218
Ancient Medicine, 55
An Enquiry Concerning Human Understanding, 154
Anesthesiology, 180
Angina, 211
Animal dissection, 96
Animal spirits, 156, 157
Anonymus Londinensis, 52
An Outline of Skepticism, 109
Anti-realist construct, 218
Anti-reductionistic, 229
Apollo, 4, 25
Apollodorus, 29
A posteriori, 148, 150, 155
A priori, 148, 150, 155, 196
A priori arguments, 141
Archidoxa, 129
Aristarchus of Samos, 135
Aristotelian, 188
Aristotelian/Galenic, 158, 159
Aristotelian time, 151, 152
Aristotle, 10, 11, 17–19, 53, 77, 81–84, 88, 92, 96, 101, 103, 111, 121, 125, 134, 139, 144, 145, 149, 153, 155, 163, 164, 173, 176, 192, 196, 197, 222, 228
Aristotle, *de caelo*, 8, 9
Aristotle discusses procedure for dissection, 97
Aristotle, *Generation of Animals*, 15

Aristotle, *Metaphysics*, 8, 12, 22
Aristotle, *On Respiration*, 16
Aristotle, *On Sensation*, 22
Aristotle, *On the Heavens*, 13
Aristotle's Categories, 125
Aristotle's four causes, 126
Aristotle's *Nicomachean Ethics and Politics*, 65
Aristotle's *Posterior Analytics*, 135
Arius Didymus, 22
Arterial and venous systems, 155
Arteries, 158
Arthur William Hope Adkins, 65
Astrophysics, 191
Athenaeus, 112
Athenaeus of Attaleis, 102
Atom, 188
Atomism, 218
A Treatise of Human Nature, 154
A universal law of everything, 223
Autopsia, 107
Avicenna, 125
Axiomatic-structured account, 226
Axiomatic system, 190
Axiomatic theory, 197
Ayer, A.J., 207

B
Bacon, F., 176
Bacteriophages, 209
Balance (symmetry), 53, 92, 185
Barber Paradox, 198
Baryons and Mesons, 219
Basi, A., 178
Bates, H.W., 166
Beck, C., 211
Belief, 73
Biblical Creationist Model, 168
Big and little Nature, 223
Big and little Nature in medicine, 207, 210–212
Big Nature, 190–192, 196, 208, 217, 221–225, 230, 232, 243
Big Nature in General Science, 222–226
Big Nature in medicine, 212–214
Big Nature's relativity, 205
Bile, serum, and phlegm, 78
Black body radiation, 201
Black box, 153
Black holes, 220, 221
Blood, 78, 92, 130, 155
Blood and blood vessels, 77

Bloodletting, 20, 156
Blood-model-A (BM-A), 155
Blood model-B (BM-B), 157
Blood temperature, 92
Bohr model of the atom, 218
Bohr, N., 188, 201, 204
Bohr's model of the atom, 201
Book of Three Principles, 128
Born, M., 204, 223
Bottom => up, 229
Brain, 92
Breast cancer, 211
Breath, 77
Bubonic plague, 234

C
Caliph Al-Hakam II, 124
Caliph Al-Muktafi, 123
Caliphates, 123
Cancer, 232
Carbolic acid, 184
Cardiology, 211
Carnap, R., 206, 207, 228
Carotid artery, 92
Carrel, A., 211
Cartesian Vortex Theory, 218
Causal momentum, 229
Causation, 229
Cauterizing wounds, 6
Censorinus, 15
Central limit theorem, 177
Certainty, 226
Chargaff, E., 209
Chargaff's rule, 209
Chase, M., 209
Cholera, 234
Christianismi Restitutio, 129
Christianity, 138
Cleanthes, 22
Clitoris, 133
Clone complex organs, 230
Cloning islet cells, 230
Cloning to alter germline cells, 231
Cnidian, 60
Cnidian Sentences, 59
Cohen, T., 227
Coherence, 131, 154, 159, 189, 197–199, 205, 215, 217, 226, 244
Coherent, 165, 223
Cold with the feminine, 18
Columbus, R., 127
Commentary on Hippocrates' On Humors, 9

Commentary on Hippocrates' On the Nature of Man, 8
Complementarity, 202
Complete, 165, 223
Complete and coherent, 159
Completeness, 131, 154, 167, 173, 189, 197, 198, 205, 215, 217, 226, 244
Completeness and elegance, 48
Completeness, coherence, simplicity, and elegance, 161, 185
Conception theory, 100
Condensed black holes, 224
Confirmation, 219
Contrary to Nature, 78
Coordinated public health response, 239
Copenhagen Interpretation of Quantum Mechanics, 205
Copernican hypothesis, 139
Copernicus, 137
Copernicus' book *De revolutionibus orbium coelestium*, 135
Coronary disease, 232
Cosmology, 222
Counterfactual confirmation, 227
COVID-19, 6, 241
COVID-19 pandemic, 214
Cowpox and smallpox, 178
Crick, F., 209
Crispr, 243
Crispr technology, 231
Cure/Healing (book), 125
Curie, M., 188, 189
Cushing, H., 211
Cyclic nature to the universe, 222
Cystic Fibrosis, 230

D
Dandy, W., 211
Dark energy, 220
Dark matter, 220
Darwin, C., 164–167, 169–171, 173
Date of Easter, 134
David Hume's mitigated skepticism, 243
Davidson, D., 203, 207
de Broglie, L., 201
Deferent, 135
de Fermat, P., 177
De humani corporis fabrica libri septem, 131
de Laplace, P., 177
De mineralibus, 129
Democritus, 187, 218
De Natura Rerum, 129
De re anatomica, 132

De revolutionibus, 143
Descartes, R., 139, 141, 142, 144, 148, 159, 226
De venarum ostiolis, 157
Devolution, 164
DeWitt, B.S., 224
Diagnosis, 50, 53–62
Diagnosis, prognosis, and treatment, 57
Diet/regimen, 132
Dihybrid crosses, 174
Diocles, 19
Diogenes Laertius, 12, 21, 29, 30
Dioscorides, 123
Disease, 6, 57, 111, 112, 176, 230
Disease inoculation, 185
Dissection on corpses, 133
Divided Line, 76
DNA, 209
Dogmatic, 109
Dogmatic practitioners, 101
Dogmatists, 99, 101, 102, 105, 111, 115
Dogmatist school of medicine, 54
Donagan, A., 227

E
Earth, 20
Ebola, 6
Ebola: 1976, 2014, 2019, 238
Edelstein, L., 65
Eidos, 164
Einstein, A., 192, 195, 197, 222–225, 227
Einstein-Maxwell, 221
Electricity and magnetism, 223
Electromagnetic, 188
Electromagnetism, 188, 218, 222, 223
Elegance, 154, 159, 185, 189, 198, 205, 215, 217, 226, 244
Elegant, 173, 198, 223
Elements of earth, 77
$E = MC^2$, 193
Empedocles, 11, 12, 14, 15, 17–21, 75, 88, 164, 218, 222
Empeiria, 84
Empiricism, 207
Empiricists, 99, 108–111, 113, 116, 125
Empirics, 115, 116
Environmental fitness pressure, 170
Epicycle, 135
Epidemics, 58, 60, 65, 109
Epidemics I, 54, 55
Epidemics III, 54, 55, 75, 185
Epidemiology, 180
Epidemiology of infectious disease, 237

Epimenides Paradox, 198
Epistasis, 231
Epistemic background in diagnosis and treatment, 60
Epistemological and metaphysical realism, 23
Epistemological anti-realists, 27
Epistemological certainty, 60
Epistemology, 20
Erasistratean, 113
Erasistratus, 109
Eristic, 78
Etymologicum Magnum, 25
Euclid, 190
Euclidean, 188, 195
Euclidean space, 151, 152, 191
Euclid's *Elements*, 154
Eudemus in Proculus, *Commentary on Euclid's* Elements, 30
Eusebius, 22
Evolutionary theory, 185
Exercise, 79
Experimental drugs for treatment, 239
Eye, 20

F
Fabricius, 155, 157
Falsification, 219
Faraday, M., 188
Feedback loops, 229
Feyerabend, 227
Field theory, 223
Fifth force, 220
Fire, 19, 20
Fitness, 167
"Fix" the defective gene, 230
Force, 188
Foucault, M., 26
Four areas of medicine, 127
Four humors, 52
Fragile X Syndrome, 230
Francis Collins, 230
Francis, D.P., 237
Franklin, R., 209
Frege, G., 196, 199
Froben, J., 128
Fully balanced, 92

G
Galapagos Islands, 168
Galen, 8, 19, 26, 63, 79, 91, 100, 109–113, 115, 116, 121, 123, 126, 155, 164, 207

Galenic, 156
Galenic humor theory, 176
Galenic texts, 111–116
Galen set out four natural faculties, 101
Galileo, 143, 155
Geist, 164
General relativity, 195, 221–225
General theory skepticism, 106
Generation of Animals, 173
Genetically-caused diseases, 230
Genetic engineering, 230, 231
Genetic epistasis, 229
Genos, 164
Geons-approach, 225
Germs, 185
Germ theory, 176–185
Gödel, K., 26
Gödel's Proof, 199
God figure, 163
Gods, 11
Gravitation, 222
Graviton, 225
Gravity, 145, 146, 195, 196, 218, 223

H
Hamiltonian Mechanics, 224
Harmony, 26
Harmony and symmetry, 76
Hartshorne, 227
Harvey, W., 155–157, 159
Hasenöhrl, F., 201
Hawking, S., 221–223
Health, 25, 53, 57, 111, 112, 184
Hearing, 23
Heart, 17–19, 92, 155, 156
Hegel, G.W.F., 26, 164
Heisenberg, W., 201, 204
Heraclides of Tarentum, 110
Heraclitus, 20–27
Heredity, 209
Heritable, 165
Herophilus, 107, 109
Hershey, A., 209
Hesiod, 164
Hinduism, 222
Hippocrates, 27, 48, 113
Hippocratic, 18, 48, 75, 112, 176
Hippocratic Corpus, 52, 64
Hippocratic Oath, 63, 66, 109, 233
Hippocratic work *Airs, Waters, and Places*, 180
Hippocratic writer of *The Sacred Disease*, 242

Hippocratic writers, 3, 54, 55, 77, 95, 100, 121, 123, 142
Hippocratic writers on Epidemics, 124
Hippocratic writings, 58
Hippolytus, 21
Hippolytus, *Refutation of All Heresies*, 8, 24
HIV/AIDS public health, 237
HIV: 1981 onwards, 236
Holism, 229
Homeric Allegories, 22
Hooke, R., 145
Hot, cold, wet, and dry, 56, 176
Hot with the masculine, 18
Hull, 227
Human dissection of corpses, 96
Human dissections, 127
Human Genome Project, 230
Hume, D., 154, 169
Huntington's Disease, 230

I
Ibn al-Nafis, 122, 123, 126
Ibn Sina, 123
Ichor, 104
Iliad, 4, 5
Immunization via inoculation, 182
Infection control, 239
Infectious disease, 5
Inflammation, 183
Inflation, 220
International public health infrastructure, 239
Intuition, 27
Inverse square law, 146
Isaac Newton's *The Principia: Mathematical Principles of Natural Philosophy*, 144
Islam, 138

J
Jenner, E., 178
Jesus, 138
Jiankui, H., 232
Juan de Quintana, 129
Judaism, 138

K
Kant, I., 146, 148, 149, 151, 153, 154, 188, 190–192, 197
Kant's Transcendental Aesthetic, 204
Kepler, 145

Kettlewell, H.B.D., 173
Killingray, D., 212
Kitab al-Shifa, 125
"Knockout" factors, 230
Kuhn, T.S., 227

L
Lamarck, J.B., 165, 166
Lavoisier, 151
Law-like nomic structure, 23
Law of Excluded Middle, 203
Law of Non-Contradiction, 203
Leptons and Hadrons, 218
Leucippus, 187, 218
Levene, P., 209
Lewis, D., 203
Life history of a star, 221
Light as electromagnetic radiation, 189
Linguistic expressions and meaning, 207
Linnaean Classification System, 165
Linus Pauling, 209
Lister, J.J., 163, 180, 183
Lister's antiseptic approach, 184
Little Nature, 217, 221, 223–225, 230, 243
Little Nature in Medicine: Infectious disease, 210
Little Nature's uncertainty principle, 205
Liver, 155
Lobachevsky, N., 191
Logical empiricist, 226, 229
Logical empiricist tradition, 228
Logicism, 199
Logicist thesis, 196–199, 227
Logos, 21, 23, 26
Loop gravity, 225
Loop quantum gravity, 223–226
Love, 13, 164
Love and Strife, 14
Lyell, C., 168

M
Magna Moralia, 29
Magnetism-field theory, 189
Maker/artifact dynamic, 137
Malthus, T.R., 166, 167, 169, 170
Marcus Aurelius, 100
Mass and force, 146
Material explanation (ME), 47, 86
Maxwell, J.C., 188, 223
Maxwell's equations, 193
McCumber, J., 228

ME accounts and focus on desired outcomes, 95
Medical diagnosis, 164
Medical surgery, 185
Mendel, G., 166, 173, 208
Mendelian Laws of Genetics, 174
Methodics, 114
Methodism, 109
Methodists, 99, 102, 105, 113, 116
Method of agreement, 181
Methodology of logical empiricism, 227
Methodus mendendi, 116
Miasma, 50
Miescher, F., 208
Mill, J.S., 177
Mill's method of difference, 180, 209
Mill's method of inductive agreement, 180
Mill's method of inductive difference, 181
Minucius Felix, 9
Modern entropy, 222
Mono-hybrid crosses, 174
Moore, 226
Mueller, I., 74, 227

N

Natural selection, 171
Nature, 28
$Nature_1$, 153, 167, 173, 184, 189, 197, 203, 226
$Nature_{1\&2}$, 141, 154
$Nature_2$, 144, 146, 154, 159, 163, 167, 168, 184, 192, 196, 205, 215, 218, 226
$Nature_3$, 154, 155, 184, 185, 197, 218
Nature (*kata phusin*), 111
Nature as big and as little, 205
Nature does nothing in vain, 101, 136, 163
Necessity, 90
Nerves, 132
Nerve-theory, 143
Neurosurgery, 211
Newton, I., 143, 145, 146, 148, 151, 159, 192, 195–197, 224
Newtonian, 154
Newtonian mechanics, 218
Newtonian physics, 190, 191, 231
Newton's second law, 202
$n/Nature_1$, 159, 163, 164, 167, 171, 189, 208
$n/Nature_2$, 147, 148, 222, 243
$n/Nature_3$, 147, 148, 243
No-hair theorem, 221
Non-reductionistic approach, 229
Normative professional code, 67
Nuclein, 208

O

"On a Heuristic Viewpoint Concerning the Production and Transformation of Light", 193
Oncology, 211
On Medical Experience, 110
On Nature, 12
On the Anatomy of Muscles, 100
"On the Electrodynamics of Moving Bodies", 193
"On the Motion Required by the Molecular Kinetic Theory of Heat—of Small Particles Suspended in a Stationary Liquid", 193
On the Natural Faculties, 113
On the Nature of Man, 57
On the Opinions of Hippocrates and Plato, 101
On the Origin of the Species, 168
On the Use of Parts, 113
On the Use of the Parts, 101, 111
Ontology, 207
Oppenheim, P., 206, 228
Opus Paramirum, 128, 129
Ordered pairs, 226, 228
Organ disorders, 230
Owen, G.E.L., 70
Owen, R., 166
Owen, R.D., 166

P

Pandemics, 242–243
Paracelsus, 127
Parallel postulate, 190
Paramedian epistemological antirealism, 26
Parkinson's Disease, 230
Parmenides, 14, 21
Particularity, 23, 58
Parts of Animals, 10, 97
Pascal, B., 177
Pasteurization, 181
Pasteur, L., 178, 181
Pasteur vaccine approach, 182
Pauli's exclusion principle, 204
Pauli, W., 203
Pedanius Dioscorides, 111
Pepsis, 52, 93, 94
Peritoma, 53
Pharmacology, 132
Phenotype via surgery, 208
Philosophers of biology, 229
Philosophy of biology, 228
Philosophy of literature, 229

Phusis$_1$, 6, 7, 9, 11, 14, 18, 19, 23, 27, 31, 35–37, 41, 47, 78, 93, 95, 102, 105, 106, 112, 113, 116, 117, 126, 131
Phusis$_1$ and *phusis*$_2$, 90
Phusis$_{1\ or\ 2}$, 47
Phusis$_{1+2}$, 102, 105, 135
Phusis$_2$, 5, 6, 9, 10, 14, 15, 19, 23, 26–28, 31, 32, 35, 38, 55, 88, 95, 104, 107, 108, 110, 113, 116, 123, 125, 130
Phusis$_3$, 27, 28, 32, 37, 40–43, 55, 58, 73, 95, 106, 107, 110, 114, 116, 125
Phusis$_{3/2}$, 36
Physics IV, 196
Pions, 219
Planck Black Body Radiation Law, 199
Planck, M., 188, 199, 204
Planck's quantum theory, 201
Plato, 20, 25, 28, 73, 74, 77, 79, 84, 101, 111, 121, 144
Plato, *Cratylus*, 22
Platonic, 125
Plato's Divided Line, 71
Plato's eristic method, 75
Plato's General Philosophy of Science, 70–75
Plato's theory of forms, 83
Plato, *Timaeus*, 8
Pleasure, 76
Plotinus, *Enneads*, 25
Plutarch, *On the Generation of the Soul* in Plato's *Timaeus*, 25
Pneuma, 20, 50, 127, 132, 156
Pneuma-enriched blood, 90
Pneuma theory, 89
Pneumatists, 99, 102, 111, 116
Polonium, 189
Polybus, 21
Pope Gregory XIII, 135, 138
Popper, K., 220, 243
Posidonius, 102
Posterior Analytics, 82–85, 176
Post-exposure prophylaxis (PEP), 236
Post-modernism, 229
Pre-exposure prophylaxis (PREP), 236
Pre-Socratics, 73, 74
Principia Ethica, 226
Principia Mathematica, 198, 226
Principle of non-contradiction, 205
Principle of precautionary reason, 232
Principle of sufficient reason, 72
Prior Analytics, 82, 83
Probability theory, 203
Process and Reality, 227
Proclus, *Commentary on Euclid*, 29
Prognosis, 53–62, 164

Prognostic, 57, 58
Pseudo-Aristotle, 29, 31
Pseudo-Aristotle, *On the World*, 24
Pseudo-Galen, 9, 10
Psuche, 20, 48–52, 76
Ptolemaic, 137, 145
Ptolemy's *Almagest*, 135
Public health, 180, 182, 233
Public health and pandemics, 212–214
Public health for populations, 208
Pulmonary capillaries, 127
Pulmonary circulation, 126
Putnam, H., 206, 228
Pythagoras, 21, 28–32

Q

Quantum gravity, 222, 224, 225
Quantum mechanics, 202, 204, 221, 224, 225
Quantum mechanism, 223
Quantum theory, 200
Quarks, 219
Quine, 207, 227

R

Ramsey sentences, 206, 228
Real, 145
Realdus Columbo, 132
Red blood cell in mammalian blood, 163
Reductionism, 199, 206
Reductionistic bent, 229
Reductionistic strategy, 187
Regimen, 52
Regimen in Acute Diseases, 61, 62
Regimen in Acute Diseases and in Diseases I, 59
Regimen in Health, 62
Regulation of hot and cold, 102
Reiman, 195
Replacement species, 168
Republic, 28
Reversibility, 221
Richard's Paradox, 199
Riemann, B., 191
Rigid designation of identity, 24
Roger Bacon, 135
Role of blood, 89
R-type into and S-type, 209
Russell, A., 165
Russell, B., 198, 226, 227
Russell's Paradox, 198
Rutherford, E., 188, 200, 201

S
Salk and Sabin polio vaccines, 210
SARS: 2003/MERS: 2012, 240
Save the phenomena, 218
Schrödinger, E.R.A., 201, 202
Schrödinger's cat, 205
Select pandemics, 233–242
Self-referential paradoxes, 198
Self-referential propositions, 26
Semantics, 207
Semmelweis, I., 179
Sepsis, 183
Servetus, M., 127, 129
Setting fractures, 6
Sextus Empiricus, 113
Sextus Empiricus, *Against the Logicians*, 21, 30
Similarity, 108, 109
Simple, 173, 223
Simple drugs, 126
Simpler, 185, 218
Simplicity, 61, 136, 137, 154, 159, 185, 189, 198, 205, 215, 217, 222, 226, 244
Simplicius, 12
Simplicius, *Commentary on Aristotle's Physics*, 13
Single, absolute truth, 138
Skepsis$_3$, 116
Skepticism$_3$, 107
Skin, 92
Skin breathing, 19
Sleeping sickness, 234
Small pox, 234
Smith, A., 169
Snow, J., 179, 180
Socrates, 125
Somatic *pneuma*, 19
Soranus, 104
Space, 145
Space and time, 190, 204
Space being non-Euclidean, 191
Space is that it may possess positive curvature, 191
Space of assumed connections, 225
Spacetime, 196, 225
Space, time, and causality, 221
Spanish Flu, 212
Spanish Flu: 1918–1919, 234
Special and general relativity, 193–196
Special relativity, 193
Spina Bifida, 230
Spin networks, 225
Standard accounts, 218, 224

Steady-state cosmological theory, 164
Sterilizing tissue with radon, 189
Stoicism, 112
Strife, 13, 164
Strings, 225
String theory, 223, 225–226
Strong and weak atomic forces, 223
Struggle for existence, 170
Suicide, 65
Super-symmetry, 225
Surgery, 132, 232
Survival competition, 167
Survival of the fittest, 171
Swainson, W., 166
Symmetrical, 109
Symmetrical (elegance), 111
Symmetrical whole, 75
Symmetry, 26, 136, 137
Symmetry between the uniform parts, 91
Symptoms, 111
Synthetic *a posteriori*, 150
Synthetic *a priori*, 150, 151
Synthetic *a priori* proposition, 151
Synthetic propositions, 148

T
Table of the Categories, 152
Tay-Sachs, 230
Technology and medicine, 230–243
Technology and the Mission of Medicine, 243
TE element, 91
TE-leaning, 95
Teleological explanation (TE), 47, 87
Teleological posits, 142
Teleology, 163
Teleology (TE) and the material account (ME), 91, 98
Thales, 9–11, 21, 218
The black hole paradox, 221
The Cannon of Medicine (book), 125, 126
The Categories of the Understanding, 153
The cat thought experiment, 202
The cause of fermentation, 181
The Copenhagen Interpretation of Quantum Mechanics, 201
The Divided Line, 71, 84
The field of connections, 225
The first cloning, 231
The Floating Man, 125
The Foundations of Arithmetic: A Logical-Mathematical Inquiry into the Concept of Number, 196

Index

The gods, 6
The Great Chain of Being, 165
The Great Surgery Book, 128
The History of Animals, 97
The Law of Dominance and Uniformity, 174
The Law of Independent Assortment, 175
The Law of Segregation, 174
Themistius, *Oration*, 22
The Nature of Man, 56
The new big and little Nature, 207
The Oath, 62–67
Theophrastus, *On Sensations*, 17
The Origin of the Species, 169–176
Theory of types, 199
The Parts of Animals, 87
The pulse, 110
The Sacred Disease, 51
The Scale of Nature, 165
The schematism of the categories, 153–158
Thessalus, 112–114
The standard account, 225
The strong force, 218
The Theory of Everything, 224
The Transcendental Dialectic, 154
The Uncertainty Principle, 204
The unit of selection debate, 229
The unity of science, 206
The weak force, 218
Thomson, J.J., 200, 201
Three-body problem, 231
Timaeus, 69–79
Timaeus 65b 4 Plato, 76
Time, 145, 192
Time and space in the transcendental aesthetic, 152
Time-line for evolution, 167
Top => down, 229
"Top-down" causation, 229
Toulmin, S., 227
Transcendental Aesthetic, 191
Transcendental Analytic, 154
Treating patients, 210–212
Treatise on Man, 142
Treatment, 53–62, 164, 230
Treatment of somatic symptoms, 231
Trophe, 52, 53
20th century medicine, 208–210
Two New Sciences, 143

U
Uncertainty, 224
Uniformtarianism principle, 168

Unifying entity of big and little Nature, 225
Unifying Existing Accounts, 224
Unity of Science as a Working Hypothesis, 228
Unordered pairs, 227, 228
Uranium, 189

V
Value-neutral, 228
Veins, 157, 158
Veins and arteries, 89
Venerable Bede, 134
Vesalius, A., 127, 130
Vibrational states, 225
Vineberg, A., 212
Vision, 23
Vivisection, 133
von Humboldt, A., 166

W
Wallace, A.R., 166, 169
War wounds, 6
Water, 20
Watson, J., 209
Wave mechanics, 201
Wet and dry, 116
Wheeler/DeWitt, 225
Wheeler, J.A., 221, 224
Whitehead, A.N., 198, 226, 227
Wilkins, M., 209
William Harvey's *De motu cordis*, 154
William Paley's *Evidence of Christianity*, 167
Willis, T., 143
Wimsatt, W., 227
Worked, 198
World Health Organization (WHO), 240
Wounds, 5

X
Xenophanes, 21
X-rays, 189

Y
Yellow fever, 234

Z
Zeno of Elea, 26, 145, 192, 194
Zeus, 5

GPSR Compliance

The European Union's (EU) General Product Safety Regulation (GPSR) is a set of rules that requires consumer products to be safe and our obligations to ensure this.

If you have any concerns about our products, you can contact us on

ProductSafety@springernature.com

In case Publisher is established outside the EU, the EU authorized representative is:

Springer Nature Customer Service Center GmbH
Europaplatz 3
69115 Heidelberg, Germany

www.ingramcontent.com/pod-product-compliance
Lightning Source LLC
Chambersburg PA
CBHW052245150925
32653CB00003B/60